Essential MATHEMATICS STAGE 9

FOR CAMBRIDGE SECONDARY 1

Patrick Kivlin, Sue Pemberton, Paul Winters

Oxford excellence for Cambridge Secondary 1

OXFORD UNIVERSITY PRESS

OXFORD
UNIVERSITY PRESS

Great Clarendon Street, Oxford, OX2 6DP, United Kingdom

Oxford University Press is a department of the University of Oxford.
It furthers the University's objective of excellence in research, scholarship,
and education by publishing worldwide. Oxford is a registered trade mark of
Oxford University Press in the UK and in certain other countries

Text © Patrick Kivlin, Sue Pemberton and Paul Winters 2014
Illustrations © Oxford University Press 2014

IGCSE® is the registered trademark of Cambridge International Examinations.

The moral rights of the authors have been asserted

First published in 2014

All rights reserved. No part of this publication may be reproduced, stored in
a retrieval system, or transmitted, in any form or by any means, without the
prior permission in writing of Oxford University Press, or as expressly permitted
by law, by licence or under terms agreed with the appropriate reprographics
rights organization. Enquiries concerning reproduction outside the scope of the
above should be sent to the Rights Department, Oxford University Press, at the
address above.

You must not circulate this work in any other form
and you must impose this same condition on any acquirer

British Library Cataloguing in Publication Data
Data available

978-1-4085-1989-9

10 9 8 7 6 5 4 3

Printed in Great Britain by CPI Group (UK) Ltd., Croydon CR0 4YY

Acknowledgements

Page make-up: OKS Prepress, India
Illustrations: OKS Prepress, India

The publishers would like to thank the following for permissions to
use their photographs:

Cover: Grafissimo/iStockphoto;

P1 top: Joe_Potato/iStockphoto; **p1 bottom:** Photobvious/i
Stockphoto; **p18:** Juulijs/Fotolia; **p34:** brytta/iStockphoto; **p49:** stockcam/
iStockphoto; **p59:** catenarymedia/iStockphoto; **p71:** traveler1116/iStockphoto;
p84: DNY59/iStockphoto; **p87 top:** Difydave/iStockphoto; **p87 bottom:**
karandaev/iStockphoto; **p88:** TPopova/iStockphoto; **p91:** scibak/iStockphoto;
p107: Branislav/iStockphoto; **p124:** williammpark/iStockphoto; **p133:**
pagadesign/iStockphoto; **p133 bottom:** viennetta/iStockphoto; **p150:**
duncan1890/iStockphoto; **p151:** GeorgePeters/iStockphoto; **p161:** wynnter/
iStockphoto; **p:171** leminuit/iStockphoto; **p173:** valdis torms/Shutterstock;
p183: holgs/iStockphoto; **p208:** Solkol/Wikimedia Commons; **p223:** S-e-v-e-r-e/
iStockphoto; **p238:** Yellowj/Shutterstock; **p248:** JM-Design/Shutterstock; **p250
top:** Natykach Nataliia/Shutterstock; **p250 middle:** humbak/Shutterstock; **p252:**
JacobH/iStockphoto; **p260:** courtneyk/iStockphoto; **p263:** robeo/iStockphoto;
p264: william87/iStockphoto; **p271:** JoeStark/iStockphoto; **p285:** LifesizeImages/
iStockphoto; **p299:** jgroup/iStockphoto; **p310:** gniedzieska/iStockphoto;
p314: Hal_P/Shutterstock; **p325:** gchutka/iStockphoto

Although we have made every effort to trace and contact all
copyright holders before publication this has not been possible in all
cases. If notified, the publisher will rectify any errors or omissions at
the earliest opportunity.

Links to third party websites are provided by Oxford in good faith
and for information only. Oxford disclaims any responsibility for
the materials contained in any third party website referenced in
this work.

Contents

	Introduction	iv
1	Integers, powers and roots	1
2	Expressions	18
3	Shapes and geometric reasoning 1	34
4	Fractions	49
5	Decimals	59
6	Processing, interpreting and discussing data	71
7	Length, mass and capacity	81
8	Equations and inequalities	91
9	Shapes and geometric reasoning 2	107
10	Presenting, interpreting and discussing data	133
11	Area, perimeter and volume	150
12	Formulae	171
13	Position and movement	183
14	Sequences	208
15	Probability	223
16	Functions and graphs	238
17	Fractions, decimals and percentages	252
18	Planning and collecting data	260
19	Ratio and proportion	271
20	Time and rates of change	285
21	Pythagoras' theorem	299
22	Trigonometry (extension work)	310
23	Matrices and transformations (extension work)	325
	Glossary	335
	Index	338

Introduction

Welcome to *Mathematics for Cambridge Secondary 1!* This student book has been written for the Cambridge International Examinations Secondary 1 Mathematics Curriculum Framework and provides complete coverage of Stage 9. Created specifically for international students and teachers by a dedicated and experienced author team, this book covers all areas in the curriculum: number, algebra, geometry measure, handling data and problem solving.

The following features have been designed to guide you through the content of the book with ease:

Learning outcomes

The learning outcomes give you an idea of what you will be covering, and what you should understand by the end of the chapter.

Worked examples

Worked examples to illustrate and expand the content. Work through these yourself and then compare your answers with the solutions.

Problem solving questions: Help develop knowledge and skills by requiring creative or methodical approaches, often in a real-life context.

Extension questions: Provide you with further challenge beyond the standard questions found in the book.

Hints: Useful tips for you to remember whilst learning the maths.

Key words: The first time key words appear in the book, they are highlighted in **bold red** text. A definition of each key word can be found in the glossary at the back of the book.

Extension Chapters: Chapters 22 and 23 are included in the Cambridge IGCSE curriculum. This additional content is included to challenge more able students.

This book is part of a series of six books and three teacher-support CD-ROMs. There are three student textbooks covering stages 7, 8 and 9 and three workbooks written to closely match the textbooks, as well as a teacher's CD-ROM for each stage.

The accompanying **Workbooks** provide extensive opportunities for you to practise your skills and apply your knowledge, both for homework and in the classroom.

The **teacher's CD-ROMs** include a wealth of interactive activities and supplementary worksheets, as well as the answers to questions in the books.

1 Integers, powers and roots

> **Learning outcomes**
> - Add, subtract, multiply and divide directed numbers.
> - Estimate square roots and cube roots.
> - Use positive, negative and zero indices and the index laws for multiplication and division of positive integer powers.
> - Use the order of operations, including brackets and powers.

Calculating by hand

Before the invention of calculators many calculations were done by hand.

There were some calculating tools that could be used.

One of these was the slide rule.

The pictures show a circular slide rule and a straight slide rule.

A slide rule was used for long multiplication and division sums.

It was based on the use of powers and indices.

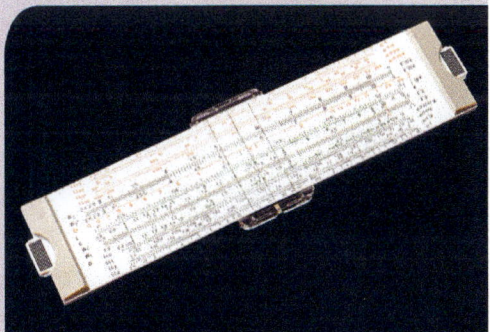

1.1 Review of directed numbers

Addition and subtraction

Positive and negative numbers are known as directed numbers because the + or − sign shows a direction along a number line.

Addition or subtraction of directed numbers can be thought of as movements along a number line.

This diagram shows the movement for (+4) + (−6)

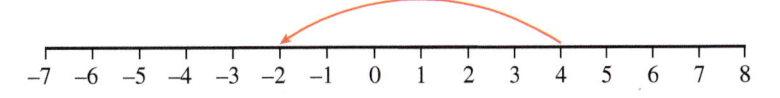

The red arrow starts at +4 and moves 6 steps to the left.

For small **integers** (whole numbers) it is easy to think of moves along the number line.

For other numbers you use the rules for combining signs.

They are:

+ + becomes +	e.g. −4 + (+6) becomes −4 + 6 = 2
+ − becomes −	e.g. −4 + (−6) becomes −4 − 6 = −10
− + becomes −	e.g. −4 − (+6) becomes −4 − 6 = −10
− − becomes +	e.g. −4 − (−6) becomes −4 + 6 = 2

Worked example 1

Work out the following:

a (+7.5) − (+12.2) **b** (−8.1) + (+14.8) **c** (−23.5) − (−17.8)

a (+7.5) − (+12.2)

= +7.5 − 12.2   − + becomes −

It is hard to think of moving 12.2 steps to the left of 7.5 on a number line.

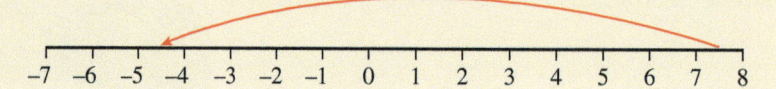

−7 −6 −5 −4 −3 −2 −1 0 1 2 3 4 5 6 7 8

Instead you can think of 7.5 − 12.2 as being the inverse or opposite of 12.2 − 7.5

Work this out normally using a calculation if necessary.

12.2 − 7.5 = 4.7

So 7.5 − 12.2 = −4.7

$$\begin{array}{r} 1\overset{1}{\cancel{2}}.\overset{\,}{2} \\ -\ 7.5 \\ \hline 4.7 \end{array}$$

b (−8.1) + (+14.8)

= −8.1 + 14.8   + + becomes +

= 14.8 − 8.1

= 6.7

c (−23.5) − (−17.8)

= −23.5 + 17.8   − − becomes +

= − (23.5 − 17.8)   using the inverse calculation

= −5.7

Multiplication and division

Here are the rules for multiplication and division.

positive and positive = positive

positive and negative = negative

negative and positive = negative

negative and negative = positive

This is often written as

$+ \; + = +$

$+ \; - = -$

$- \; + = -$

$- \; - = +$

> The basic rule is the same for both multiplication and division.
> If the two signs are the same the answer will be positive.
> If the two signs are different the answer will be negative.

Worked example 2

Work out: **a** $-7 \times +12$ **b** -23.6×-9 **c** $-63 \div 9$ **d** $220.8 \div -12$

a The signs are different so the answer will be negative.

You know that $7 \times 12 = 84$

Therefore $-7 \times 12 = -84$

b The signs are the same so the answer will be positive.

Work out $23.6 \times 9 = 212.4$ > You may need to write a calculation for this.

Therefore $-23.6 \times -9 = 212.4$

$$\begin{array}{r} 23.6 \\ \times \quad 9 \\ \hline 212.4 \\ 3\;5 \end{array}$$

c The signs are different so the answer will be negative.

You know that $63 \div 9 = 7$

Therefore $-63 \div 9 = -7$

d The signs are different so the answer will be negative.

Work out $220.8 \div 12 = 18.4$

Therefore $220.8 \div -12 = -18.4$

$$12 \overline{)220.^{10}8}^{\;\;18.4}$$

Mathematics for Cambridge Secondary 1

Worked example 3

Work out: **a** $-4 \times +8 \times -10$ **b** $(-6)^2$

a When there are more than two numbers work through them two at a time.

It may help to bracket them in pairs.

$-4 \times +8 \times -10$
$= (-4 \times +8) \times -10$
$= -32 \times -10$
$= 320$

if there is an even number of negative signs, the answer will be positive

if there is an odd number of negative signs, the answer will be negative

b $(-6)^2 = (-6) \times (-6) = 36$

It is worth noting that when a negative number is squared, the answer is positive.

Exercise 1.1

Work out:

1. $(-2.5) + (+8.5)$
2. $3 + (-9)$
3. $(-6.4) - (+9)$
4. $6.2 - (-8.5)$
5. $(-13) + (-14)$
6. $(-8) + (+28)$
7. $(-5.8) - (-7.8)$
8. $0.5 + (+1.5)$
9. $(-3.9) + (+4.9)$
10. $(-5.2) + (-4.8)$
11. $7.6 - (+5.3)$
12. $3 - (+13.6)$
13. $15.3 + (-12.2)$
14. $(-12.8) - (+13.8)$
15. $(-21) - (-26.9)$
16. $(-13.7) + (-16.8)$
17. $-9 \times +8$
18. -7×-6
19. $+9.2 \times -5$
20. -4.3×-8
21. $+16.3 \times +5$
22. $-17.2 \times +3$
23. $+25.8 \times -2$
24. -16.4×-10
25. $-72 \div 8$
26. $-48 \div -6$
27. $128 \div -8$
28. $-354 \div -6$
29. $18.7 \div -11$
30. $-650 \div -10$
31. $-65.1 \div 7$
32. $-89.5 \div -5$
33. $(-3)^2$
34. $(-8)^2$
35. $(-10)^2$
36. $-5 \times -4 \times 6$
37. $-4 \times -2 \times -7$
38. $9 \times -3 \times -10$
39. $-3 \times 8 \times -4 \times -10$
40. $(-2)^3$ (Work this out as $-2 \times -2 \times -2$)

1.2 Square and cube roots

Square roots

The integers 1, 4, 9, 16, 25, 36, 49, … are **square numbers**. A square number is a number formed by multiplying any integer by itself.

Each of them has a positive integer square root and a negative integer square root.

For example the square roots of 36 are 6 and −6.

This is because $6^2 = 36$ and $(-6)^2 = 36$

> **Worked example 1**
>
> Write down the square roots of:
>
> **a** 81   **b** 169   **c** 400   **d** 225
>
> **a** $9^2 = 81$ so the square roots of 81 are 9 or −9
> **b** $13^2 = 169$ so the square roots of 169 are 13 or −13
> **c** $20^2 = 400$ so the square roots of 400 are 20 or −20
> **d** $15^2 = 225$ so the square roots of 225 are 15 or −15

You should know the square numbers up to 400.

The square root of any other number such as $\sqrt{20}$ is not an integer.

You need to be able to estimate the answer.

To do this you look for square numbers larger and smaller than the number.

Think of the list of square numbers.

$$20$$
$$1 \quad 4 \quad 9 \quad 16 \quad 25 \quad 36 \quad 49 \quad 64 \quad 81 \quad 100 \ \ldots$$

20 is bigger than 16 and smaller than 25.

$\sqrt{20}$ is bigger than $\sqrt{16}$ and smaller than $\sqrt{25}$.

$\sqrt{20}$ is bigger than 4 and less than 5.

20 is about halfway between 16 and 25 so the square root of 20 is about 4.5 or −4.5

> Don't forget the negative square root.

Using a calculator gives $\sqrt{20} = 4.47$ to 2 decimal places so 4.5 is a good estimate.

Mathematics for Cambridge Secondary 1

Worked example 2

Find an estimate for the square roots of these numbers.

a 40   **b** 55   **c** 160

Use a calculator to check the answer.

a 40 is between 36 and 49.

$\sqrt{40}$ is between 6 and 7.

40 is nearer to 36 than to 49.

This diagram of a number line may help. The square roots of 40 are about 6.3 or −6.3

```
36  37  38  39  40  41  42  43  44  45  46  47  48  49
6                    ↑                                7
                About 6.3
```

On the calculator $\sqrt{40}$ = 6.325 to 3 decimal places.

b 55 is between 49 and 64.

$\sqrt{55}$ is between 7 and 8.

55 is slightly nearer to 49 than 64.

The square roots of 55 are about 7.4 or −7.4

On the calculator $\sqrt{55}$ = 7.416 to 3 decimal places.

c 160 is between 144 and 169.

$\sqrt{160}$ is between 12 and 13.

160 is nearer to 169 than to 144.

The square roots of 160 are about 12.7 or −12.7

On the calculator $\sqrt{160}$ = 12.649 to 3 decimal places.

Cube roots

Estimating cube roots uses the same method.

One difference is that negative numbers have a negative cube root.

$\sqrt[3]{-125} = -5$ because $(-5)^3 = -5 \times -5 \times -5 = -125$ *There is an odd number of negative signs so the answer will be negative.*

The cube numbers from 1^3 to 10^3 and from $(-1)^3$ to $(-10)^3$ are:

$1 = 1^3 = 1 \times 1 \times 1$ $-1 = (-1)^3 = -1 \times -1 \times -1$

$8 = 2^3 = 2 \times 2 \times 2$ $-8 = (-2)^3 = -2 \times -2 \times -2$

$27 = 3^3 = 3 \times 3 \times 3$ $-27 = (-3)^3 = -3 \times -3 \times -3$

$64 = 4^3 = 4 \times 4 \times 4$ $-64 = (-4)^3 = -4 \times -4 \times -4$

$125 = 5^3 = 5 \times 5 \times 5$

$216 = 6^3 = 6 \times 6 \times 6$

$343 = 7^3 = 7 \times 7 \times 7$

$512 = 8^3 = 8 \times 8 \times 8$

$729 = 9^3 = 9 \times 9 \times 9$

$1000 = 10^3 = 10 \times 10 \times 10$

$-125 = (-5)^3 = -5 \times -5 \times -5$

$-216 = (-6)^3 = -6 \times -6 \times -6$

$-343 = (-7)^3 = -7 \times -7 \times -7$

$-512 = (-8)^3 = -8 \times -8 \times -8$

$-729 = (-9)^3 = -9 \times -9 \times -9$

$-1000 = (-10)^3 = -10 \times -10 \times -10$

You should learn 1^3, 2^3, 3^3, 4^3, 5^3 and 10^3.

Worked example 3

Estimate the following:

a $\sqrt[3]{20}$ **b** $\sqrt[3]{-140}$ **c** $\sqrt[3]{750}$

a 20 is between 8 and 27.

$\sqrt[3]{20}$ is between 2 and 3.

20 is nearer to 27 than 8.

$\sqrt[3]{20}$ is about 2.7

b -140 is between -125 and -216.

$\sqrt[3]{-140}$ is between -5 and -6.

-140 is nearer to -125 than -216.

$\sqrt[3]{-140}$ is about -5.2

c 750 is between 729 and 1000.

$\sqrt[3]{750}$ is between 9 and 10.

750 is a lot nearer to 729 than 1000.

$\sqrt[3]{750}$ is about 9.1

Exercise 1.2

1 Copy and complete this working for estimating $\sqrt{70}$.

70 is between 64 and ………

$\sqrt{70}$ is between …….. and ……..

70 is nearer to ……… than ……..

$\sqrt{70}$ is about ……………

2 Estimate the following square roots. Use a calculator to check the answers.

a $\sqrt{5}$ **b** $\sqrt{15}$ **c** $\sqrt{45}$ **d** $\sqrt{90}$ **e** $\sqrt{300}$

f $\sqrt{200}$ **g** $\sqrt{380}$ **h** $\sqrt{265}$ **i** $\sqrt{180}$ **j** $\sqrt{125}$

Mathematics for Cambridge Secondary 1

3 a Work out the values of 21^2 and 22^2.

   b Use your answers to part **a** to estimate $\sqrt{450}$.

   c Use a calculator to check the answer.

4 Write down all the cube numbers between 1 and 1000.

5 Use your answer to question **4** to estimate the following cube roots.

   Use a calculator to check the answers.

   a $\sqrt[3]{4}$ **b** $\sqrt[3]{30}$ **c** $\sqrt[3]{90}$ **d** $\sqrt[3]{200}$ **e** $\sqrt[3]{-50}$

   f $\sqrt[3]{-500}$ **g** $\sqrt[3]{950}$ **h** $\sqrt[3]{-120}$ **i** $\sqrt[3]{-10}$ **j** $\sqrt[3]{350}$

6 Here is a number pattern using square numbers.

   $1^2 - 0^2 = 1 + 0 = 1$

   $2^2 - 1^2 = 2 + 1 = 3$

   $3^2 - 2^2 = 3 + 2 = 5$

   $4^2 - 3^2 = \_\_ + \_\_ = 7$

   $5^2 - 4^2 = \_\_ + \_\_ = \_\_$

   $6^2 - 5^2 = \_\_ + \_\_ = \_\_$

   a Copy the pattern and fill in the missing numbers.

   b Complete the next two lines of the pattern.

   c Work out: **i** $10^2 - 9^2$ **ii** $16^2 - 15^2$

   > Look at the number pattern and use it to work these out.

7 Tom and George were talking about their homework.

   Tom said that $\sqrt{-100} = -10$ because $\sqrt{100} = 10$

   George said that $\sqrt{-100}$ cannot be worked out because negative numbers have no square root.

   Who is right?

   > Try working out $(-10)^2$. What happens when you multiply two negative numbers together?

1.3 Powers and indices

Index notation

In the number 5^3 the small number ^{3} is called the **index** or **power**. A power tells you how many times a number is multiplied by itself.

The 5 is called the **base number**.

Indices is the plural of index.

The index shows how many of the base numbers are multiplied together.

$4^6 = 4 \times 4 \times 4 \times 4 \times 4 \times 4 = 4096$

8

1 Integers, powers and roots

There are six 4's all multiplied together. This is worked out one step at a time.

$4 \times 4 = 16 \quad 16 \times 4 = 64 \quad 64 \times 4 = 256 \quad 256 \times 4 = 1024 \quad 1024 \times 4 = 4096$

Here are some more examples:

$9^4 = 9 \times 9 \times 9 \times 9 = 6561$

$3^7 = 3 \times 3 \times 3 \times 3 \times 3 \times 3 \times 3 = 2187$

$7^1 = 7$

In the last example the index 1 is not needed.

Any number can be thought of as having an index 1.

$x^1 = x$

Worked example 1

Write these using indices.

a $5 \times 5 \times 5 \times 5 \times 5 \times 5 \times 5 \times 5$

b $10 \times 10 \times 10 \times 10 \times 10 \times 10$

c $3 \times 3 \times 4 \times 4 \times 4 \times 4 \times 4$

a There are eight 5's multiplied together.

$5 \times 5 \times 5 \times 5 \times 5 \times 5 \times 5 \times 5 = 5^8$

b There are six 10's.

$10 \times 10 \times 10 \times 10 \times 10 \times 10 = 10^6$

c Here there are two 3's and five 4's.

The base numbers are different so they are done separately.

$3 \times 3 \times 4 \times 4 \times 4 \times 4 \times 4 = 3^2 \times 4^5$

Worked example 2

Evaluate the following:

a 2^6 **b** 6^3 **c** 8^7

a $2^6 = 2 \times 2 \times 2 \times 2 \times 2 \times 2 = 64$

b $6^3 = 6 \times 6 \times 6 = 216$

c Parts **a** and **b** could easily be worked out in your head.

Part **c** may need a calculator. Your calculator has a button marked x^y or $x^\square$.

Use it to show that $8^7 = 2\,097\,152$

9

The index laws

Look at the multiplication $6^4 \times 6^3$

It can be written out as $6^4 \times 6^3 = (6 \times 6 \times 6 \times 6) \times (6 \times 6 \times 6) = 6 \times 6 \times 6 \times 6 \times 6 \times 6 \times 6 = 6^7$

Four 6's   Three 6's   Seven 6's altogether

In short $6^4 \times 6^3 = 6^{4+3} = 6^7$

This can be written as an index law

$$x^a \times x^b = x^{a+b}$$

When multiplying powers you add the index numbers.

Now look at the division $7^5 \div 7^3$

Writing this in fraction form gives $7^5 \div 7^3 = \dfrac{7 \times 7 \times 7 \times 7 \times 7}{7 \times 7 \times 7} = \dfrac{7 \times 7 \times \cancel{7} \times \cancel{7} \times \cancel{7}}{\cancel{7} \times \cancel{7} \times \cancel{7}}$

$$= 7 \times 7 = 7^2$$

After cancelling there are $5 - 3 = 2$ sevens left.

In short $7^5 \div 7^3 = 7^{5-3} = 7^2$

This can be written as another index law

$$x^a \div x^b = x^{a-b}$$

When dividing powers you subtract the index numbers.

This result leads to two new rules.

Look at the division $6^5 \div 6^5$

Using the division rule for indices the answer is $6^5 \div 6^5 = 6^{5-5} = 6^0$

But $6^5 \div 6^5 = \dfrac{6 \times 6 \times 6 \times 6 \times 6}{6 \times 6 \times 6 \times 6 \times 6} = 1$

But two answers to the same question must be equal.

Therefore $6^0 = 1$

This is true for any division of the form $x^a \div x^a$ and leads to the result

$$x^0 = 1$$

Any number raised to the power 0 equals 1.

Look at the power $(4^2)^5$.

This means $4^2 \times 4^2 \times 4^2 \times 4^2 \times 4^2 = 4^{10}$

Note that $10 = 2 \times 5$

Now look at $(6^3)^4 = 6^3 \times 6^3 \times 6^3 \times 6^3 = 6^{12}$

In this case note that $12 = 3 \times 4$

This gives another index law

$$(x^a)^b = x^{a \times b}$$

When you raise a power to a power you multiply the index numbers together.

Worked example 3

Simplify the following using index notation.

a 7×7^3 **b** $8^9 \div 8^5$ **c** $(3^2)^7$ **d** $5^3 \times (5^4)^3 \div 5^7$

a To multiply the powers you add the index numbers.

Remember that $7 = 7^1$

$7^1 \times 7^3 = 7^4$

b To divide the powers you subtract the index numbers.

$8^9 \div 8^5 = 8^4$

c To raise a power to a power you multiply the index numbers.

$(3^2)^7 = 3^{14}$

d This is a combination of addition, subtraction and multiplication of powers.

$5^3 \times (5^4)^3 \div 5^7$

$= 5^3 \times 5^{12} \div 5^7$

$= 5^{3+12-7} = 5^8$

Worked example 4

Work out the value of the following:

a $4^6 \div 4^4$ **b** 25^0 **c** $7^4 \times 7^5 \div 7^8$

a $4^6 \div 4^4 = 4^2 = 16$ note that it is easier to simplify the indices before working out the power

b $25^0 = 1$ any number raised to the power 0 equals 1

c $7^4 \times 7^5 \div 7^8 = 7^{4+5-8} = 7^1 = 7$

Exercise 1.3

1 Simplify these expressions using index numbers.

a $2 \times 2 \times 2 \times 2$
b $5 \times 5 \times 5$
c $8 \times 8 \times 8 \times 8 \times 8 \times 8$
d $7 \times 7 \times 7 \times 7 \times 7 \times 7 \times 7 \times 7 \times 7$
e $11 \times 11 \times 11 \times 11$
f $3 \times 3 \times 3 \times 3 \times 3 \times 3 \times 3 \times 3 \times 3$
g $3 \times 3 \times 3 \times 3 \times 4 \times 4$
h $2 \times 2 \times 2 \times 9 \times 9 \times 9$
i $2 \times 2 \times 2 \times 2 \times 2 \times 2 \times 7$
j $x \times y \times y \times y$
k $a \times a \times b \times b \times b \times c \times c$
l $p \times q \times r \times p \times q \times q \times r \times r$

2 Find the value of the following:
   a 2^4 b 3^3 c 5^4 d 10^5 e 9^2
   f 1^8 g 1^{10} h 8^0 i 12^1 j 15^0

3 Simplify these expressions. Leave the answers in index form.
   a $3^4 \times 3^5$ b $4^7 \times 4^2$ c $5^9 \times 5^2$ d 3×3^4
   e $7^4 \div 7^2$ f $6^8 \div 6^5$ g $10^{11} \div 10^4$ h $5^9 \div 5^8$
   i $(2^3)^3$ j $(3^4)^2$ k $(10^2)^4$ l $(7^4)^6$

4 Simplify these expressions. Leave the answers in index form.
   a $8^2 \times 8^3 \times 8^4$ b $5^9 \times 5^2 \div 5^6$ c $(8^2)^3 \div 8^6$
   d $(5^3)^4 \times 5$ e $7^3 \times 7 \div 7^2$ f $(9^4)^5 \div (9^3)^6$
   g $(5^7)^3 \times 5^2 \div 5^8$ h $3^5 \times 3^3 \div 3^8$ i $2 \times 2^3 \div 2^4$
   j $x^5 \times x^4 \times x^2$ k $y^8 \times y^4 \div y^{10}$ l $(t^3)^2 \times t^6 \div t^5$

5 Simplify these expressions and then work out the value.
   a $2^3 \times 2^4 \div 2^5$ b $(6^3)^4 \div 6^{10}$ c $(5^3)^4 \div (5^2)^6$
   d $7^5 \div 7^4$ e $10^7 \times 10 \div 10^6$ f $3^5 \times (3^3)^3 \div 3^{11}$

6 Solve these equations to find the value of x.
   a $9^x = 1$ b $(5^x)^3 = 5^6$ c $(8^4)^x = 8^{20}$
   d $(10^x)^4 = (10^6)^2$ e $(3^x)^6 = 3^6$

> For parts **b**, **c**, **d** and **e** write both sides as a single power. For example in part **b** write $5^{3x} = 5^6$ so $3x = 6$.

1.4 Negative indices

Look at the division $4^2 \div 4^3$

Using the index laws $4^2 \div 4^3 = 4^{-1}$

But division also gives $4^2 \div 4^3 = \dfrac{\cancel{4} \times \cancel{4}}{4 \times \cancel{4} \times \cancel{4}} = \dfrac{1}{4}$

This shows that $4^{-1} = \dfrac{1}{4}$

Now look at the division $2 \div 2^4$

The index laws give the result $2 \div 2^4 = 2^{-3}$

Division gives $2 \div 2^4 = \dfrac{\cancel{2}}{2 \times 2 \times 2 \times \cancel{2}} = \dfrac{1}{2^3}$

This shows that $2^{-3} = \dfrac{1}{2^3}$

These are two examples of the general rule

$$x^{-a} = \dfrac{1}{x^a}$$

1 Integers, powers and roots

Worked example 1

Write as a fraction:

a 3^{-2} **b** $10^3 \times 10^{-5}$

a $3^{-2} = \frac{1}{3^2} = \frac{1}{9}$

b $10^3 \times 10^{-5} = 10^{3-5}$

$= 10^{-2}$

$= \frac{1}{10^2}$

$= \frac{1}{100}$

Worked example 2

Simplify these expressions. Leave the answers in index form.

a $6^3 \times 6^{-5} \times 6^4$ **b** $(3^4)^{-1} \div (3^{-2})^3$

a $6^3 \times 6^{-5} \times 6^4$

$= 6^{3 + (-5) + 4}$

$= 6^2$

b $(3^4)^{-1} \div (3^{-2})^3$

$= 3^{(-4) - (-6)}$

Remember the rules for adding and subtracting directed numbers.

$= 3^{-4+6}$

$= 3^2$

Exercise 1.4

1 Write as a fraction:

a 2^{-4} **b** 3^{-1} **c** 5^{-3} **d** 10^{-2}

e 8^{-2} **f** 7^{-2} **g** 3^{-3} **h** 2^{-3}

i 15^{-2} **j** 10^{-3}

2 Find the value of these as a fraction or a whole number.

a $5^2 \times 5^{-3}$ **b** 6×6^{-1} **c** $2^4 \times 2^{-3}$ **d** $10^{-2} \times 10$

e $7^4 \times 7^{-5}$ **f** $3^4 \times 3^{-6} \times 3^2$ **g** $2^9 \div 2^{10}$ **h** $3 \div 3^{-1}$

i $6^2 \div 6^3$ **j** $10^4 \div 10^3$ **k** $2^5 \times 2^{-1} \div 2^2$ **l** $4^2 \times 4^{-5} \times 4$

13

3 Simplify these expressions. Leave the answers in index form.

    a $4^3 \times 4^{-7}$ **b** $5^{-2} \times 5^5$ **c** $8^2 \div 8^4$ **d** $6^7 \div 6^{-3}$

    e $(3^2)^{-1} \times 3^{-4}$ **f** $2^5 \div (2^2)^{-3}$ **g** $2 \times 2^{-5} \div 2^3$ **h** $3^4 \div 3^{-5} \times (3^2)^2$

    i $x^5 \times x^{-5}$ **j** $x^4 \times x^{-6}$ **k** $y^2 \div y^5$ **l** $x^7 \div x^{-2} \times x^{-4}$

4 Find the value of:

    a 9×3^{-3} **b** 8×2^{-4} **c** $10^{-3} \times 100$ **d** $16 \div 2^4$

    e $4^3 \div 2^5$ **f** $9^2 \times 3^{-3}$ **g** $4^{-3} \times 8^2$ **h** $8^2 \div 2^3 \times 4^{-2}$

5 Find the value of x in these equations.

    a $(3^2)^x = \frac{1}{9}$ *Write the right-hand side as a power of 3. $\frac{1}{9} = \frac{1}{3^2} = 3^{-2}$ so $3^{2x} = 3^{-2}$*

    b $(2^x)^2 = \frac{1}{16}$ **c** $5^x \times 5^6 = 25$ **d** $7^4 \times 7^x = 49$

    e $(2^x)^x = 16$ note that there are two possible answers to **e**

1.5 Order of operations

The order in which calculations are done is important.

In your Stage 8 Student Book you learned the word **BIDMAS** which tells you the order of importance.

                  Brackets

                  Indices or powers

                  Division
                  Multiplication

                  Addition
                  Subtraction

Notice how they are spaced.

This is because division and multiplication have equal importance.

Addition and subtraction have equal importance as well.

Worked example 1

Calculate: $24 + 4^2 \times (9 - 7) - 27 \div 3$

$24 + 4^2 \times \mathbf{(9 - 7)} - 27 \div 3$	brackets come first
$= 24 + \mathbf{4^2} \times 2 - 27 \div 3$	indices are cleared next
$= 24 + \mathbf{16 \times 2} - \mathbf{27 \div 3}$	division and multiplication come next
$= 24 + 32 - 9$	finally, the addition and subtraction
$= 47$	

Worked example 2

Put brackets in to make this equation correct.

$21 - 12 \div 6 - 2^2 = 23$

You may need to try brackets in several different places.

a With no brackets:

$21 - 12 \div 6 - 2^2$

$= 21 - 12 \div 6 - 4$

$= 21 - 2 - 4 = 15$ this is not correct

b $(21 - 12) \div 6 - 2^2$

$= 9 \div 6 - 2^2$

$= 9 \div 6 - 4$

$= 1\frac{1}{2} - 4 = -2\frac{1}{2}$ this is not correct

c $21 - 12 \div (6 - 2^2)$

$= 21 - 12 \div 2$

$= 21 - 6 = 15$ this is not correct

d $21 - (12 \div 6 - 2^2)$

$= 21 - (12 \div 6 - 4)$

$= 21 - (2 - 4)$

$= 21 - (-2) = 23$ this is correct

The last one gives the correct answer so this is the required position for the brackets.

Exercise 1.5

Work out:

1. $17 - 5 \times 3$
2. $32 \div 8 + 3^2$
3. $69 - 52 \times 2$
4. $7 + 3 \times 4$
5. $3 \times 5 + 6 \times 2$
6. $9 \times 4 - 20 \div 5$
7. $6 \times (3 + 2)$
8. $12 \div (8 - 4)$
9. $9 \times (2 + 3) \times 10$
10. $3^2 - 2^2 \times 2$
11. $(3^2 - 2^2) \times 2$
12. $60 \div 2^2 - 8$
13. $36 - 5^2 + 2 \times 3$
14. $3 \times (4^2 - 3^2)$
15. $(5 + 2 \times 4)^2$
16. $2^3 + 3^3 \times 2$
17. $17 + 3^2 \times (6 - 2)$
18. $12^2 - (2 + 7 \times 4)$
19. $36 - 3^3 \div (12 - 9) - 2^3$
20. $(1 + 2)^3 + 4 \times 5 - 6$

Put brackets in (if necessary) to make these correct.

21. $12 + 3 \times 2 = 30$
22. $5 \times 8 - 3 = 25$
23. $32 \div 8 - 4 = 8$
24. $36 \div 3 + 6 \times 2 = 8$
25. $5^2 - 7 \times 2 = 36$
26. $9 - 7 \times 3^2 = 18$
27. $2^3 + 1 \times 3^2 + 1 = 90$
28. $6 + 2 \times 5^2 = 56$

Mathematics for Cambridge Secondary 1

29 Ronald wrote down an equation but forgot to write the brackets.

$$5 + 12 \times 3 - 1 = 2 + 3 \times 6 - 1$$

Work out where to put one pair of brackets on each side to make the equation correct.

> You may need to try using brackets in several different places before you find the right answer.

Review

1 Work out:
- **a** $(-4) + (-5)$
- **b** $(-3) - 4$
- **c** $8 + (-3)$
- **d** -7×-4
- **e** $36 \div (-3)$
- **f** -3.2×5
- **g** -2.4×-4
- **h** $-63 \div -7$

2 Write down the value of:
- **a** 6^2
- **b** $(-8)^2$
- **c** 15^2
- **d** $(-17)^2$
- **e** 1^8
- **f** 12^0
- **g** 5^{-2}
- **h** 6^{-1}

3 Write down the value of:
- **a** $\sqrt{361}$
- **b** $\sqrt{121}$
- **c** $\sqrt[3]{64}$
- **d** $\sqrt[3]{27}$

4 Estimate:
- **a** $\sqrt{28}$
- **b** $\sqrt[3]{40}$
- **c** $\sqrt{300}$
- **d** $\sqrt[3]{85}$

5 Simplify these expressions using index form.
- **a** $7 \times 7 \times 7 \times 7$
- **b** $12 \times 12 \times 12$
- **c** $5 \times 5 \times 5 \times 3 \times 3$
- **d** $4 \times 4 \times 4 \times 6 \times 6 \times 6 \times 6$

6 Simplify these expressions. Leave the answer in index form.
- **a** $4^2 \times 4^4$
- **b** $5^8 \div 5^3$
- **c** $6^2 \times 6^4 \times 6$
- **d** $(10^2)^3$
- **e** $(5^3)^4$
- **f** $x^4 \times x^3 \div x$
- **g** $y^5 \div y^3 \times y^2$
- **h** $x^4 \times x^3 \div x^{-1}$

7 Simplify these expressions and then work out the value.
- **a** $8^{-5} \times 8^7$
- **b** $(5^2)^3 \div 5^4$
- **c** $9 \div 3^2$
- **d** 16×2^{-3}

8 Work out:
- **a** $5 + 4 \times 8$
- **b** $43 - 5 \times 6 + 2$
- **c** $12 \times (3 + 2)$
- **d** $100 \div 5^2 + 7$
- **e** $3^3 + (12 - 2 \times 4)$
- **f** $(-3)^2 + 2^2 \times 5^2$
- **g** $12 + 2^3 \div 4 - 3$
- **h** $6^2 + 4 \times (5 + 2^2)$

9 Put brackets in to make these equations correct.
- **a** $3^2 \times 4 + 1 = 45$
- **b** $144 \div 8 + 2^2 \times 2 = 24$
- **c** $3 + 2 \times 5^2 - 9 = 80$

Examination-style questions

1 Work out: **a** 4^3 **b** $\sqrt{324}$ [2]

2 Fill in the missing index number. $6 \times 6 \times 6 \times 6 \times 6 = 6^{\square}$ [1]

3 Find the value of $5^3 + 13^2$ [2]

4 Work out:

a $(4 + 17) \times 5 - 3$

b $4^2 + 48 \div (4^2 - 4)$

5 Here is a list of numbers.

2, 6, 7, 9, 16, 24, 30, 60

Fill in the blanks with numbers from the list.

a _____ and _____ are factors of 12. [2]

b _____ and _____ are multiples of 15. [2]

c _____ and _____ are prime numbers. [2]

d _____ and _____ are square numbers. [2]

6 Work out the value of:

a $\sqrt[3]{125}$ **b** 8^0 [2]

7 Work out the value of:

a 2^{-1} **b** 2^4 [2]

8 Put a pair of brackets in to make this equation correct.

$7 \times 2 + 1 - 4 = 17$ [1]

9 Simplify these expressions. Leave the answers in index form.

a $x^3 \times x^{-5} \div x^4$ [2]

b $(y^3)^5 \div y^4$ [2]

10 Choose the correct word from this list to complete the statements.

cube   **square root**   **square**   **cube root**

a The _____ of 14 is 196 [1]

b The _____ of 125 is 5 [1]

2 Expressions

Learning outcomes

- Know the origins of the word *algebra* and its links to the work of the Arab mathematician Al'Khwarizmi.
- Add and subtract simple algebraic expressions.
- Use index notation for positive integer powers; apply the index laws for multiplication and division to simple algebraic expressions.
- Simplify algebraic expressions by taking out single-term common factors.
- Substitute positive and negative numbers into expressions.
- Expand the product of two linear expressions of the form $x \pm n$ and simplify the corresponding quadratic expression.
- Construct algebraic expressions.

Al'Khwarizmi

Muhammad ibn Musa Al'Khwarizmi is a famous Arabic mathematician who was born around 783.

He wrote numerous books. In one of his books called 'The Compedious Book on Calculation by Completion and Balancing' he used the word 'al-jabra', which means subtracting a number from both sides of an equation.

The word 'al-jabra' is the origin of the word 'algebra' that we use today.

A unique Arabic copy of this book is kept in Oxford and a Latin translation of the book is kept in Cambridge.

The illustration shows a stamp, which says his name and also '1200 years'. It was produced in 1983 to commemorate 1200 years since his birth.

2.1 Multiplying and dividing algebraic expressions

You can write $a \times a \times a \times a$ as a^4 ← **index** or **power** / **base number**

and $y \times y \times y \times y \times y \times y \times y$ as y^7

$3x \times 2x \times 5x$ can be written as $3 \times x \times 2 \times x \times 5 \times x = 3 \times 2 \times 5 \times x \times x \times x$
$$= 30 \times x^3$$
$$= 30x^3$$

$\dfrac{2x}{6}$ can be written as $\dfrac{\cancel{2} \times x}{\cancel{2} \times 3}$ which simplifies to $\dfrac{x}{3}$

Worked example 1

Simplify these expressions.   **a** $5a \times a \times 2a$ **b** $2x \times 3y \times x \times 4y$

a $5a \times a \times 2a = 5 \times a \times a \times 2 \times a$
$= 5 \times 2 \times a \times a \times a$
$= 10a^3$

b $2x \times 3y \times x \times 4y = 2 \times x \times 3 \times y \times x \times 4 \times y$
$= 2 \times 3 \times 4 \times x \times x \times y \times y$
$= 24x^2y^2$

Worked example 2

Simplify these expressions.   **a** $\dfrac{12x}{3y}$ **b** $\dfrac{5x^2}{10xy}$

a $\dfrac{12x}{3y} = \dfrac{\cancel{3} \times 4 \times x}{\cancel{3} \times y}$ divide numerator and denominator by 3
$= \dfrac{4x}{y}$

b $\dfrac{5x^2}{10xy} = \dfrac{\cancel{5} \times \cancel{x} \times x}{2 \times \cancel{5} \times \cancel{x} \times y}$ divide numerator and denominator by 5 and by x
$= \dfrac{x}{2y}$

Exercise 2.1

1 Simplify:

   a $3x \times 4x$ **b** $5x \times x$ **c** $2y \times y \times 3y$
   d $2a \times 3a \times 4a$ **e** $2b \times 3b \times b \times 4b \times b$ **f** $4p \times p \times 3p \times p \times p$

2 Simplify:

   a $5a \times a \times 2b$ **b** $2p \times 3q \times p \times 4q \times p$ **c** $7x \times 3y \times 2x$
   d $2x \times 3y \times 4y \times 5x$ **e** $6a \times b \times a \times 7ab$ **f** $x \times 8xy \times 2xy \times 3x$

3 Simplify:

   a $\dfrac{5x}{10}$ **b** $\dfrac{6y}{8}$ **c** $\dfrac{8a}{12}$ **d** $\dfrac{14x}{21}$ **e** $\dfrac{6x}{4}$
   f $\dfrac{5x}{x^2}$ **g** $\dfrac{3x^2}{12x}$ **h** $\dfrac{12xy}{8x}$ **i** $\dfrac{6y}{y^2}$ **j** $\dfrac{2xy}{y^2}$
   k $\dfrac{7x}{14y}$ **l** $\dfrac{9xy}{3xyz}$ **m** $\dfrac{8ab^2}{b}$ **n** $\dfrac{2xy}{6x^2y^2}$ **o** $\dfrac{2a^2b}{5ab^2}$

Mathematics for Cambridge Secondary 1

4 Write in the form 2^x. (For example $16 = 2^4$)

    a 32   **b** 8   **c** 4   **d** 64   **e** 256

5 Write in the form x^y. (For example $125 = 5^3$)

    a 49   **b** 27   **c** 10 000   **d** 169   **e** 243

6 Solve the equations.

The first one has been done for you.

    a $2^x = 8$ write 8 as a power of 2

         $2^x = 2^3$ the base numbers are the same so the indices must be the same

         $x = 3$

    b $2^x = 64$ **c** $3^x = 9$ **d** $10^x = 100$ **e** $5^x = 125$ **f** $2^x = 256$

    g $4^x = 64$ **h** $2^x = 512$ **i** $3^x = 27$ **j** $2^x = 16$ **k** $3^x = 81$

7 Solve the equations.

The first one has been done for you.

    a $5^{x+1} = 125$ write 125 as a power of 5

         $5^{x+1} = 5^3$ the base numbers are the same so the indices must be the same

         $x + 1 = 3$ subtract 1 from both sides

         $x = 2$

    b $2^{2x} = 64$ **c** $2^{x+1} = 16$ **d** $2^{x-1} = 8$ **e** $5^{x-2} = 125$ **f** $3^{x-1} = 81$

    g $2^{x-2} = 128$ **h** $2^{2x+1} = 32$ **i** $5^{x-1} = 125$ **j** $3^{2x+1} = 27$ **k** $10^{3x} = 1\,000\,000$

2.2 The index laws

The following three rules are known as the laws of indices. They were introduced in Chapter 1. There are more laws of indices that you will learn for your IGCSE.

Multiplying

Consider $2^5 \times 2^3 = (2 \times 2 \times 2 \times 2 \times 2) \times (2 \times 2 \times 2) = 2^8$

If you look carefully you can see a rule. $2^5 \times 2^3 = 2^{5+3} = 2^8$

$$\boxed{\text{RULE 1} \quad x^a \times x^b = x^{a+b}}$$

Dividing

Consider $4^5 \div 4^3 = \dfrac{\cancel{4} \times \cancel{4} \times \cancel{4} \times 4 \times 4}{\cancel{4} \times \cancel{4} \times \cancel{4}} = 4^2$

If you look carefully you can see a second rule. $4^5 \div 4^3 = 4^{5-3} = 4^2$

$$\boxed{\text{RULE 2} \quad x^a \div x^b = x^{a-b}}$$

2 Expressions

Raising to a power

Consider $(5^2)^3 = (5 \times 5) \times (5 \times 5) \times (5 \times 5) = 5^6$

If you look carefully you can see a third rule. $(5^2)^3 = 5^{2 \times 3} = 5^6$

> **RULE 3** $(x^a)^b = x^{a \times b}$

Worked example 1

Simplify: **a** $y^3 \times y^4$ **b** $a^8 \div a^3$ **c** $(f^3)^4$

a $y^3 \times y^4 = y^{3+4}$ add the indices
$= y^7$

b $a^8 \div a^3 = a^{8-3}$ subtract the indices
$= a^5$

c $(f^3)^4 = f^{3 \times 4}$ multiply the indices
$= f^{12}$

Worked example 2

Simplify: **a** $5xy^2 \times 2x^2y^5 \times 7x^3y^4$ **b** $\dfrac{2x^7y^8}{6x^3y^2}$

a $5xy^2 \times 2x^2y^5 \times 7x^3y^4 = 5 \times 2 \times 7 \times x^1 \times x^2 \times x^3 \times y^2 \times y^5 \times y^4$
$\qquad\qquad\qquad\qquad\qquad\quad = 70 \times x^{1+2+3} \times y^{2+5+4}$
$\qquad\qquad\qquad\qquad\qquad\quad = 70x^6y^{11}$

b $\dfrac{2x^7y^8}{6x^3y^2} = \dfrac{2}{6} \times (x^7 \div x^3) \times (y^8 \div y^2)$
$\qquad\quad = \dfrac{1}{3} \times x^{7-3} \times y^{8-2}$
$\qquad\quad = \dfrac{x^4y^6}{3}$

Exercise 2.2

Simplify the expressions.

1   **a** $x^3 \times x^2$ **b** $y^2 \times y^4$ **c** $a^8 \times a$ **d** $p^4 \times p^7$
     e $f \times f^3$ **f** $y^3 \times y^3$ **g** $q^6 \times q^3$ **h** $d^4 \times d^8$

2   **a** $2x^3 \times 3x^4$ **b** $4y^5 \times 2y^3$ **c** $3a^4 \times 5a^2$ **d** $4f^2 \times f^5$
     e $5y^3 \times 5y^3$ **f** $4d \times 3d^5$ **g** $7e^8 \times 4e^6$ **h** $9c^4 \times 4c^9$

3   **a** $x^3 \times x^5 \times x^7$ **b** $y^6 \times y \times y^2$ **c** $2a^3 \times 5a^2 \times 3a^4$ **d** $8d^4 \times d \times 2d^7$

4 a $x^7 \div x^2$ **b** $x^8 \div x^4$ **c** $a^8 \div a^5$ **d** $p^6 \div p$

5 a $\dfrac{16x^7}{4x^5}$ **b** $\dfrac{8x^6}{16x^2}$ **c** $\dfrac{3a^5}{12a^2}$ **d** $\dfrac{2y^8}{8y^2}$

  e $\dfrac{5y^7}{15y^4}$ **f** $\dfrac{12w^6}{18w^2}$ **g** $\dfrac{25a^8}{50a^3}$ **h** $\dfrac{18g^5}{27g^2}$

6 a $\dfrac{x^4 \times x^3}{x^5}$ **b** $\dfrac{y^8 \times y^8}{y^7}$ **c** $\dfrac{x^{10}}{x^2 \times x^4}$ **d** $\dfrac{a^4 \times a^7}{a^2 \times a^3}$

  e $\dfrac{2x^2 \times 3x^5}{4x^3}$ **f** $\dfrac{3y^4 \times 6y^3}{2y^5}$ **g** $\dfrac{8x^9}{2x^3 \times 2x^4}$ **h** $\dfrac{2a^5 \times 5a^2}{3a \times 4a^4}$

7 a $(y^3)^2$ **b** $(x^4)^3$ **c** $(x^2)^6$ **d** $(a^4)^4$ $(y^3)^2$ means $y^3 \times y^3$

8 a $(2x^3)^2$ **b** $(3a^2)^4$ **c** $(10y^4)^3$ **d** $\left(\dfrac{h^3}{2}\right)^2$ $(2x^3)^2$ means $2x^3 \times 2x^3$

2.3 Factorising

In your Stage 8 Student Book you learned how to **expand brackets**.

Expanding $2(x + 5)$ gives $2x + 10$

This can be found using the following two methods:

Method 1

By using a diagram.

The area of a rectangle that has sides of 2 and $x + 5$ can be found by splitting the rectangle up into two parts as shown below.

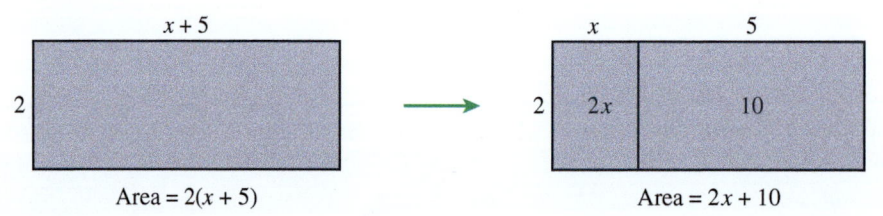

Area = $2(x + 5)$ Area = $2x + 10$

So $2(x + 5) = 2x + 10$

Method 2

By multiplying out the brackets.

$2(x + 5) = 2 \times x + 2 \times 5$
$ = 2x + 10$

You must multiply each term in the bracket by 2.

Factorising is the reverse of expanding brackets.

So factorising $2x + 10$ gives $2(x + 5)$

When you are asked to factorise an expression it is important that you factorise it fully.

For example $12x + 8 = 2(6x + 4)$ is only partially factorised.

The correct answer should be $12x + 8 = 4(3x + 2)$

2 Expressions

Worked example

Factorise: **a** $30x + 12$ **b** $3x^2 - 9xy$ **c** $6xy^2 + 8x^2y^3$

a $30x + 12$
$= 6(5x + 2)$

6 is the highest common factor (HCF) of the two terms
$30x = 6 \times 5x$ and $12 = 6 \times 2$
it is always sensible to check your answer by expanding the brackets
check: $6(5x + 2) = 6 \times 5x + 6 \times 2 = 30x + 12$

b $3x^2 - 9xy$
$= 3x(x - 3y)$

$3x$ is the HCF of the two terms
$3x^2 = 3x \times x$ and $9xy = 3x \times 3y$
check: $3x(x - 3y) = 3x \times x - 3x \times 3y = 3x^2 - 9xy$

c $6xy^2 + 8x^2y^3$
$= 2xy^2(3 + 4xy)$

$2xy^2$ is the HCF of the two terms
$6xy^2 = 2xy^2 \times 3$ and $8x^2y^3 = 2xy^2 \times 4xy$
check: $2xy^2(3 + 4xy) = 2xy^2 \times 3 + 2xy^2 \times 4xy = 6xy^2 + 8x^2y^3$

Exercise 2.3

1 Expand the brackets.

   a $8(x + 7)$ **b** $3(5 - y)$ **c** $7(9y - 2)$ **d** $x(x - 2)$
   e $y(3y + 8)$ **f** $2x(5 - 4x)$ **g** $2xy(3 - y)$ **h** $5x^2(3 - 2x)$

2 Copy and complete.

   a $8x + 10 = 2(4x + \square)$ **b** $6x - 9 = 3(2x - \square)$ **c** $21x + 14 = 7(3x + \square)$
   d $18x + 27 = 9(\square + 3)$ **e** $24y + 40 = \square(3y + \square)$ **f** $54 - 42x = \square(9 - \square)$

3 Kheirunnisa is asked to factorise some expressions for homework.

   She has made some mistakes.

   a $12y - 8 = 4(3y - 4)$

   b $30x + 15 = 5(6x + 3)$

   c $4x - 10 = 4(x - 2.5)$

   Copy out each question and correct her mistakes.

In questions **4** to **8**, factorise the expressions.

4 a $3x + 6$ **b** $7y - 14$ **c** $8x + 24$ **d** $9y + 12$
   e $3x - 15$ **f** $16y - 12$ **g** $30x + 14$ **h** $20 - 16x$
   i $42 + 21y$ **j** $60 - 45x$ **k** $38 - 19y$ **l** $144 - 108x$

Mathematics for Cambridge Secondary 1

5 a $x^2 + 4x$ b $y^2 - 5y$ c $x^2 - 10x$ d $2a^2 - 4a$
   e $x^2 + 9x$ f $8x - 3x^2$ g $10y - 2y^2$ h $16x - 12x^2$
   i $14x - 7x^2$ j $6x + 4x^2$ k $12y^2 - 9y$ l $72x^2 - 54x$

6 a $6\pi + 2$ b $24 - 16\pi$ c $\pi r + 2r$ d $\pi r^2 + 2\pi rh$

7 a $2x + 4y + 6z$ b $9x^2 + 6x + 3$ c $8a - 12b + 30c$ d $12x^3 + 6x^2 - 3x$

8 a $xy + x^2$ b $2x^2 + 6xy$ c $8x^2y - 4xy$ d $10x^3y^2 - 15x^3y$
   e $16xy^3 - 24x^3y$ f $24a^2b + 16ab$ g $45x^5y - 54x^2y$ h $8x^6y^4 - 6x^5y^2$

9 You can simplify algebraic fractions by factorising and then cancelling factors in the numerator and denominator.

For example:

$$\frac{6x + 18}{24}$$
$$= \frac{6 \times (x + 3)}{6 \times 4}$$
$$= \frac{x + 3}{4}$$

and

$$\frac{x^2 - 2x}{xy}$$
$$= \frac{x \times (x - 2)}{x \times y}$$
$$= \frac{x - 2}{y}$$

Simplify the following fractions.

a $\dfrac{4x + 2}{10}$ b $\dfrac{9x - 6}{12}$ c $\dfrac{6x + 12}{8}$ d $\dfrac{24 - 16y}{20}$

e $\dfrac{5a + 5b}{5c}$ f $\dfrac{8x - 4y}{16z}$ g $\dfrac{3x + 3}{x^2 + x}$ h $\dfrac{5x}{x^2 - 5x}$

2.4 Adding and subtracting algebraic fractions

The rules for adding and subtracting algebraic fractions are exactly the same as the rules for numerical fractions.

Case 1

The denominators are the same:

$$\frac{2}{7} + \frac{3}{7} = \frac{5}{7}$$

Similarly $\dfrac{2x}{7} + \dfrac{3x}{7} = \dfrac{5x}{7}$

Case 2

The denominators are different:

$$\frac{2}{3} + \frac{1}{4} = \frac{8}{12} + \frac{3}{12}$$
$$= \frac{11}{12}$$

> You must change the fractions so that they have the same common denominator, which in this case is 12.

Similarly $\dfrac{2x}{3} + \dfrac{x}{4} = \dfrac{8x}{12} + \dfrac{3x}{12}$
$$= \frac{11x}{12}$$

24

2 Expressions

Worked example

Write these as single fractions.

a $\dfrac{9}{y} - \dfrac{2}{y}$ **b** $\dfrac{5x}{4} - \dfrac{2x}{5}$ **c** $\dfrac{2}{3x} + \dfrac{1}{6x}$ **d** $\dfrac{3}{x} + \dfrac{2}{y}$

a $\dfrac{9}{y} - \dfrac{2}{y} = \dfrac{7}{y}$ the denominators are the same

b $\dfrac{5x}{4} - \dfrac{2x}{5} = \dfrac{25x}{20} - \dfrac{8x}{20}$ you must change the fractions so that they have the same common denominator, which in this case is 20

$\qquad = \dfrac{17x}{20}$

c $\dfrac{2}{3x} + \dfrac{1}{6x} = \dfrac{4}{6x} + \dfrac{1}{6x}$ you must change the fractions so that they have the same common denominator, which in this case is $6x$

$\qquad = \dfrac{5}{6x}$

d $\dfrac{3}{x} + \dfrac{2}{y} = \dfrac{3y}{xy} + \dfrac{2x}{xy}$ you must change the fractions so that they have the same common denominator, which in this case is xy

$\qquad = \dfrac{3y + 2x}{xy}$

Exercise 2.4

1 Write these as a single fraction.

     a $\dfrac{x}{3} + \dfrac{x}{3}$ **b** $\dfrac{2x}{5} + \dfrac{x}{5}$ **c** $\dfrac{x}{7} + \dfrac{3x}{7}$ **d** $\dfrac{x}{5} + \dfrac{1}{5}$ **e** $\dfrac{2x}{9} + \dfrac{2x}{9}$

     f $\dfrac{x}{8} - \dfrac{y}{8}$ **g** $\dfrac{5x}{7} - \dfrac{2x}{7}$ **h** $\dfrac{x}{7} - \dfrac{2}{7}$ **i** $\dfrac{5}{x} + \dfrac{2}{x}$ **j** $\dfrac{7}{y} - \dfrac{3}{y}$

     k $\dfrac{x}{y} + \dfrac{2}{y}$ **l** $\dfrac{a}{x} - \dfrac{b}{x}$ **m** $\dfrac{x}{4} + \dfrac{x}{4}$ **n** $\dfrac{5x}{6} - \dfrac{x}{6}$ **o** $\dfrac{8}{9xy} - \dfrac{2}{9xy}$

2 Leo writes:

$\dfrac{x}{8} + \dfrac{2x}{3} = \dfrac{3x}{11}$

Explain why he is wrong and work out the correct answer.

3 Write these as a single fraction.

     a $\dfrac{x}{2} + \dfrac{x}{4}$ **b** $\dfrac{x}{4} + \dfrac{x}{5}$ **c** $\dfrac{y}{3} - \dfrac{y}{4}$ **d** $\dfrac{x}{2} + \dfrac{x}{8}$ **e** $\dfrac{2x}{5} + \dfrac{x}{10}$

     f $\dfrac{5x}{8} - \dfrac{x}{2}$ **g** $\dfrac{2x}{5} - \dfrac{x}{4}$ **h** $\dfrac{5y}{7} - \dfrac{y}{2}$ **i** $\dfrac{3x}{8} + \dfrac{x}{3}$ **j** $\dfrac{2y}{3} + \dfrac{y}{6}$

4 Find the odd one out.

| $\dfrac{x}{6} + \dfrac{5x}{12}$ | $\dfrac{5x}{6} - \dfrac{x}{4}$ | $\dfrac{x}{4} + \dfrac{x}{3}$ | $\dfrac{x}{6} + \dfrac{3x}{4}$ | $\dfrac{2x}{3} - \dfrac{x}{12}$ |

In questions **5** to **7** write as single fractions.

5 a $\dfrac{x}{2} + \dfrac{x}{3} + \dfrac{x}{4}$ **b** $\dfrac{x}{3} + \dfrac{2x}{5} + \dfrac{x}{4}$ **c** $\dfrac{x}{2} + \dfrac{x}{10} + \dfrac{2x}{5}$ **d** $\dfrac{5x}{6} + \dfrac{x}{4} - \dfrac{2x}{3}$

25

6 a $\dfrac{1}{2x} + \dfrac{3}{x}$ b $\dfrac{2}{3y} + \dfrac{1}{y}$ c $\dfrac{3}{4x} + \dfrac{1}{2x}$ d $\dfrac{3}{xy} + \dfrac{2}{y}$

7 a $\dfrac{x}{4} + \dfrac{x+1}{2}$ b $\dfrac{x+2}{3} + \dfrac{x+7}{4}$ c $\dfrac{x-8}{4} + \dfrac{x+2}{5}$ d $\dfrac{x-4}{3} - \dfrac{x-5}{7}$

2.5 Substitution into expressions

This section revises the work that you did in Stage 8 on substituting positive and negative numbers into expressions.

Substitution into an expression means replacing the letters in an expression by the given numbers.

When a calculation involves more than one operation it is important that you do the operations in the correct order.

1 Work out the **Brackets** first.
2 Work out the **Indices** or powers next.
3 Work out the **Division** and **Multiplication** next.
4 Work out the **Addition** and **Subtraction** last.

> **BIDMAS**
> Brackets
> Indices or powers
> Division
> Multiplication
> Addition
> Subtraction

Worked example

If $x = 3$ and $y = -4$, find the values of:

a $5x + 3y$ b $x^2 + y^2$ c $2x^2 + y$ d $4xy + 3y^2$

a $5x + 3y = (5 \times 3) + (3 \times -4)$
$= 15 + -12$
$= 3$

b $x^2 + y^2 = 3^2 + (-4)^2$
$= 9 + 16$
$= 25$

c $2x^2 + y = 2 \times 3^2 + -4$
$= 18 - 4$
$= 14$

d $4xy + 3y^2 = (4 \times 3 \times -4) + 3 \times (-4)^2$
$= -48 + 48$
$= 0$

Exercise 2.5

1 If $x = 7$, find the value of:

a $5x - 3$ **b** $x^2 + 1$ **c** $8 - 2x$ **d** $3x^2$

e $2x^2 - 4$ **f** $\dfrac{x+3}{5}$ **g** $\dfrac{2x+1}{5x-5}$ **h** $\dfrac{x^2+2}{x-4}$

i x^3 **j** $(2x)^2 + 5x$ **k** $2x + \dfrac{x}{14}$ **l** $x^2 - \dfrac{21}{x}$

2 If $x = 3$ and $y = 5$, find the value of:

a $2x + 4y$ **b** $5x - 3y$ **c** $3xy$ **d** $x^2 + y^2$

e $2x^2 y$ **f** $3xy^2 - 4x$ **g** $\dfrac{x+y}{2}$ **h** $\dfrac{4xy}{5}$

i $\dfrac{x^3}{y+4}$ **j** $(2xy)^2$ **k** $(x-y)^2$ **l** $\left(\dfrac{3x}{2y}\right)^2$

3 If $a = 2$ and $b = -3$ and $c = -4$, find the value of:

a $2a + 3b$ **b** abc **c** $ab + c$ **d** $a^2 + b^2 + c^2$

e $2b^2 + a$ **f** $5ac$ **g** $b^3 + 1$ **h** $5c^2 - b$

i $(ab)^2 + c$ **j** $\dfrac{c}{a} + b$ **k** $(a + b - c)^2$ **l** $\left(\dfrac{5ab}{c+2}\right)^2$

4 Find the value of x that makes each of these expressions equal.

$x^2 + 13$ $2x^2 - 12$ $3x^2 + 4x - 17$

2.6 Expanding double brackets

$(x + 2)(x + 5)$ means $(x + 2) \times (x + 5)$

The area of a rectangle that has sides of length $x + 2$ and $x + 5$ can be found by splitting the rectangle up into four parts as shown below.

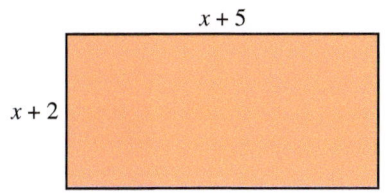

Area = $(x + 2)(x + 5)$

Area = $x^2 + 5x + 2x + 10$

So $(x + 2)(x + 5) = x^2 + 5x + 2x + 10$

$ = x^2 + 7x + 10$

You can expand the double brackets $(x + 2)(x + 5)$ without using a diagram.

To do this you must follow these steps:

Step 1:

Multiply each term in the second bracket by x.

$(x + 2)(x + 5)$ $x \times x = x^2$ and $x \times 5 = 5x$

Step 2:

Multiply each term in the second bracket by 2.

$(x + 2)(x + 5)$ $2 \times x = 2x$ and $2 \times 5 = 10$

Step 3:

Combine the results of Step 1 and Step 2.

So $(x + 2)(x + 5) = x^2 + 5x + 2x + 10$
$= x^2 + 7x + 10$

Worked example 1

Use a diagram to expand $(x + 6)(x + 11)$

	x	11
x	x^2	$11x$
6	$6x$	66

$(x + 6)(x + 11) = x^2 + 11x + 6x + 66$
$= x^2 + 17x + 66$

Worked example 2

Expand and simplify: **a** $(x + 3)(x + 4)$ **b** $(x - 7)(x + 2)$ **c** $(x - 6)(x - 7)$

a $(x + 3)(x + 4)$ $x \times x = x^2$ $x \times 4 = 4x$ $3 \times x = 3x$ $3 \times 4 = 12$
$= x^2 + 4x + 3x + 12$
$= x^2 + 7x + 12$

b $(x - 7)(x + 2)$ $x \times x = x^2$ $x \times 2 = 2x$ $-7 \times x = -7x$ $-7 \times 2 = -14$
$= x^2 + 2x - 7x - 14$
$= x^2 - 5x - 14$

c $(x - 6)(x - 7)$ $x \times x = x^2$ $x \times -7 = -7x$ $-6 \times x = -6x$ $-6 \times -7 = 42$
$= x^2 - 7x - 6x + 42$ be careful with double negatives: $-6 \times -7 = +42$
$= x^2 - 13x + 42$

Exercise 2.6

1 Use a diagram to help you expand these double brackets.

   a $(x + 1)(x + 3)$ **b** $(x + 8)(x + 2)$ **c** $(x + 5)(x + 5)$ **d** $(x + 6)(x + 4)$

   e $(x + 8)(x + 1)$ **f** $(x + 7)(x + 9)$ **g** $(x + 4)(x + 8)$ **h** $(x + 2)(x + 10)$

2 Expand these double brackets without using a diagram.

   a $(x + 2)(x + 4)$ **b** $(x + 5)(x + 1)$ **c** $(x + 4)(x + 4)$ **d** $(x + 7)(x + 3)$

   e $(x + 9)(x + 8)$ **f** $(x + 6)(x + 2)$ **g** $(x + 7)(x + 8)$ **h** $(x + 11)(x + 4)$

3 Expand and simplify:

   a $(x + 3)^2$ **b** $(x + 1)^2$ **c** $(x + 7)^2$ $(x + 3)^2$ means $(x + 3)(x + 3)$

   d $(x + 2)^2$ **e** $(x + 9)^2$ **f** $(x + 5)^2$

4 Correct the mistakes that Amelia has made in her expanding brackets homework.

   a $(x + 3)(x - 6) = x^2 - 6x + 3x - 3 = x^2 - 3x - 3$

   b $(x - 4)(x - 5) = x^2 - 5x - 4x - 20 = x^2 - 9x - 20$

   c $(x - 6)^2 = x^2 + 36$

In questions **5** to **8**, expand the brackets and simplify.

5 **a** $(x - 3)(x + 4)$ **b** $(x + 5)(x - 7)$ **c** $(x - 6)(x + 8)$ **d** $(x + 2)(x - 9)$

   e $(x + 3)(x - 7)$ **f** $(x - 6)(x + 7)$ **g** $(x + 9)(x - 1)$ **h** $(x + 4)(x - 5)$

   i $(x + 4)(x - 2)$ **j** $(x - 9)(x + 8)$ **k** $(x + 10)(x - 2)$ **l** $(x + 5)(x - 7)$

6 **a** $(x - 3)(x + 3)$ **b** $(x + 5)(x - 5)$ **c** $(x - 6)(x + 6)$ **d** $(x + 2)(x - 2)$

   e $(x + 4)(x - 4)$ **f** $(x - 1)(x + 1)$ **g** $(x + 15)(x - 15)$ **h** $(x + 9)(x - 9)$

7 **a** $(x - 5)(x - 4)$ **b** $(x - 8)(x - 7)$ **c** $(x - 6)(x - 8)$ **d** $(x - 2)(x - 3)$

   e $(x - 3)(x - 7)$ **f** $(x - 6)(x - 9)$ **g** $(x - 11)(x - 2)$ **h** $(x - 8)(x - 10)$

8 **a** $(x - 1)^2$ **b** $(x - 7)^2$ **c** $(x - 4)^2$ **d** $(x - 13)^2$

9

| A $x - 3$ | B $x + 1$ | C $x + 3$ | D $x + 2$ | E $x - 1$ |

Which two cards when multiplied together give:

   a $x^2 - x - 6$ **b** $x^2 + 4x + 3$ **c** $x^2 + x - 2$

   d $x^2 - 1$ **e** $x^2 - 2x - 3$ **f** $x^2 + 5x + 6$

   g $x^2 + 2x - 3$ **h** $x^2 + 3x + 2$ **i** $x^2 - 9$

10 Expand the brackets and simplify:

   a $(x + 4)(x + 5) + (x + 3)(x + 9)$ **b** $(x + 7)(x + 1) + (x + 2)(x + 5)$

   c $x(x + 2) + (x + 8)(x + 1)$ **d** $(x + 9)(x + 2) - (x + 4)(x + 6)$

   e $(x + 4)^2 + (x + 3)^2$ **f** $(x + 8)^2 - (x - 3)^2$

Mathematics for Cambridge Secondary 1

11 Solve these equations:

    a $x(x + 8) = (x + 3)(x + 4)$ **b** $(x + 4)(x + 2) = (x + 3)(x + 1)$

    c $(x + 2)^2 = (x - 3)^2$ **d** $(x + 4)(x + 2) = (x - 3)(x - 4)$

12 Show that $n^2 + (n + 1)^2 + (n + 2)^2 = 3n^2 + 6n + 5$

13 Expand and simplify:

    a $(2x + 3)(x + 5)$ **b** $(3x + 1)(x - 7)$ **c** $(x - 6)(2x + 4)$ **d** $(5x + 3)(x - 2)$

    e $(2x + 1)(3x - 5)$ **f** $(4x - 3)(2x + 1)$ **g** $(8x + 5)(3x - 4)$ **h** $(5x - 4)(x - 5)$

    i $(3x - 4)(x - 2)$ **j** $(7x - 1)(3x - 2)$ **k** $(5x + 7)^2$ **l** $(4x - 3)^2$

14 Write as a single fraction:

    a $\dfrac{2}{x} + \dfrac{5}{x + 7}$ **b** $\dfrac{2}{x + 3} + \dfrac{5}{x + 1}$ **c** $\dfrac{4}{x - 2} + \dfrac{3}{x + 5}$ **d** $\dfrac{5}{x - 3} - \dfrac{2}{x + 4}$

2.7 Constructing expressions

Worked example 1

Write down an expression for the area of the rectangle.

Area = base × height

$= (x + 8)(x - 3)$

$= x^2 - 3x + 8x - 24$

$= x^2 + 5x - 24$

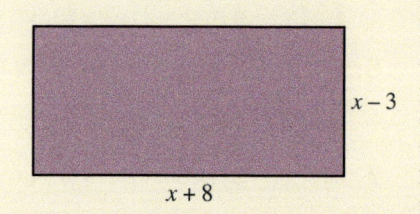

Worked example 2

Write down an expression for:

a the length *EF*

b the length *AF*

c the perimeter of the shape *ABCDEF*.

a $EF = AB - DC = (2x + 5) - (x + 7)$

        $= 2x + 5 - x - 7$

        $= x - 2$

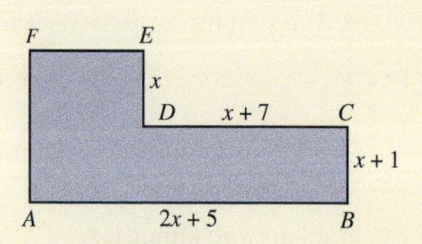

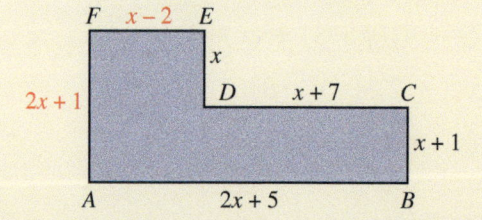

30

b $AF = BC + DE = x + 1 + x$
$= 2x + 1$

c Perimeter $= AB + BC + CD + DE + EF + FA$
$= (2x + 5) + (x + 1) + (x + 7) + x + (x - 2) + (2x + 1)$
$= 8x + 12$

Exercise 2.7

1 Write expressions for the perimeter of each of these shapes.

Simplify your answers.

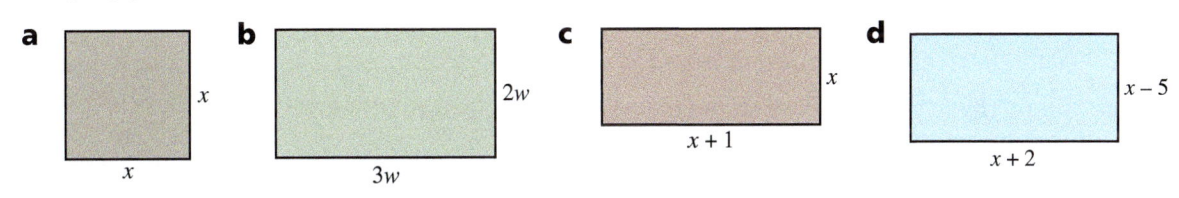

a square with sides x, x

b rectangle with sides $2w$, $3w$

c rectangle with sides x, $x + 1$

d rectangle with sides $x - 5$, $x + 2$

2 Write expressions for the area of each of the shapes in question **1**.

Simplify your answers.

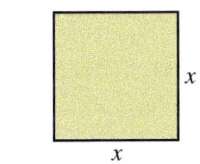

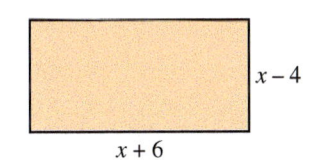

3 square with side x; rectangle with sides $x - 4$, $x + 6$

The square and the rectangle have the same area. Find the value of x.

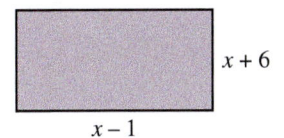

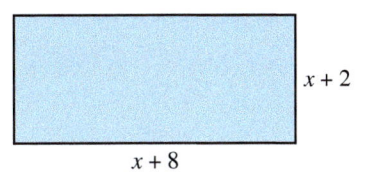

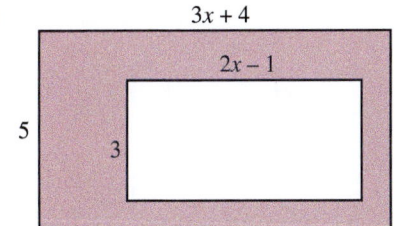

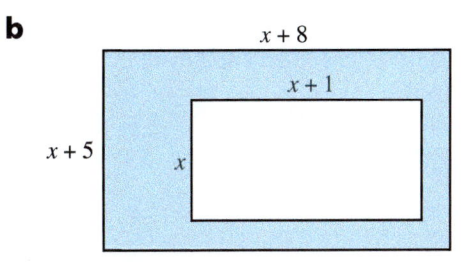

4 rectangle with sides $x + 6$, $x - 1$; rectangle with sides $x + 2$, $x + 8$

Write an expression for the **total area** of these two rectangles.

Simplify your expression.

5 Write an expression for the shaded area in each of these shapes.

Simplify your expressions.

a outer rectangle $3x + 4$ by 5, inner rectangle $2x - 1$ by 3

b outer rectangle $x + 8$ by $x + 5$, inner rectangle $x + 1$ by x

31

Mathematics for Cambridge Secondary 1

6 Write down an expression for:

  a the length of DC

  b the length of DE

  c the perimeter of the shape ABCDEF

  d the area of the shape ABCDEF.

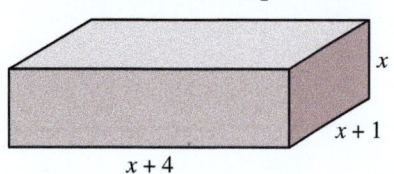

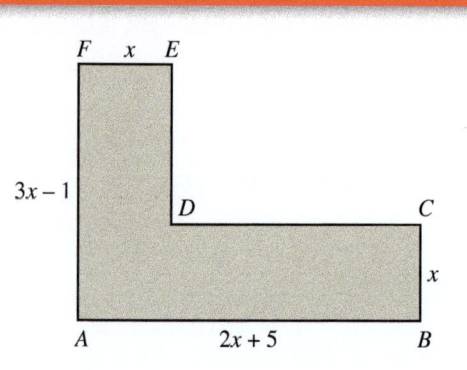

7 Write down and simplify an expression for the surface area of the cuboid.

8 a Deepa cycles x km at an average speed of 20 km/h.

  Write down an expression, in terms of x, for the time it takes Deepa to complete this journey.

  b Deepa then cycles another $2x$ km at an average speed of 30 km/h.

  Write down an expression, in terms of x, for the total time taken, in hours, by Deepa to complete the two stages of her journey and show that it simplifies to $\frac{7x}{60}$.

Review

1 Simplify:

  a $7x \times x$ **b** $5a \times 2b \times 3a$ **c** $2xy \times 3x \times 5xy \times x$

2 Simplify:

  a $\frac{9x}{12}$ **b** $\frac{3xy}{x^2}$ **c** $\frac{6x}{10y}$ **d** $\frac{4a^2b}{6ab^2}$

3 Solve the equation: $2^x = 32$

4 Simplify:

  a $x^4 \times x^5$ **b** $2y^3 \times 5y$ **c** $8a^4 \times 3a^7$ **d** $y^4 \times y \times y^8$

  e $x^9 \div x^3$ **f** $\frac{35x^8}{40x^2}$ **g** $\frac{x^{15}}{x^4 \times x^2}$ **h** $(x^5)^4$

5 Factorise:

  a $20x + 16$ **b** $y^2 - 4y$ **c** $3x^2 - 6x$ **d** $6x + 4y + 8z$

6 Write as a single fraction:

  a $\frac{x}{5} + \frac{x}{5}$ **b** $\frac{x}{6} + \frac{x}{3}$ **c** $\frac{7x}{10} - \frac{x}{4}$ **d** $\frac{5}{3x} + \frac{1}{3x}$

7 If $a = 5$ and $b = -2$, find the value of:

  a $a^2 + b^2$ **b** $5a - 6b$ **c** $\frac{4a}{b}$ **d** $2a^2 + b$

8 Expand and simplify:

  a $(x+7)(x+2)$ **b** $(x-6)(x+3)$ **c** $(x+1)(x-2)$

  d $(x-5)(x-9)$ **e** $(x+4)^2$ **f** $(x-2)^2$

Examination-style questions

1. A pen costs 75 cents.

   Write an expression for the total cost, in cents, of x pens. [1]

2. Simplify:

   $5x + 7y + 3x - 4y$ [1]

3. Find the value of:

   $7x - 3y$ when $x = 2$ and $y = -4$ [2]

4. A book costs $7 and a magazine costs $3.

   Write an expression for the total cost, in dollars, of b books and m magazines. [2]

5. Expand: $4(x - 6)$ [1]

6. Expand and simplify:

   $7(3x - 2) - 2(4x - 3)$ [2]

7. The total mass of x sweets is 250 g.

   Write down an expression for the mass of one sweet. [1]

8. Expand and simplify: $(x + 8)(x - 5)$ [2]

9. Simplify:

   a $m^5 \times m^3$ [1]

   b $p^8 \div p^2$ [1]

   c $2x^3y^4 \times 3x^2y^3$ [2]

10. Simplify: $\dfrac{x}{3} + \dfrac{2x}{9}$ [2]

11. Factorise

    a $8x + 24$ [1]

    b $x^2 - 7x$ [1]

3 Shapes and geometric reasoning 1

> **Learning outcomes**
>
> - Prove that the sum of the exterior angles of any polygon is 360°.
> - Prove and use the formula for the sum of the interior angles of any polygon.
> - Calculate the interior angle or exterior angle of any regular polygon.
> - Solve problems using properties of angles, of parallel and intersecting lines, and of triangles, other polygons and circles, justifying inferences and explaining reasoning with diagrams and text.

Mathematics in design

This picture shows part of a wall at the Reales Alcazares in Seville, Spain.

This old royal palace is now a World Heritage Site.

It contains some fine examples of how Islamic builders and architects used mathematical shapes to inspire tile designs.

Other Moorish palaces, such as the Alhambra in Granada, Spain, also have many beautiful examples of Islamic art.

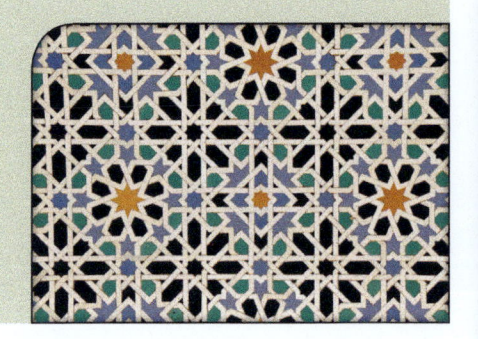

3.1 Polygons

Interior angles

A **polygon** is a two-dimensional shape with straight sides.

Polygons can be **convex** or **concave**.

Convex polygons

Concave polygons

3 Shapes and geometric reasoning 1

Concave polygons have one **vertex** which points into the shape.

Here are some examples of polygons with their names.

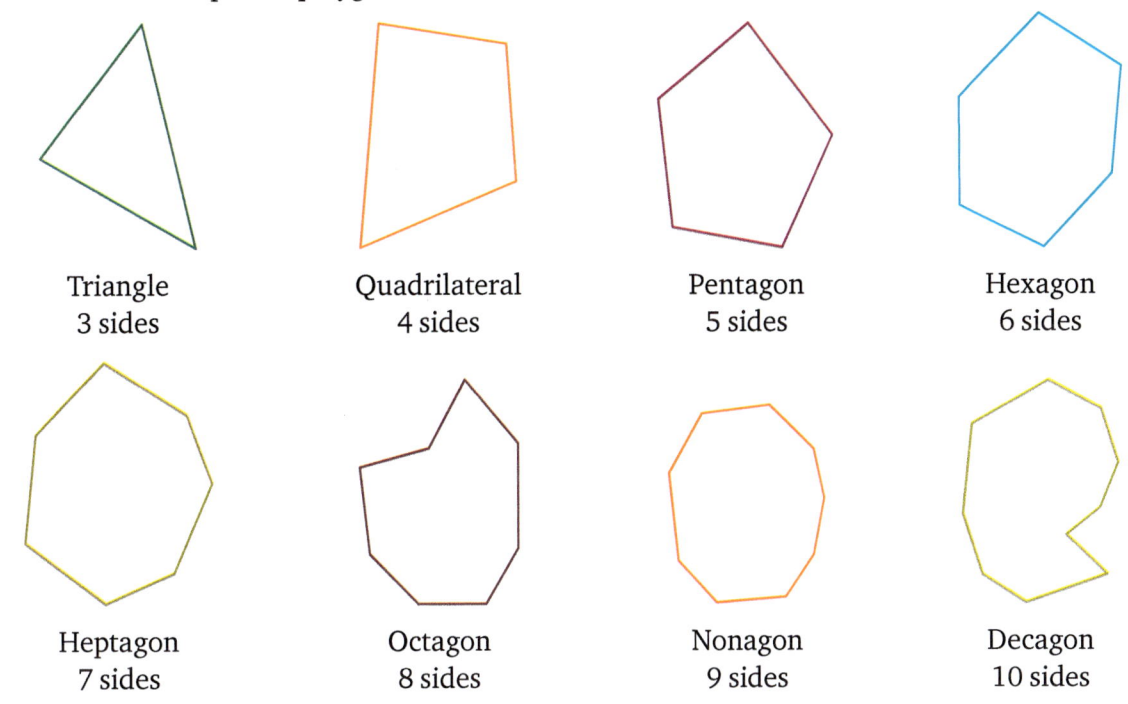

| Triangle | Quadrilateral | Pentagon | Hexagon |
| 3 sides | 4 sides | 5 sides | 6 sides |

| Heptagon | Octagon | Nonagon | Decagon |
| 7 sides | 8 sides | 9 sides | 10 sides |

The **interior angles** of a polygon are the angles inside the shape.

The interior angles of this hexagon are $a°$, $b°$, $c°$, $d°$, $e°$ and $f°$.

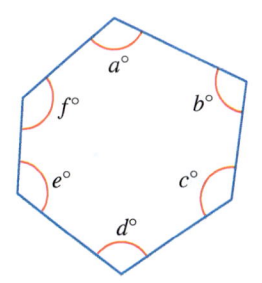

By drawing **diagonals** the hexagon can be split into four triangles.

The interior angles of each triangle add up to 180°.

The interior angles of the hexagon add up to $4 \times 180° = 720°$

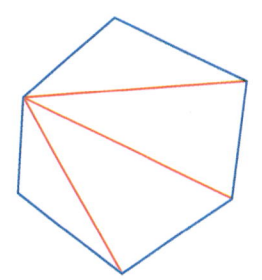

Worked example 1

By drawing diagonals find the sum of the interior angles of this pentagon.

35

Mathematics for Cambridge Secondary 1

There are several ways of drawing diagonals to split the pentagon.

In each case there are three triangles.

$3 \times 180° = 540°$

The sum of the interior angles of the pentagon is 540°.

Worked example 2

Use the result from Worked example 1 to find the missing angle in this pentagon.

The four given angles add up to $48° + 110° + 90° + 39° = 287°$

The sum of the interior angles of a pentagon is 540°.

The missing angle is $540° - 287° = 253°$

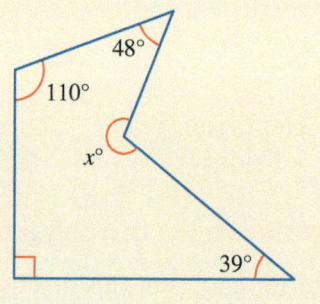

> Check that the answer is sensible. This angle is bigger than 180° and less than 270°. The answer 253° is realistic.

Exterior angles

The **exterior angles** of a convex polygon are formed by extending the side at each vertex.

The exterior angles of a pentagon are shown here.

At each vertex the interior angle and the exterior angle lie along a straight line.

At each vertex: interior angle + exterior angle = 180°

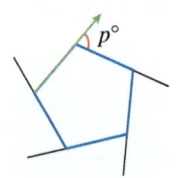

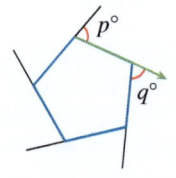

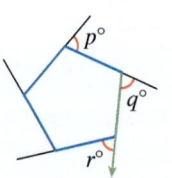

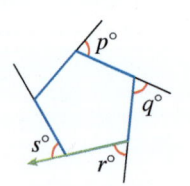

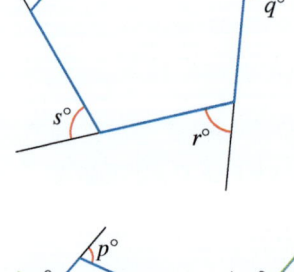

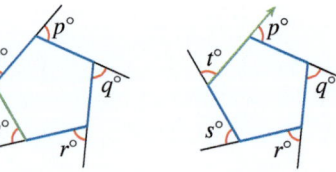

Imagine an arrow along one side of the pentagon.

It rotates through angle $p°$ to lie on the next side.

36

3 Shapes and geometric reasoning 1

Then it rotates through angle $q°$, then $r°$, then $s°$ and then $t°$.

It is now back to where it started. It has turned through 360°.

This shows that **the exterior angles of a convex polygon add up to 360°**.

Worked example 3

Find the size of the angles marked $x°$ in this diagram.

The sum of the exterior angles is 360°.

$$55 + 65 + x + x + 75 + x = 360$$
$$195 + 3x = 360$$
$$3x = 165$$
$$x = 55$$

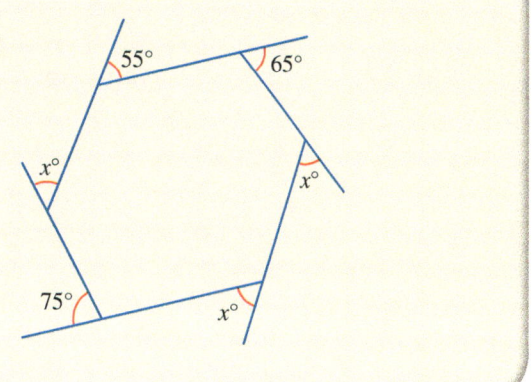

Exercise 3.1

1 a Draw an octagon.

  b Draw diagonals to split the octagon into triangles.

  c Write down how many triangles there are.

  d Find the sum of the interior angles of an octagon.

2 Copy this table.

Number of sides	Name of polygon	Number of triangles	Sum of interior angles
3	Triangle	1	180°
4			
5			
6	Hexagon	4	720°
7			
8			
9			
10			
12	Dodecagon		

  a Fill in the names of the polygons. Some are done for you.

  b Draw diagrams of each polygon and split them into triangles. Write the number of triangles into the table.

  c Work out the sum of the interior angles and fill in the final column.

3 Use the answers from question **2** to find the missing angles.

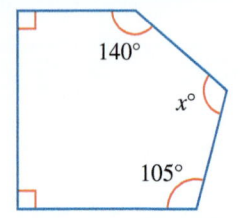

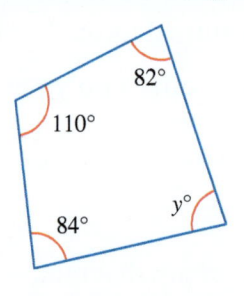

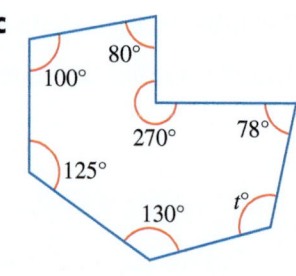

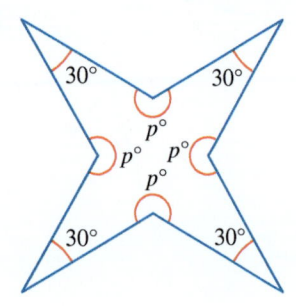

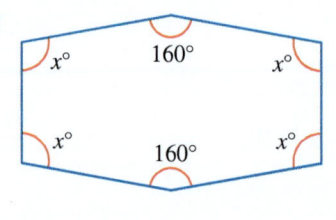

a [pentagon with angles 140°, 105°, two right angles, and $x°$]

b [quadrilateral with angles 82°, 110°, 84°, and $y°$]

c [polygon with angles 80°, 100°, 270°, 78°, $t°$, 130°, 125°]

d [star shape with four 30° points and four $p°$ angles in center]

e [hexagon with angles 160°, 160°, and four $x°$ angles]

f [regular pentagon with five $a°$ angles]

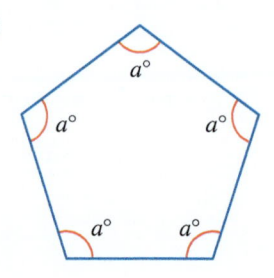

4 Find the missing angles.

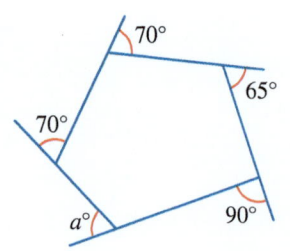

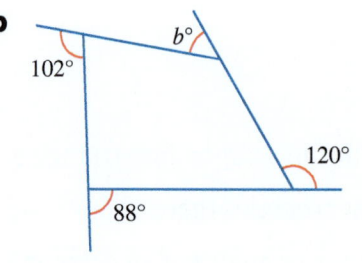

a [pentagon with angles 70°, 65°, 90°, $a°$, 70°]

b [quadrilateral with angles 102°, $b°$, 120°, 88°]

c [hexagon with six $c°$ angles]

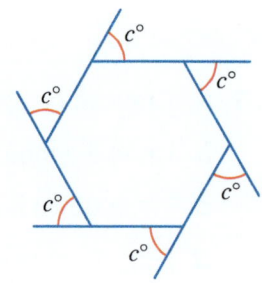

5 Here are four statements about concave polygons.

Decide which are true and which are false.

Draw diagrams to explain your answers.

You may need to try some sketches to see if these can be done.

a Concave polygons have at least one interior angle greater than 180°.

b Concave polygons cannot be divided into triangles.

c It is impossible to draw a concave triangle.

d It is impossible to draw a concave quadrilateral.

6 Here is a convex pentagon.

Work out the value of x.

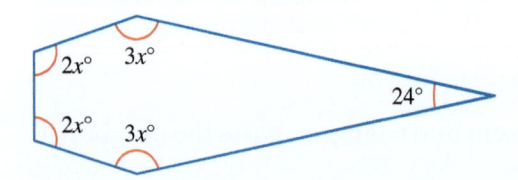

[pentagon with angles $2x°$, $3x°$, 24°, $3x°$, $2x°$]

Make an equation using the individual angles and the sum of the angles in a pentagon.

3.2 More polygons

Formula for the sum of interior angles

In the table in Exercise 3.1 the number of triangles is always two less than the number of sides.

If a polygon has n sides, it can be divided into $n - 2$ triangles.

The sum of the interior angles is $(n - 2) \times 180°$ or $180(n - 2)°$

This gives a general formula:

> the sum of the interior angles of a polygon with n sides is $180(n - 2)°$

The sum of exterior angles

A polygon has n sides.

At each vertex the interior angle + exterior angle = $180°$

There are x vertices so the total of all interior angles + exterior angles = $180° \times n$ or $180n°$

But the sum of the interior angles is $180(n - 2)°$

So the sum of the exterior angles = $180n - 180(n - 2)$

$$= 180n - 180n + 360$$
$$= 360$$

This proves that the sum of the exterior angles is $360°$.

> ### Worked example 1
>
> Find the sum of the interior angles of a polygon with 20 sides.
>
> Substitute $n = 20$ into the formula $180(n - 2)°$.
>
> $180 \times (20 - 2) = 180 \times 18$
> $ = 3240$
>
> The sum of the interior angles is $3240°$.

Regular polygons

All of the sides of a **regular polygon** are equal.

All of the angles of a regular polygon are equal.

Here are some regular polygons.

| Equilateral triangle | Square | Regular pentagon | Regular hexagon | Regular octagon |

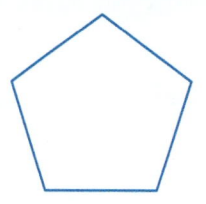

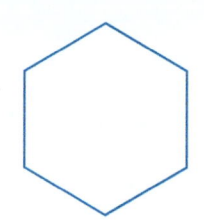

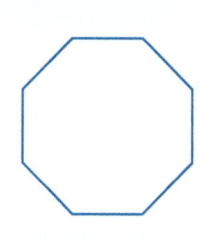

Exterior angles

The exterior angles of a regular polygon add up to 360°.

All the exterior angles are equal.

Each exterior angle is equal to 360° divided by the number of sides.

$$\text{Each exterior angle of a regular polygon with } n \text{ sides} = \frac{360°}{n}$$

Interior angles

There are two methods for finding the interior angle of a regular polygon.

1 Using the formula for the sum of the interior angles.

   The sum of the interior angles of a polygon with n sides is $180(n - 2)°$

   All of the interior angles are equal.

   $$\text{Each interior angle of a regular polygon with } n \text{ sides} = \frac{180(n - 2)°}{n}$$

2 Using the exterior angles.

   Each exterior angle $= \frac{360°}{n}$

   The exterior angle and interior angle at each vertex add up to 180°.

   $$\text{Each interior angle of a regular polygon with } n \text{ sides} = 180° - \frac{360°}{n}$$

Worked example 2

A regular polygon has 15 sides.

a Find the size of each interior angle.

b Work out the size of each exterior angle.

c Check that they add up to 180°.

a Substitute $n = 15$ into $\frac{180(n - 2)}{n}$

$= \frac{180 \times (15 - 2)}{15}$

$= \frac{180 \times 13}{15}$

$= \frac{2340}{15}$

$= 156$

Each interior angle is 156°.

b Substitute $n = 15$ into $\frac{360}{n}$
$= \frac{360}{15}$
$= 24$
Each exterior angle is 24°.

c Check that $156° + 24° = 180°$ ✓

Exercise 3.2

1 These are regular polygons. Find the size of the angles marked with letters.

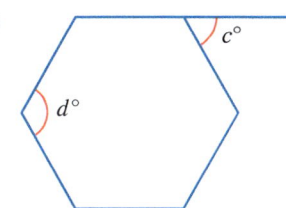

a (pentagon with angles $a°$ and $b°$)   b (hexagon with angles $c°$ and $d°$)   c (octagon with angles $f°$ and $e°$)

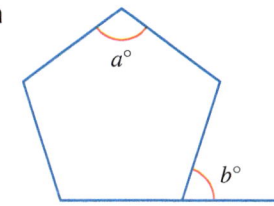

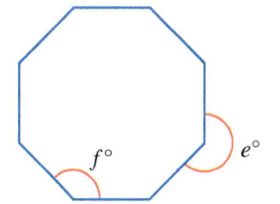

2 Find the size of **i** each interior angle and **ii** each exterior angle of regular polygons with:

  a 9 sides
  b 12 sides
  c 20 sides
  d 7 sides.

3 Copy and complete this table. Some of the angles are already shown.

Number of sides	3	4	5	6	7	8	9	10	12	15	20
Exterior angle	120°									24°	
Interior angle										156°	

4 Use your table in question **3** to help answer this question.

  a What happens to the exterior angle as the number of sides increases?

  b What happens to the interior angle as the number of sides increases?

  c Work out the interior angle of a regular polygon with 100 sides to test your theory.

5 These diagrams each show a vertex of a regular polygon.

  Work out how many sides each one has.

  a (vertex showing 160°)   b (vertex showing 165°)

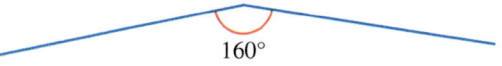

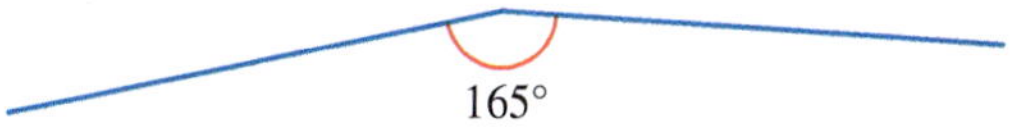

> You will find it helpful to work out the exterior angle first.

6 There are two formulae for working out the interior angle, x, of a regular polygon.
One of them is $x = \dfrac{180(n-2)°}{n}$

Multiply out the brackets and simplify this equation.

Show that it is equivalent to $x = 180° - \dfrac{360°}{n}$

> When you have multiplied out the brackets you need to write it as two separate fractions.

3.3 Angles and lines

Review of parallel lines

When a straight line crosses two parallel lines there are two intersections or cross-over points.

Corresponding angles are in the same position at each intersection.

These are some examples of pairs of corresponding angles.

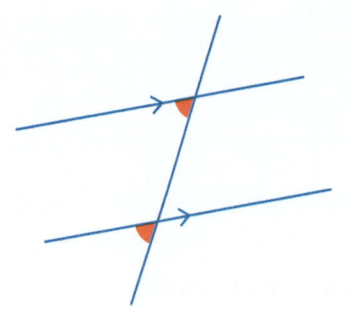

Corresponding angles are equal.

Alternate angles are in the opposite position at each intersection.

Here are some examples of pairs of alternate angles.

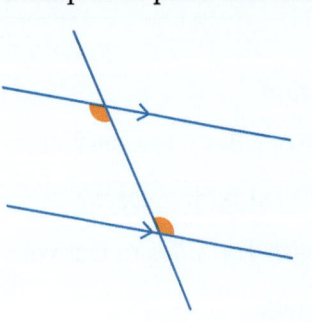

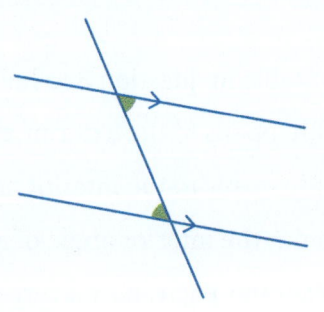

Alternate angles are equal.

You may be asked to state the facts that you are using to answer a question.

Worked example 1

Find the size of the angles marked with letters in this diagram.

Give reasons for your answers.

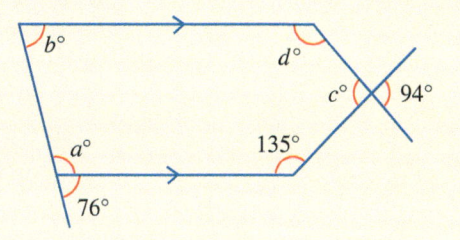

$a = 180 - 76$
$ = 104$ angles on a straight line add up to 180°

$b = 76$ corresponding angles

$c = 94$ vertically opposite angles

> Remember that when two straight lines cross there are two pairs of opposite angles at the vertex, known as vertically opposite angles. Each pair of vertically opposite angles are equal.

$d = 540 - (104 + 76 + 135 + 94)$
$ = 540 - 409$
$ = 131$ interior angles of a pentagon add up to 540°

Worked example 2

A is the centre of a circle.

B and C are points on the circumference.

ACD is a straight line.

Angle $BAC = 80°$

a Work out angle ABC.

b Work out angle BCD.

Give reasons for your answers.

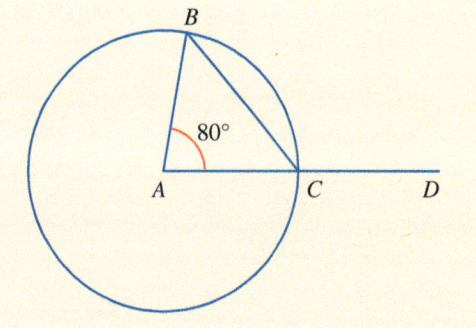

a $AB = AC$ they are each a radius of the circle

Triangle ABC is an isosceles triangle.

Angle $ABC = (180° - 80°) \div 2$ angles in an isosceles triangle add up to 180°
$ = 50°$

b   Angle $ACB = 50°$ triangle ABC is isosceles

Angle $BCD = 180° - 50°$ angles on a straight line add up to 180°
$ = 130°$

Mathematics for Cambridge Secondary 1

Exercise 3.3

1 Find the angles marked with letters.

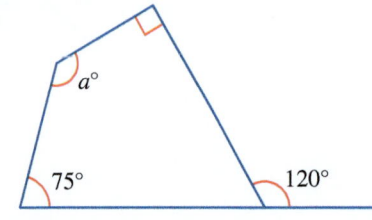

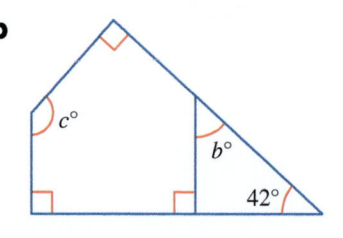

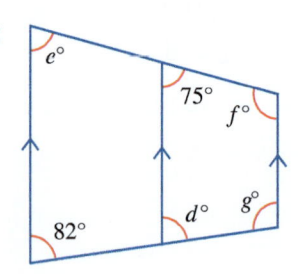

a (angles: $a°$, $75°$, $120°$, right angle)

b (angles: $c°$, $b°$, $42°$, right angles)

c (angles: $e°$, $75°$, $f°$, $d°$, $g°$, $82°$)

2 ABC is a triangle.

PQ is a line parallel to AB.

Angle ABC = 40°

Angle CPQ = 80°

Find the angles marked $w°$, $x°$, $y°$ and $z°$.

Give reasons for your answers.

3 ABCDE is a pentagon.

AM is a line of symmetry.

Angle ABC = 85°

Angle CDE = 130°

Find:

  a angle AED

  b angle BCD

  c angle BAE

  d angle AMC.

4 ABCDEF is a regular hexagon.

  a Work out angle ABC.

A diagonal has been drawn between A and C.

  b What type of triangle is ABC?

  c Work out angle BAC.

  d Work out angle CAF.

Another diagonal has been drawn between C and F.

  e Work out angle AFC.

  f What type of triangle is ACF?

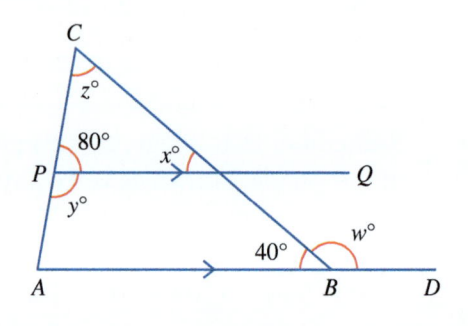

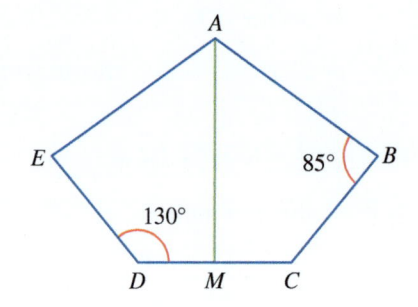

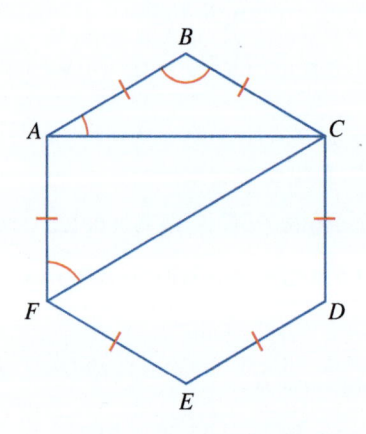

3 Shapes and geometric reasoning 1

5 The diagram shows a circle with centre O.

ABC is a triangle with A, B and C on the circumference.

Angle $AOB = 128°$

Angle $AOC = 110°$

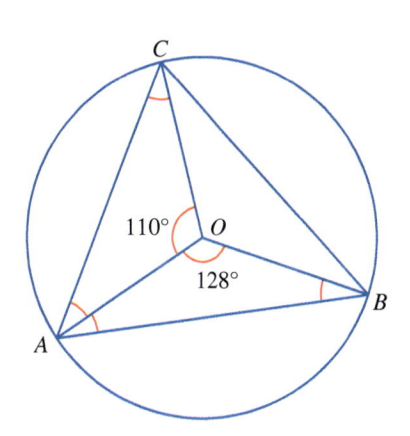

a Work out angle BOC. Give a reason for your answer.

b Work out angle OCA. Give a reason for your answer.

c Work out angle OCB. Give a reason for your answer.

d Work out angle ACB.

e Show that angle AOB is twice the size of angle ACB.

> O is the centre of the circle so $OA = OB = OC$. Use isosceles triangles for this question.

> This is a particular case of a general result. The angle at the circumference is twice the angle at the centre.

Review

1 a Draw a pentagon.

b Draw diagonals to divide the pentagon into triangles.

c Work out the sum of the interior angles of a pentagon.

2 a Write down the mathematical name for this shape.

b Work out the sum of the interior angles of the shape.

c Use the answer to **b** to work out the size of angle $x°$.

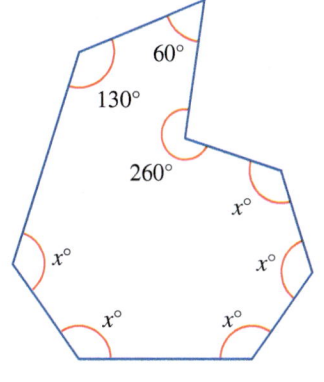

3 Find the size of the angles marked with letters.

a

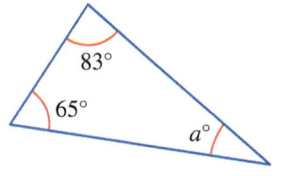

b

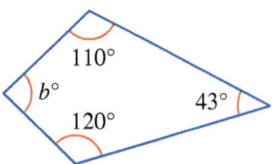

4 a Copy and complete this sentence.

The exterior angles of a polygon add up to _____.

b Find the size of angle $a°$ in the diagram.

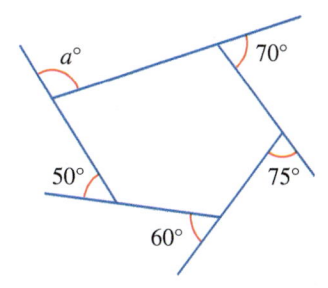

45

5 Here are four polygons.

A B C D

Copy these statements and fill in the missing letters.

a ____ and ____ are pentagons.

b ____ and ____ are concave polygons.

c ____ and ____ are regular polygons.

d ____ and ____ are hexagons.

6 Write down the mathematical name for a regular quadrilateral.

7 Work out the size of each exterior angle of a polygon with:

a 6 sides   **b** 9 sides   **c** 12 sides   **d** 24 sides.

8 Work out the size of each interior angle of the polygons in question 7.

9 The diagram shows part of a regular polygon.

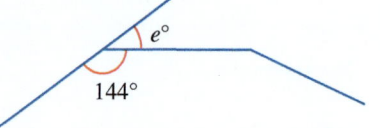

a Work out the size of angle $e°$.

b How many sides does the complete polygon have?

10 In the diagram AB is parallel to PQ.

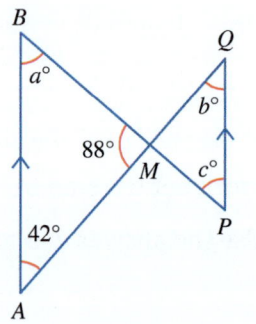

Work out the size of:

a angle $a°$

b angle $b°$

c angle $c°$.

11 A, B and C are points on the circumference of a circle.

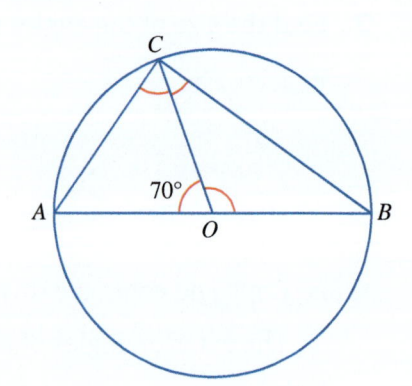

O is the centre of the circle.

Angle AOC = 70°

a What type of triangle is AOC?

b Work out the size of angle ACO.

c Write down the size of angle COB.

d Work out the size of angle BCO.

e Work out the size of angle ACB.

12 The diagram shows a circle like the one in question **11**.

In this case angle $AOC = 40°$

Repeat questions **a** to **e** from question **11**.

What do you notice about angle ACB in each case?

> These are examples of a general rule.
>
> If triangle ABC has vertices on the circumference of a circle and AB is a diameter then angle ACB is a right angle.

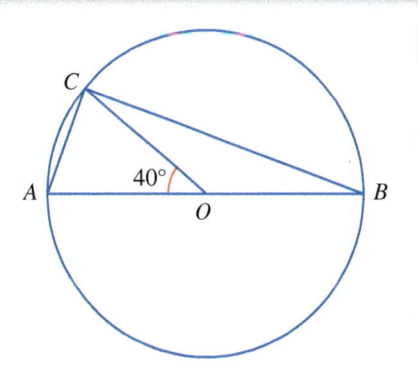

Examination-style questions

1   **a** Write down the size of angle p in degrees.

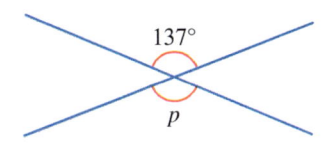

[1]

   b Work out the size of angle q in degrees

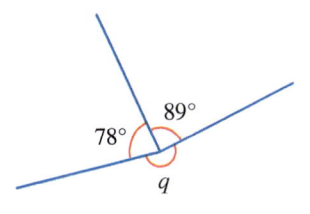

[2]

2   **a** What type of angle is x?

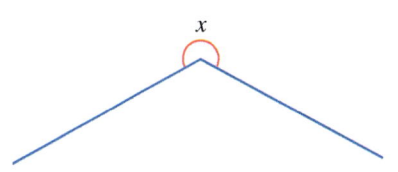

[1]

   b Copy and complete.

   $t =$ _____ because _____

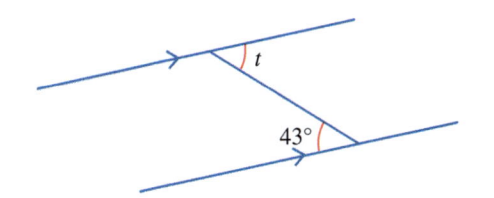

[2]

Mathematics for Cambridge Secondary 1

3

a What is the name for this shape? [1]

b Work out the sum of the interior angles. [2]

4 a Copy and complete.

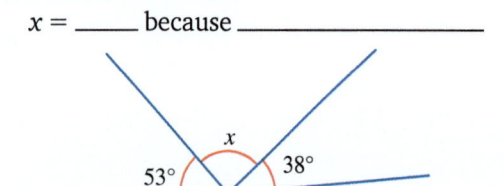

$x =$ _____ because _____

(53°, x, 38°)

b Copy and complete. [2]

$y =$ _____ because _____

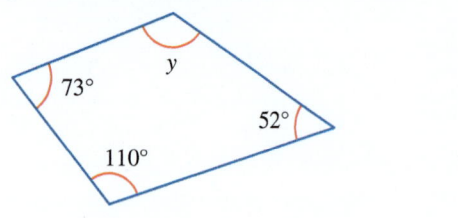

(73°, y, 52°, 110°)

[2]

5 The diagram shows part of a regular polygon.

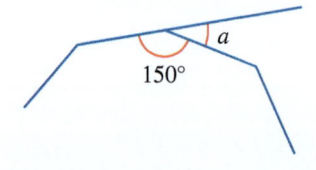

(150°, a)

a Write down the size of angle a. [1]

b Work out the number of sides in the polygon. [2]

6 In the diagram ACE is a triangle.

BD is parallel to AE.

Angle BAE = 66° and angle AED = 72°

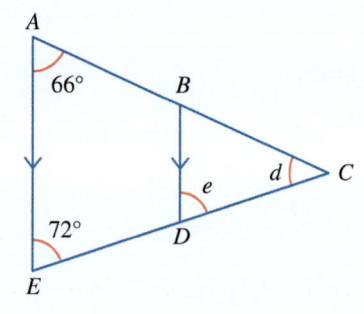

Work out, giving reasons for your answers:

a angle d [2]

b angle e. [2]

4 Fractions

> **Learning outcomes**
> - Write a fraction in its simplest form.
> - Add and subtract fractions.
> - Multiply and divide fractions.
> - Cancelling common factors before multiplying or dividing.

Who used fractions before we did?

The ancient Egyptians used fractions almost 4000 years ago. They used pictures to represent fractions and these were difficult to work with.

The Romans used fractions based on weight and twelfths. Their fractions used words and were also difficult to work with.

The Babylonians used fractions. Their system was based on 60. They had no zero and some numbers could be interpreted in different ways.

About 1500 years ago a system of numbers was developed in India. The Indian numerals spread to Arabia through trading and became the numbers we use today.

Indian fractions were written in a very similar way to those we use now. One number was written over another, but they had no horizontal line in the middle. The Arabs added the horizontal line.

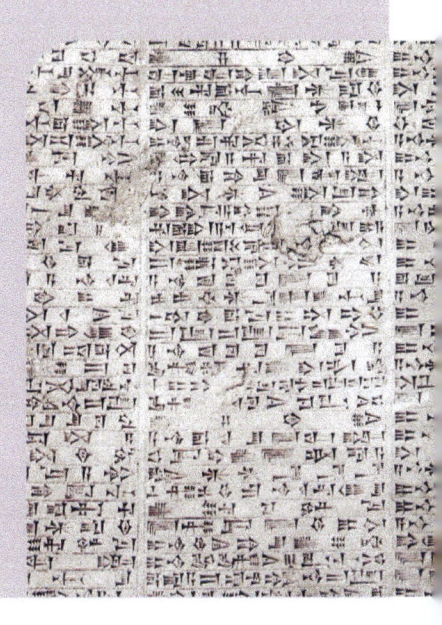

4.1 Addition and subtraction

Equivalent fractions

The fractions $\frac{1}{4}, \frac{2}{8}, \frac{3}{12}, \frac{4}{16}$ are equivalent to each other. They are called **equivalent fractions**.

You can multiply the **numerator** and **denominator** by the same number to obtain an equivalent fraction. The numerator is the top number and the denominator is the bottom number.

You can also divide the numerator and denominator by the same number to obtain an equivalent fraction. You need to find common factors of the numerator and denominator.

This is called **simplifying**. Sometimes it is called **cancelling**.

When a fraction cannot be simplified any further it is said to be in its **simplest form**.

Mathematics for Cambridge Secondary 1

Worked example 1

Simplify: $\dfrac{28}{32}$

Give your answer in its simplest form.

$$\dfrac{28}{32} \xrightarrow{\div 2} \dfrac{14}{16} \xrightarrow{\div 2} \dfrac{7}{8}$$

This could be done in one step by dividing the numerator and denominator by 4.

$$\dfrac{28}{32} = \dfrac{7}{8}$$

This cannot be simplified further. It is in its simplest form.

Addition of fractions

Fractions need to have the same denominator in order to be added together.

The method is:

    Find the lowest common multiple of the denominators.

    Write each fraction with this denominator.

    Add the two fractions.

    Simplify the answer if possible.

When adding **mixed numbers** one way is to add the integer and fraction parts separately. It can also be done by changing the mixed number to improper fractions.

> Remember that a mixed number is a number that has a whole number part and a fraction part.

Worked example 2

Work out: $\dfrac{5}{6} + \dfrac{7}{8}$

The lowest common multiple of 6 and 8 is 24.

$$\dfrac{5}{6} \xrightarrow{\times 4} \dfrac{20}{24} \qquad \dfrac{7}{8} \xrightarrow{\times 3} \dfrac{21}{24}$$

$$\dfrac{5}{6} + \dfrac{7}{8} = \dfrac{20}{24} + \dfrac{21}{24}$$

$$= \dfrac{20 + 21}{24}$$

$$= \dfrac{41}{24}$$

$$= 1\dfrac{17}{24}$$

4 Fractions

Worked example 3

Work out: $1\frac{5}{6} + 2\frac{7}{8}$

$1\frac{5}{6} + 2\frac{7}{8} = 1 + 2 + \frac{5}{6} + \frac{7}{8}$ — separate the integer part from the fraction part

$= 3 + \frac{20}{24} + \frac{21}{24}$ — add the integer parts, write the fraction parts with the same denominator

$= 3 + \frac{41}{24}$ — add the fraction parts together

$= 3 + 1 + \frac{17}{24}$ — write any improper fraction as a proper fraction

$= 4\frac{17}{24}$

Subtraction of fractions

This is done in almost exactly the same way as for addition of fractions.

The method is:

- Find the lowest common multiple of the denominators.
- Write each fraction with this denominator.
- Subtract the two numerators.
- Simplify the answer if possible.

When subtracting mixed numbers one way is to write them as **improper fractions**.

> Remember that a fraction is an improper fraction if the numerator is larger than the denominator.

Worked example 4

Work out: $1\frac{1}{2} - \frac{4}{5}$

$1\frac{1}{2} - \frac{4}{5} = \frac{3}{2} - \frac{4}{5}$ — write mixed numbers as improper fractions

$\frac{3}{2} \stackrel{\times 5}{=} \frac{15}{10} \qquad \frac{4}{5} \stackrel{\times 2}{=} \frac{8}{10}$ — the lowest common multiple of 2 and 5 is 10

$\frac{3}{2} - \frac{4}{5} = \frac{15}{10} - \frac{8}{10}$

$= \frac{15 - 8}{10}$

$= \frac{7}{10}$ — subtract the numerators

Mathematics for Cambridge Secondary 1

Worked example 5

Work out: $3\frac{1}{4} - 1\frac{2}{5}$

$3\frac{1}{4} - 1\frac{2}{5} = \frac{13}{4} - \frac{7}{5}$ write mixed numbers as improper fractions

$\frac{13}{4} \overset{\times 5}{=} \frac{65}{20}$ $\frac{7}{5} \overset{\times 4}{=} \frac{28}{20}$ the lowest common multiple of 4 and 5 is 20

$\frac{13}{4} - \frac{7}{5} = \frac{65}{20} - \frac{28}{20}$

$= \frac{65 - 28}{20}$

$= \frac{37}{20}$ subtract the numerators

$= 1\frac{17}{20}$ finish by writing any improper fractions as mixed numbers

Exercise 4.1

1 Write these fractions in their simplest form.

    a $\frac{4}{24}$ **b** $\frac{16}{52}$ **c** $\frac{12}{99}$ **d** $\frac{18}{32}$

2 Work these out. Give your answers in their simplest form.

    a $\frac{3}{8} + \frac{3}{4}$ **b** $\frac{2}{5} + \frac{3}{7}$ **c** $\frac{3}{8} + \frac{1}{6}$ **d** $1\frac{3}{4} + 1\frac{2}{3}$

3 Work these out. Give your answers in their simplest form.

    a $\frac{7}{8} - \frac{5}{16}$ **b** $\frac{2}{3} - \frac{4}{7}$ **c** $2\frac{3}{8} - \frac{2}{3}$ **d** $2\frac{4}{9} - \frac{7}{12}$

4 Work these out. Give your answers in their simplest form.

    a $\frac{11}{18} + \frac{7}{24}$ **b** $\frac{13}{36} + \frac{11}{20}$ **c** $\frac{7}{24} - \frac{5}{30}$ **d** $\frac{8}{15} - \frac{7}{27}$

5 Work these out. Give your answers in their simplest form.

    a $3\frac{5}{12} + 4\frac{11}{18}$ **b** $7\frac{17}{24} + 3\frac{15}{32}$ **c** $5\frac{7}{18} - 2\frac{11}{12}$ **d** $4\frac{11}{24} - 1\frac{13}{18}$

6 Claude cycles from Carnac to Auray. He calls in at Plouharnel and Ploemel on the way.

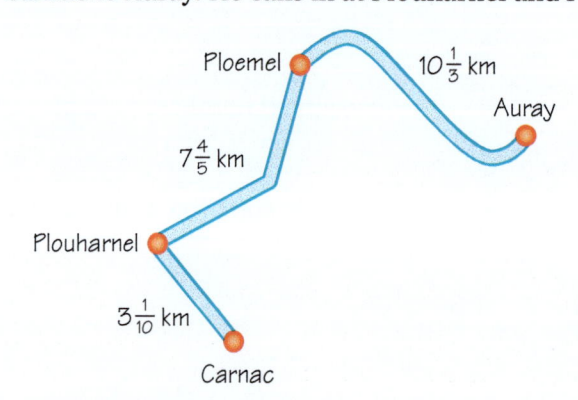

Ploemel to Auray: $10\frac{1}{3}$ km

Plouharnel to Ploemel: $7\frac{4}{5}$ km

Carnac to Plouharnel: $3\frac{1}{10}$ km

How far does Claude cycle altogether? Give your answer as a fraction in its simplest form.

4 Fractions

7 A full jar of jam has a total mass of $\frac{9}{20}$ kg.

The jam has a mass of $\frac{7}{24}$ kg.

What is the mass of the jar?

8 Work out the missing fraction in each of these.

> Write the fractions as improper fractions to start with.

a $2\frac{5}{12} + ? = 4\frac{5}{16}$ **b** $5\frac{7}{12} - ? = \frac{11}{18}$

9 a The numbers $\frac{1}{2}$, a, $\frac{3}{4}$ are written in increasing order of size.

The value of a is exactly halfway between $\frac{1}{2}$ and $\frac{3}{4}$.

> Use equivalent fractions to help.

Find the value of a.

b The numbers $\frac{1}{2}$, m, n, $\frac{3}{4}$ are written in increasing order of size.

The differences between the numbers are all equal.

Find the values of m and n.

10 Work these out. Give your answers in their simplest form.

a $1\frac{2}{5} + \frac{7}{20} + \frac{7}{30}$ **b** $1\frac{4}{5} + \frac{1}{6} + \frac{7}{12}$ **c** $1\frac{3}{5} + 2\frac{5}{6} + \frac{8}{15}$

d $4\frac{1}{4} + \frac{3}{5} - \frac{7}{20}$ **e** $3\frac{2}{3} + 1\frac{5}{6} - \frac{14}{15}$ **f** $4\frac{3}{8} + \frac{7}{12} - \frac{5}{6}$

11 A rectangular photograph is $6\frac{1}{4}$ cm long by $8\frac{5}{8}$ cm wide.

a What is the perimeter of the photograph?

b The photograph is to be mounted on card leaving a border $\frac{2}{3}$ cm wide.

What is the perimeter of the mount?

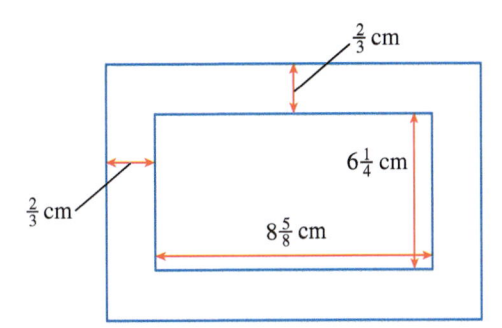

4.2 Multiplying fractions

Multiplying an integer by a fraction

Remember an **integer** is a whole number.

$3 \times \frac{4}{5}$ means the same as $\frac{4}{5} + \frac{4}{5} + \frac{4}{5}$ which is $\frac{4+4+4}{5} = \frac{3 \times 4}{5} = \frac{12}{5}$

To multiply an integer by a fraction you multiply the numerator by the integer. This gives you the numerator of your answer. The denominator does not change.

To work out a fraction of an amount you multiply the amount by the fraction. Remember 'of' means 'multiply' so $\frac{2}{3}$ of 36 means $\frac{2}{3} \times 36$

53

Mathematics for Cambridge Secondary 1

Worked example 1

Work out: **a** $\frac{5}{8} \times 36$ **b** $\frac{7}{12}$ of 18

a $\frac{5}{8} \times 36 = \frac{180}{8}$
$= \frac{45}{2}$
$= 22\frac{1}{2}$

b $\frac{7}{12} \times 18 = \frac{126}{12}$
$= \frac{21}{2}$
$= 10\frac{1}{2}$

Multiplying first can involve large numbers. To avoid this you can cancel first, then multiply.

a $\frac{5}{\underset{2}{8}} \times \overset{9}{36} = \frac{45}{2}$ cancel by 4
$= 22\frac{1}{2}$

b $\frac{7}{\underset{2}{12}} \times \overset{3}{18} = \frac{21}{2}$ cancel by 6
$= 10\frac{1}{2}$

Multiplying a fraction by a fraction

To multiply a fraction by a fraction you multiply the numerators together, and multiply the denominators together. To help keep the numbers small you can cancel before you multiply, as in Worked example 2.

When you multiply mixed numbers you **must** write them as improper fractions first.

Worked example 2

Work these out. Give your answers in their simplest form.

a $\frac{2}{3} \times \frac{5}{8}$ **b** $\frac{7}{12} \times \frac{3}{28}$

a $\frac{\overset{1}{2}}{3} \times \frac{5}{\underset{4}{8}} = \frac{5}{12}$ cancel by 2

b $\frac{\overset{1}{7}}{\underset{4}{12}} \times \frac{\overset{1}{3}}{\underset{4}{28}} = \frac{1}{16}$ cancel by 7, and cancel by 3

54

4 Fractions

Worked example 3

Work these out. Give your answers in their simplest form.

Write any improper fractions as mixed numbers.

a $2\frac{5}{8} \times 2\frac{2}{7}$ **b** $3\frac{3}{7} \times 2\frac{1}{10}$

a $2\frac{5}{8} \times 2\frac{2}{7} = \frac{21^3}{8_1} \times \frac{16^2}{7_1} = \frac{6}{1} = 6$

b $3\frac{3}{7} \times 2\frac{1}{10} = \frac{{}^{12}24}{7_1} \times \frac{21^3}{10_5} = \frac{36}{5} = 7\frac{1}{5}$

Exercise 4.2

1 Work these out. Give your answers in their simplest form.

Write any improper fractions as mixed numbers.

- **a** $\frac{2}{3}$ of 36
- **b** $\frac{3}{8}$ of 24
- **c** $\frac{2}{5}$ of 24
- **d** $\frac{5}{12}$ of 16
- **e** $\frac{3}{4} \times 24$
- **f** $\frac{7}{8} \times 20$
- **g** $\frac{7}{15} \times 24$
- **h** $\frac{5}{12} \times 28$

2 Work these out. Give your answers in their simplest form.

Write any improper fractions as mixed numbers.

- **a** $\frac{3}{5} \times \frac{7}{9}$
- **b** $\frac{4}{7} \times \frac{7}{9}$
- **c** $\frac{8}{9} \times \frac{3}{5}$
- **d** $\frac{6}{11} \times \frac{4}{9}$

3 Work these out. Give your answers in their simplest form.

Write any improper fractions as mixed numbers.

- **a** $1\frac{1}{9} \times \frac{3}{5}$
- **b** $2\frac{4}{5} \times \frac{4}{7}$
- **c** $\frac{25}{32} \times 3\frac{1}{5}$
- **d** $1\frac{1}{8} \times \frac{2}{3}$
- **e** $1\frac{7}{9} \times \frac{3}{8}$
- **f** $\frac{4}{5} \times 2\frac{1}{12}$
- **g** $\frac{2}{5} \times 1\frac{7}{8}$
- **h** $2\frac{1}{10} \times \frac{5}{7}$
- **i** $1\frac{2}{3} \times \frac{9}{10}$
- **j** $1\frac{1}{9} \times 2\frac{2}{5}$
- **k** $1\frac{7}{8} \times 1\frac{1}{5}$
- **l** $2\frac{2}{3} \times 2\frac{1}{4}$
- **m** $2\frac{2}{9} \times \frac{6}{25}$
- **n** $1\frac{1}{10} \times 1\frac{11}{25}$
- **o** $4\frac{1}{6} \times 1\frac{13}{35}$
- **p** $1\frac{11}{12} \times 1\frac{1}{5}$

4 Greta travels 24 km to work every day.

She cycles $\frac{3}{5}$ of the way and walks the rest.

How much further does she cycle than walk?

5 A box contains 6 identical paperback books each with a mass of $\frac{3}{10}$ kg, and 5 identical hardback books each with a mass of $\frac{4}{5}$ kg. The empty box has a mass of $2\frac{3}{8}$ kg. What is the total mass of the box with all the books?

> Find the total mass of the paperback books and the total mass of the hardback books to start with.

6 Work these out. Give your answers in their simplest form.

Write any improper fractions as mixed numbers.

- **a** $1\frac{4}{5} \times \frac{2}{3} \times 2\frac{1}{2}$
- **b** $2\frac{1}{4} \times 2\frac{1}{3} \times 2\frac{2}{3}$

55

4.3 Dividing fractions

Dividing an integer by a fraction

To divide an integer by a fraction you multiply by the reciprocal of the fraction.

That is, you turn the fraction 'upside down' and multiply by it.

> **Worked example 1**
>
> Work out: $4 \div \frac{2}{3}$
>
> $$4 \div \frac{2}{3} = \overset{2}{\cancel{4}} \times \frac{3}{\cancel{2}_1}$$
> $$= \frac{6}{1}$$
> $$= 6$$

Dividing a fraction by a fraction

To divide a fraction by a fraction you multiply the first fraction by the reciprocal of the **second** fraction.

That is, you turn the **second** fraction 'upside down' and multiply by it.

Remember to write any mixed numbers as improper fractions before you start.

> **Worked example 2**
>
> Work out: $2\frac{6}{7} \div 1\frac{1}{4}$
>
> $$2\frac{6}{7} \div 1\frac{1}{4} = \frac{20}{7} \div \frac{5}{4}$$
> $$= \frac{\overset{4}{\cancel{20}}}{7} \times \frac{4}{\cancel{5}_1}$$
> $$= \frac{16}{7}$$
> $$= 2\frac{2}{7}$$

Exercise 4.3

1 Work these out.

    a $4 \div \frac{1}{2}$ **b** $3 \div \frac{3}{4}$ **c** $6 \div \frac{3}{8}$ **d** $8 \div \frac{2}{5}$

2 Work these out. Give your answers in their simplest form.

Write any improper fractions as mixed numbers.

    a $\frac{4}{5} \div \frac{2}{3}$ **b** $\frac{7}{10} \div \frac{3}{5}$ **c** $\frac{12}{19} \div \frac{6}{7}$ **d** $\frac{16}{25} \div \frac{8}{15}$

3 Work these out. Give your answers in their simplest form.

Write any improper fractions as mixed numbers.

a $1\frac{7}{9} \div 2\frac{2}{3}$
b $\frac{8}{15} \div \frac{24}{25}$
c $\frac{5}{18} \div 2\frac{1}{12}$
d $1\frac{1}{15} \div 1\frac{3}{5}$

e $1\frac{7}{8} \div 2\frac{1}{4}$
f $1\frac{1}{8} \div 1\frac{1}{2}$
g $1\frac{1}{20} \div 2\frac{4}{5}$
h $1\frac{7}{8} \div \frac{3}{4}$

i $3\frac{3}{5} \div \frac{9}{10}$
j $\frac{4}{9} \div 2\frac{2}{3}$
k $2\frac{1}{10} \div 2\frac{4}{5}$
l $2\frac{2}{5} \div \frac{8}{15}$

m $2\frac{1}{4} \div \frac{3}{10}$
n $1\frac{7}{8} \div \frac{5}{12}$
o $2\frac{2}{3} \div 2\frac{2}{9}$
p $1\frac{11}{25} \div 1\frac{1}{15}$

4 An empty case has a mass of $3\frac{3}{8}$ kg.

A toolkit has a mass of $1\frac{1}{8}$ kg.

> Remember to only include the mass of the case once.

The case is safe to carry provided its total mass, including contents, is less than 25 kg.

What is the largest number of toolkits that can safely be put into the case?

5 Work these out. Simplify your answers as much as possible.

a $2\frac{2}{5} \times \frac{5}{9} \div 1\frac{3}{5}$
b $1\frac{5}{9} \times \frac{5}{7} \div 1\frac{1}{24}$

Review

1 Work these out. Give your answers in their simplest form.

a $\frac{13}{20} + \frac{11}{18}$
b $\frac{5}{12} - \frac{3}{20}$
c $3\frac{25}{32} + 4\frac{7}{24}$
d $4\frac{7}{27} - 2\frac{11}{18}$

2 Lauren walks $\frac{4}{5}$ km to the bus stop.

Then she travels $4\frac{2}{3}$ km on the bus.

Finally she walks $\frac{7}{12}$ km to school.

How far does she travel in total?

3 Use the lengths shown in the diagram to find:

a the overall length of the key

b the length of the handle.

(Diagram shows: $?$, $4\frac{9}{10}$ cm, $5\frac{1}{5}$ cm, $\frac{11}{12}$ cm, $\frac{8}{15}$ cm)

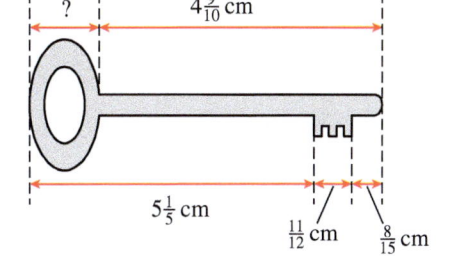

4 Work these out. Give your answers in their simplest form.

Write any improper fractions as mixed numbers.

a $\frac{4}{15} \times \frac{9}{16}$
b $\frac{2}{9} \times \frac{3}{8}$
c $\frac{4}{25} \times \frac{15}{16}$

d $2\frac{7}{10} \times 2\frac{7}{9}$
e $4\frac{2}{3} \times 2\frac{1}{28}$
f $2\frac{2}{9} \times 1\frac{11}{25}$

5 Work these out. Give your answers in their simplest form.

Write any improper fractions as mixed numbers.

a $1\frac{3}{8} \div \frac{1}{4}$
b $\frac{4}{5} \div \frac{3}{10}$
c $\frac{7}{8} \div \frac{5}{12}$

d $4\frac{2}{3} \div \frac{7}{9}$
e $2\frac{5}{8} \div 1\frac{5}{16}$
f $1\frac{1}{6} \div 2\frac{1}{3}$

57

6 A pack contains 20 identical sweets.

   The empty pack has a mass of $12\frac{1}{2}$ g.

   Each sweet has a mass of $5\frac{2}{5}$ g.

   What is the total mass of the pack and sweets?

7 The diagram shows a garden spade.

   a What is the total length of the spade?

   b What is the distance between the points at the end of each of the prongs?

   $4\frac{3}{5}$ cm

   $90\frac{3}{8}$ cm

   $24\frac{1}{3}$ cm

   $25\frac{1}{4}$ cm

Examination-style questions

1 Write $\frac{45}{54}$ as a fraction in its simplest form. [1]

2 Write $\frac{26}{7}$ as a mixed number. [1]

3 Work out $\frac{1}{6}$ of 84. [1]

4 Copy and complete this statement with one of = < >

   $\frac{15}{100}$ ☐ $\frac{1}{5}$ [1]

5 Write down all the fractions that are equivalent.

   $\frac{7}{27}$ $\frac{4}{5}$ $\frac{32}{40}$ $\frac{28}{32}$ $\frac{20}{25}$ [2]

6 Work out $2\frac{1}{3} + 1\frac{2}{5}$ [2]

7 Work out $\frac{2}{5}$ of 75 [2]

8 Work out $\frac{3}{5}$ of $\frac{4}{7}$ [2]

9 Copy the calculation below and write a fraction in the box to make it correct.

   $2\frac{1}{3} \times$ ☐ $= 5\frac{5}{6}$ [2]

5 Decimals

> **Learning outcomes**
>
> - Recognise the equivalence of 0.1, $\frac{1}{10}$, and 10^{-1}.
> - Multiply and divide whole numbers and decimals by 10 to the power of any positive or negative integer.
> - Divide by decimals.
> - Multiply by decimals.
> - Recognise the effects of multiplying and dividing by numbers between 0 and 1.
> - Round numbers to a given number of decimal places or significant figures.
> - Give solutions to problems with an appropriate degree of accuracy.

Shortcuts

Multiplication is a shortcut for addition.
$2 + 2 + 2 + 2 + 2$ can be written as 5×2, which is 10.

Powers, or indices, are a shortcut for multiplication.
$2 \times 2 \times 2 \times 2 \times 2$ can be written as 2^5, which is 32.

Many people have taken time to make calculating easier and quicker.

In 1614 John Napier, a Scottish mathematician, invented a method for making calculations easier. A development of Napier's method was still taught to students in schools until recently.

Nowadays you reach for a calculator when calculations get tough.

5.1 Integer powers of 10

Reading and writing integer powers of 10

You already know:

$10 \times 10 \times 10 = 10^3 = 1000$

$10 \times 10 \qquad\; = 10^2 = 100$

$10 \qquad\qquad\quad\; = 10^1 = 10$

Look at the list. Two things happen each time the power is reduced by 1.

1 There is one fewer '10' on the left.

2 The value on the right is divided by 10.

The pattern continues:

$1 \qquad = 10^0 = 1$

$\frac{1}{10} \qquad = 10^{-1} = \frac{1}{10} \qquad = 0.1$

$\frac{1}{10 \times 10} \qquad = 10^{-2} = \frac{1}{100} \qquad = 0.01$

$\frac{1}{10 \times 10 \times 10} = 10^{-3} = \frac{1}{1000} \qquad = 0.001$

Writing numbers as 10^2 and 10^{-3} is sometimes called writing them in index form.

Worked example 1

Write the value for each of these: **a** 10^6 **b** 10^{-5}

a $10^6 = 10 \times 10 \times 10 \times 10 \times 10 \times 10 = 1\,000\,000$ (This is one million.)

b $10^{-5} = \frac{1}{10 \times 10 \times 10 \times 10 \times 10} = \frac{1}{100\,000} = 0.00001$

Worked example 2

Write each of these in index form: **a** 10 000 **b** 0.0001

a $10\,000 = 10^4$

b $0.0001 = 10^{-4}$

Multiplying and dividing by integer powers of 10

First write the power of 10 as a number if the power is greater than 0, or as a fraction if the power is less than 0. Then multiply or divide.

You can do a quick check to see if the answer is likely to be correct.

When you multiply by a number between 0 and 1 your answer will be smaller than the one you start with.

When you divide by a number between 0 and 1 your answer will be larger than the one you start with.

Worked example 3

Work out:

a 245×10^3

b 0.25×10^{-1}

c $2.45 \div 10^{-3}$

d $230 \div 10^2$

a $245 \times 10^3 = 245 \times 1000$ write the integer power of 10 as a number (3 is greater than 0)

$ = 245\,000$ check: 10^3 is greater than 1, the answer is larger than 245 ✓

b $0.25 \times 10^{-1} = 0.25 \times \frac{1}{10}$ write the integer power of 10 as a fraction (-1 is less than 0)

$\phantom{0.25 \times 10^{-1}} = 0.025$ check: 10^{-1} is between 0 and 1, the answer is less than 0.25 ✓

c $2.45 \div 10^{-3} = 2.45 \div \frac{1}{1000}$ write the integer power of 10 as a fraction (-3 is less than 0)

$\phantom{2.45 \div 10^{-3}} = 2.45 \times 1000$ to divide by a fraction you turn it upside down and multiply

$\phantom{2.45 \div 10^{-3}} = 2450$ check: 10^{-3} is between 0 and 1, the answer is larger than 2.45 ✓

d $230 \div 10^2 = 230 \div 100$ write the integer power of 10 as a number (2 is greater than 0)

$ = 2.3$ check: 10^2 is greater than 1, the answer is less than 230 ✓

Exercise 5.1

Do not use a calculator for this exercise.

1 Write each of these in index form.

    a 10 000 000   **b** 100   **c** 100 000   **d** 0.00001   **e** 0.1   **f** 0.000001

2 Write down the value of each of the following.

    a 10^3 **b** 10^5 **c** 10^9 **d** 10^{-5} **e** 10^{-2} **f** 10^0

3 Work these out.

    a 2.3×10^2 **b** 4.5×10^4 **c** 210×10^2 **d** 5.32×10^5

    e 0.45×10^2 **f** 20.1×10^{-1} **g** 508×10^{-2} **h** 5.4×10^{-3}

4 Work these out.

    a $289 \div 10^2$ **b** $305 \div 10^{-1}$ **c** $25.8 \div 10^{-2}$ **d** $240 \div 10^3$

    e $0.25 \div 10^{-3}$ **f** $0.205 \div 10^1$ **g** $2.5 \div 10^2$ **h** $3.05 \div 10^{-2}$

5 Copy each of the following. Replace the ? with the integer power of 10 that makes the calculation correct.

> It will help to decide whether you are multiplying or dividing by a number larger or smaller than 1.

    a $40.8 \times ? = 408$ **b** $248.9 \times ? = 2.489$ **c** $48.5 \div ? = 4850$ **d** $2.8 \div ? = 0.0028$

6 Copy each of the following. Replace the ? with the value that makes the calculation correct. Each of the numbers needed has the digits 2 and 4, in that order.

    a $? \times 10^3 = 24$ **b** $? \times 10^{-2} = 24$ **c** $? \div 10^5 = 2.4$ **d** $? \div 10^{-3} = 240$

Mathematics for Cambridge Secondary 1

7 Khalil is given this calculation by his teacher.

245 ☐ ? = 2.45

He has to replace the ☐ with × or ÷

He has to replace the ? with a number.

He says there is more than one way to do this.

Find as many ways as possible to replace the ☐ and ?

5.2 Dividing by a decimal

Sometimes you may need to divide by a decimal.

You can often transform this to a division by an **integer**.

Use **equivalent fractions** to do this.

> When two fractions have the same value they are known as equivalent fractions.

Worked example

Work these out.

a $\dfrac{6}{0.2}$ **b** $\dfrac{30}{0.1}$ **c** $\dfrac{0.09}{0.01}$ **d** $2.5 \div 0.05$

a $\dfrac{6}{0.2} = \dfrac{60}{2}$ (×10 to numerator and denominator)   multiply the numerator and denominator by 10

$= 30$

b $\dfrac{30}{0.1} = \dfrac{300}{1}$ (×10 to numerator and denominator)   multiply the numerator and denominator by 10

$= 300$

c $\dfrac{0.09}{0.01} = \dfrac{9}{1}$ (×100 to numerator and denominator)   multiply the numerator and denominator by 100

$= 9$

d $2.5 \div 0.05 = \dfrac{2.5}{0.05}$ write the division as a fraction

$\dfrac{2.5}{0.05} = \dfrac{250}{5}$ (×100 to numerator and denominator)   multiply the numerator and denominator by 100

$= 50$

Exercise 5.2

Do not use a calculator for this exercise.

1 Work these out.

   a $\frac{6}{0.2}$
   b $\frac{16}{0.4}$
   c $\frac{12}{0.3}$
   d $\frac{20}{0.1}$
   e $\frac{1.4}{0.7}$
   f $\frac{0.04}{0.2}$
   g $\frac{1.8}{0.3}$
   h $\frac{80}{0.4}$

2 Work these out.

   a $\frac{1.2}{0.02}$
   b $\frac{3.6}{0.09}$
   c $\frac{16}{0.08}$
   d $\frac{0.8}{0.04}$
   e $\frac{1.6}{0.02}$
   f $\frac{4}{0.05}$
   g $\frac{0.15}{0.05}$
   h $\frac{14}{0.07}$

3 Work these out.

   a $\frac{28}{0.7}$
   b $\frac{32}{0.04}$
   c $\frac{15}{0.1}$
   d $\frac{49}{0.01}$
   e $\frac{120}{0.04}$
   f $\frac{90}{0.9}$
   g $\frac{1.6}{0.04}$
   h $\frac{0.08}{0.02}$

4 Work these out.

   a $25 \div 0.1$
   b $1.6 \div 0.4$
   c $640 \div 0.08$
   d $0.8 \div 0.02$

5 Faith says '2.5 ÷ 0.5 is 0.5'.

   > Work out the correct answer first.

   What mistake has Faith made?

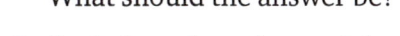

   What should the answer be?

6 Ilecia has a barrel containing 140 litres of water.

   > Work out the correct answer first.

   She says 'If I fill 70 bottles each containing 0.5 litre from the barrel it will be empty'.

   What mistake has Ilecia made?

   How many bottles can she actually fill?

7 a 0.3 kg of gold is sold for $13 971. How much is this per kg?

   b 0.4 kg of platinum is sold for $19 220. Is it more expensive to buy 1 kg of platinum or 1 kg of gold? Show how you decided.

   > Work out the cost per kg for platinum first.

8 Work these out.

   a $225 \div 1.5$
   b $125 \div 2.5$
   c $90 \div 1.5$

9 A pipe is cut into lengths 0.25 metre long.

   If the pipe is 4 metres long how many lengths can be cut?

5.3 Multiplying with decimals

You can use facts you already know to help you multiply decimals.

You can use **equivalent calculations** to help you multiply decimals.

For example, to multiply by 0.3 you can multiply by 3 and then divide the result by 10.

To multiply by 0.03 you can multiply by 3 and then divide the result by 100.

In calculations such as 300 × 0.2 you divide 300 by 10 and multiply 0.2 by 10.

This gives 30 × 2 as an equivalent calculation.

30 × 2 is easy to work out in your head as 60.

Worked example

Work out: **a** 700 × 80   **b** 400 × 0.6   **c** 0.15 × 0.06

a 700 × 80 = 7 × 100 × 8 × 10   *this is a method you can use to work these out in your head*
            = 7 × 8 × 100 × 10
            = 56 × 1000
            = 56 000

b 400 × 0.6 = 40 × 6   *this is an equivalent calculation, as it gives the same answer*
              = 240

c 0.15 × 0.06

First remove the decimal points. 15 × 6 = 90

Then count up the number of digits after the decimal points in the question.

There are four digits after the decimal points in the question.

You need four digits after the decimal point in the answer.

The answer is 0.0090

> Always take care with counting the decimal places. An answer such as 90 uses two of the four digits after the decimal point.

Exercise 5.3

Do not use a calculator for this exercise.

1 Work these out.

  a 30 × 60   **b** 40 × 30   **c** 20 × 300   **d** 500 × 30
  e 400 × 50   **f** 60 × 7000   **g** 600 × 5000   **h** 240 × 600

2 Work these out.

  a 400 × 0.2   **b** 300 × 0.8   **c** 800 × 0.7   **d** 500 × 0.4
  e 150 × 0.3   **f** 250 × 0.7   **g** 125 × 0.3   **h** 25 × 0.4

3 Work these out.

    a 0.25×0.2 **b** 0.35×0.4 **c** 0.24×0.03 **d** 0.24×0.4

    e 0.02×0.03 **f** 0.32×0.04 **g** 0.55×0.4 **h** 7.5×0.4

4 What do you notice about these?

    a 12×0.2 $48 \div 20$ 6×0.4 24×0.1 $9.6 \div 4$

    b $64 \div 20$ 32×10^{-1} 80×0.04 $96 \div 30$ $12.8 \div 4$

5 Write down some calculations that all have the answer 4.8

6 Three numbers multiply together to give 0.08. What can the numbers be?

> Find three numbers that multiply together to give 8, then decide where the decimal points will need to be in your numbers.

7 Work these out.

    a 1.5×1.5 **b** 1.2×1.2 **c** 1.6×0.16 **d** 14×0.14

5.4 Rounding

Decimal places

You can round numbers to a given number of **decimal places**.

Rounding to 1 decimal place gives you 1 digit after the decimal point.

Rounding to 2 decimal places gives you 2 digits after the decimal point.

You can round to 3 decimal places, 4 decimal places, and so on.

To round to 1 decimal place you need to look at the 2nd digit after the decimal point.

If it is '5 or more' you round up. If it is less than 5 you round down.

To round to 2 decimal places you need to look at the 3rd digit after the decimal point.

Instead of writing decimal places every time you can write 'd.p.' to save time.

Worked example 1

Round these number to the given number of decimal places.

a 23.956 to 1 d.p.

b 4.9762 to 2 d.p.

c 23.5804 to 3 d.p.

d 8.0325 to 3 d.p.

a 23.956 is between 23.9 and 24.0

The 2nd digit after the decimal point is '5 or more' so round up.

23.956 is 24.0 to 1 d.p.   the final '0' in 24.0 tells you the number is correct to 1 d.p.

b 4.9762 is between 4.97 and 4.98

The 3rd digit after the decimal point is '5 or more' so round up.

4.9762 is 4.98 to 2 d.p.

c 23.5804 is between 23.580 and 23.581

The 4th digit after the decimal point is not '5 or more' so round down.

23.5804 is 23.580 to 3 d.p.   the final '0' in 23.580 tells you the number is correct to 3 d.p.

d 8.0325 is between 8.032 and 8.033

The 4th digit after the decimal point is '5 or more' so round up.

8.0325 is 8.033 to 3 d.p.

Significant figures

Numbers can also be rounded to a given number of **significant figures**.

Significant figures are counted from the first non-zero digit.

For the number 342.014

3	4	2	.	0	1	4
↑	↑	↑		↑	↑	↑
1st sig fig	2nd sig fig	3rd sig fig		4th sig fig	5th sig fig	6th sig fig

The first significant figure is the 3. It is the most significant figure as it is worth 300.

For the number 0.0345.

0 . 0	3	4	5
	↑	↑	↑
	1st sig fig	2nd sig fig	3rd sig fig

The first significant figure is the 3. The zeros at the front of the number are not significant.

When you round to a given number of significant figures you look at the 'next' significant figure. If it is '5 or more' then you round up. Otherwise you round down.

Instead of writing significant figures all the time you can write 's.f.' to save time.

5 Decimals

Worked example 2

Round each of these numbers to 2 significant figures.

a 789 **b** 6.0524 **c** 0.002534 **d** 0.603

a 7   8   9
   ↑   ↑   ↑
  1st 2nd 3rd
  sig fig sig fig sig fig

The 3rd significant figure is '5 or more' so round up.

789 = 790 (2 s.f.)

> Always check that the most significant figure has the same place value after rounding as it did before rounding. Here the 7 is worth 700 before rounding. It is still worth 700 after rounding.

b 6 . 0   5 2 4
   ↑   ↑   ↑
  1st 2nd 3rd
  sig fig sig fig sig fig

The 3rd significant figure is '5 or more' so round up.

6.0524 = 6.1 (2 s.f.)

c 0 . 0 0 2   5   3   4
            ↑   ↑   ↑
           1st 2nd 3rd
           sig fig sig fig sig fig

The 3rd significant figure is not '5 or more' so round down.

0.002534 = 0.0025 (2 s.f.)

d 0 . 6   0   3
       ↑   ↑   ↑
      1st 2nd 3rd
      s.f. s.f. s.f.

The 3rd significant figure is not '5 or more' so round down.

0.603 = 0.60 (2 s.f.)

Making decisions

Sometimes you have to decide on an appropriate **degree of accuracy**.

This means you have to decide how many significant figures to give an answer to.

An answer to a problem cannot have more significant figures than the information given in the problem.

Mathematics for Cambridge Secondary 1

> **Worked example 3**
>
> Bill made 16 biscuits using a total of 125 g of flour.
>
> How much flour was in each biscuit?
>
> Give your answer to an appropriate degree of accuracy.
>
> $125 \div 16 = 7.8125$ g
>
> The flour was measured to 3 significant figures.
>
> The mass of flour in each biscuit cannot be given to more than this.
>
> 7.8 g (2 s.f.) is sensible as 2 s.f. is 1 s.f. less than the information given.

Exercise 5.4

1 Round these numbers to the given number of decimal places.

   a 12.59 to 1 d.p.   b 3.428 to 2 d.p.   c 5.98 to 1 d.p.   d 49.9752 to 1 d.p.

   e 19.95 to 1 d.p.   f 1.0508 to 3 d.p.   g 3.05 to 1 d.p.   h 5.8602 to 3 d.p.

2 Round these numbers to 2 significant figures.

   a 48.2   b 2.59   c 5.605   d 8.03

   e 40.09   f 7.98   g 29.7   h 49.09

3 Round these numbers to 1 significant figure.

   a 32   b 407   c 970   d 2348

   e 0.056   f 76.8   g 8345   h 99

4 Round these to 3 significant figures.

   a 4589   b 23.56   c 1.0435   d 0.002387

   e 0.04289   f 0.05832   g 0.0088412   h 0.05997

5 Use a calculator to work these out.

   Give your answers correct to 2 significant figures.

   a $23 \div 7$ b $6 \div 38$ c $3 \div 129$ d 520×530

   e $0.28 \div 34$ f $20 \div 3$ g $8 \div 9$ h $49 \div 11$

6 A pizza with a mass of 0.850 kg is divided equally between 6 people.

   How much does each person have? Give your answer in kg to an appropriate degree of accuracy.

7 Tomasz runs 125 metres in 22.5 seconds.

   How many metres each second is this? Give your answer to an appropriate degree of accuracy.

8 Louisa measures a distance to be 100 metres.

   Explain why you cannot tell whether it was measured to 1, 2, or 3 significant figures. Give some examples to help explain.

9 Max works out 3.5 × 3.5 = 12.25

3.25 × 3.25 = 10.5625

3.255 × 3.255 = 10.595025

> Try some numbers to check the statements.

He says 'when you multiply a 1 d.p. number by a 1 d.p. number you always get an answer with 2 d.p.'

'and when you multiply a 2 d.p. number by a 2 d.p. number you always get an answer with 4 d.p., and so on'.

Is Max correct?

10 Phil measures the width of his book. It is 23.5 cm to 1 decimal place. What is the smallest and largest width his book could be?

Review

1 Work these out.

a 4.5×10^2
b 450×10^{-1}
c 508×10^{-3}
d 0.0054×10^3
e 0.0492×10^{-1}
f $42 \div 10^2$
g $589 \div 10^1$
h $0.05 \div 10^{-3}$
i $0.0089 \div 10^{-2}$
j $0.025 \div 10^{-3}$

2 Work these out.

a $\frac{2}{0.4}$
b $\frac{16}{0.8}$
c $\frac{14.4}{0.9}$
d $\frac{1.8}{0.6}$
e $\frac{6.3}{0.07}$
f $\frac{1.5}{0.01}$
g $\frac{0.8}{0.02}$
h $\frac{0.18}{0.03}$
i $3 \div 0.1$
j $20 \div 0.01$

3 Work these out.

a 60×300
b 5000×20
c 300×0.8
d 250×0.2
e 0.35×0.2
f 0.08×0.04
g 0.24×0.6
h 6000×0.09

4 Write these numbers to the given number of decimal places.

a 2.58 to 1 d.p.
b 2.68961 to 2 d.p.
c 2.089 to 1 d.p.
d 8.99 to 1 d.p.
e 2.034 to 1 d.p.

5 Write these numbers to the given number of significant figures.

a 458 to 2 s.f.
b 2.35841 to 3 s.f.
c 0.0258 to 1 s.f.
d 99.8 to 2 s.f.
e 99.8 to 1 s.f.

Examination-style questions

1. Copy each of the following. Write the missing numbers in the spaces.

    a $17.26 \times 1000 = \ldots\ldots$ [1]

    b $3.8 \div \ldots\ldots = 0.038$ [1]

2. Calculate: $34.2 \div 6$ [1]

3. Look at this calculation:

    $37 \times 291 = 10\,767$

    Work out:

    a 3.7×2.91 [1]

    b $107.67 \div 0.37$ [1]

4. Here are some numbers.

    0.01   0.1   1   10   100

    Use four of these numbers to make these statements correct.

    You may use each number only once.

    $24 \times \ldots = 24 \div \ldots$

    $24 \div \ldots = 24 \times \ldots$ [2]

5. Look at this calculation:

    $$\frac{26.3 + 47.9}{107.4 - 3.29}$$

    a   Use your calculator to work out the answer.

    Write down all of the figures on your display. [1]

    b   Write your answer correct to 3 decimal places. [1]

6. Copy each of the following. Write numbers in the boxes to make the following calculations correct.

    a $4.68 \div 0.8 = 4.68 \times \Box \div 8$ [1]

    b $56.3 \times \Box = 56.3 \times 7 \div 100$ [1]

7. Copy each of the following. Write the missing numbers in the spaces.

    a $265 \times \ldots = 2.65$ [1]

    b $45.8 \div \ldots = 458$ [1]

8. Here is part of Nicole's maths homework:

    $263 \times 1.15 = 241$

    Nicole's answer is wrong.

    Explain how you can tell she is wrong **without** working out the correct answer. [1]

6 Processing, interpreting and discussing data

Chapter 18 covers collecting data
Chapter 10 covers presenting data

> **Learning outcomes**
> - Calculate statistics to represent data sets.
> - Select the most appropriate statistics for the problem.

Medical statistics

Florence Nightingale is well known for her work as a nurse. She introduced many modern nursing techniques.

This followed her work in the Crimea where she collected data.

This provided evidence that more troops were dying of diseases such as typhus and cholera than died in London during the plague of 1655.

There was an annual mortality rate of 60%. Between the ages of 25 and 35 the mortality rate was twice that in civilian life.

Following a report written by Florence Nightingale a department devoted to Army Medical Statistics was created.

6.1 Finding average values for large data sets

The mean

The mean is found using the formula:

$$\text{mean} = \frac{\text{the sum of all the values}}{\text{the number of values}}$$

Large data sets are usually grouped into classes.

You do not know the actual values.

You calculate an **estimate for the mean** using the mid-class values.

> **Worked example 1**
>
> The grouped frequency table shows the time taken to complete a bus journey.
>
> Calculate an estimate of the mean time for the bus journeys.
>
> | Time, t (minutes) | Frequency |
> |---|---|
> | $5 < t \leqslant 10$ | 4 |
> | $10 < t \leqslant 15$ | 18 |
> | $15 < t \leqslant 20$ | 32 |
> | $20 < t \leqslant 25$ | 14 |
> | $25 < t \leqslant 30$ | 7 |

Mathematics for Cambridge Secondary 1

Time, t (minutes)	Frequency, f	Mid-class value, m	f × m
5 < t ≤ 10	4	7.5	30
10 < t ≤ 15	18	12.5	225
15 < t ≤ 20	32	17.5	560
20 < t ≤ 25	14	22.5	315
25 < t ≤ 30	7	27.5	192.5
Total	75		Total 1322.5

Mean = $\dfrac{\text{total of } f \times m \text{ column}}{\text{total of the frequencies}} = \dfrac{1322.5}{75} = 17.6$ to 1 d.p.

The median

The median is the middle value when the data set is listed in order of size.

For grouped data, you can identify which group (or class) contains the median.

Worked example 2

Find an estimate of the median bus journey time for the data in Worked example 1.

The median is the middle value.

This is the 38th value in the table.

When the times are listed in order of size:

The first 4 times are 10 minutes or less.

The first 22 times are 15 minutes or less.

The first 54 times are less than 20 minutes.

The 38th time is in the class 15 < t ≤ 20

Time, t (minutes)	Frequency
5 < t ≤ 10	4
10 < t ≤ 15	18
15 < t ≤ 20	32
20 < t ≤ 25	14
25 < t ≤ 30	7

The mode

The mode is the value that is most common.

You can find the modal class for grouped data.

Worked example 3

Find the modal class of the bus journey times in Worked example 1.

The modal class is the class with the greatest frequency.

The modal class is 15 < t ≤ 20

Time, t (minutes)	Frequency
5 < t ≤ 10	4
10 < t ≤ 15	18
15 < t ≤ 20	32
20 < t ≤ 25	14
25 < t ≤ 30	7

Exercise 6.1

1 The tables show grouped data.

For each table find:

a the modal class

b the class which contains the median

c an estimate for the mean using mid-class values.

A

Class	0–9	10–19	20–29	30–39	40–49
Frequency	4	8	12	6	2

B

Class	10–19	20–29	30–39	40–49	50–59	60–69
Frequency	4	18	16	8	3	1

2 The table shows the heights, in cm, of a group of students.

a Calculate an estimate for the mean height of the students using mid-class values.

b Find the class which contains the median height of the students.

Height, h (cm)	Frequency
$150 < h \leqslant 160$	14
$160 < h \leqslant 170$	48
$170 < h \leqslant 180$	60
$180 < h \leqslant 190$	5

3 The table shows the masses of a group of babies.

a Calculate an estimate for the mean mass of the babies using mid-class values.

b Find the class which contains the median mass of the babies.

Mass, w (kg)	Frequency
$0 < w \leqslant 2$	8
$2 < w \leqslant 4$	40
$4 < w \leqslant 6$	29
$6 < w \leqslant 8$	2

4 The tables show the time taken by two groups of students to solve a puzzle.

Group A	
Time, t (minutes)	Frequency
$0 < t \leqslant 10$	4
$10 < t \leqslant 20$	12
$20 < t \leqslant 30$	25
$30 < t \leqslant 40$	18
$40 < t \leqslant 50$	11

Group B	
Time, t (minutes)	Frequency
$0 < t \leqslant 10$	3
$10 < t \leqslant 20$	10
$20 < t \leqslant 30$	20
$30 < t \leqslant 40$	12
$40 < t \leqslant 50$	4

a Calculate estimates for the mean time for each group.

b Which group generally solves the puzzle quickest? Give a reason for your answer.

73

5 Lanika says that girls have bigger handwriting than boys.

She asks some students to write the sentence 'The quick brown fox jumps over the lazy dog'.

Each sentence is then measured.

The results are shown in the tables.

Girls	
Length, l, (cm)	Frequency
$14 < l \leq 16$	4
$16 < l \leq 18$	8
$18 < l \leq 20$	32
$20 < l \leq 22$	16
$22 < l \leq 24$	12

Boys	
Length, l, (cm)	Frequency
$14 < l \leq 16$	4
$16 < l \leq 18$	9
$18 < l \leq 20$	29
$20 < l \leq 22$	14
$22 < l \leq 24$	4

Is Lanika correct? Show how you decide.

> Use estimates of the mean length for boys and girls to help.

6 Hassan measures the 'time waiting', t minutes, of customers at a supermarket checkout.

His results are shown in the table.

Time, t (minutes)	$0 < t \leq 1$	$1 < t \leq 2$	$2 < t \leq 3$	$3 < t \leq 4$	$4 < t \leq 5$	$5 < t \leq 6$	$6 < t \leq 7$
Frequency	8	18	30	20	10	8	6

The manager is required to employ more staff on checkouts if the average 'time waiting' is more than 3 minutes. If the average time waiting is below 3 minutes the manager does not need to employ more staff on checkouts.

a Hassan says 'the average shows that more staff are required'. Show how the mean supports his statement.

b The manager says 'the average shows that more staff are not needed'. Show how the median supports the manager's statement.

6.2 Selecting the most appropriate average

It is important to know which average to select to represent a set of data.

You want to use an average value to give a typical value.

Here are some of the advantages and disadvantages for the three averages you know how to find.

Average	Advantages	Disadvantages
Mode	Always one of the data values. The only average that can be found for data that is not numerical.	Sometimes there is no mode. There may be more than one mode.
Mean	Uses all the values in its calculation.	Can be affected by a small number of very large or very small values. Unlikely to be one of the data values. Can take time to calculate.
Median	Not affected by very large or very small values.	Can be time consuming as you have to order the data.

Worked example 1

Some houses in a village are for sale.

Here are the prices.

$90 000, $105 000, $104 000, $90 000, $92 000, $201 000, $95 000

a Find the mode, mean and median price.

b Which of these averages best represents these prices? Give a reason for your answer.

a The mode is $90 000

 The median is $95 000

 The mean is $111 000

b The mode is the lowest price, so does not represent the data well.

 The mean value is higher than all the prices except for the highest price, so does not represent the data well.

 The median best represents the prices as it is amongst the data.

Worked example 2

Toste is given $4 each week pocket money.

He does some research to find out what other children in his area receive.

He finds:

the mean is $5.40

the median is $5.20

the mode is $5.

He says 'I need an increase of $1.50 to catch up to the average pocket money each week.'

His mum says 'I'll give you an increase of $1 to bring you up to the average.'

a Give a reason why you might agree with Toste.

b Give a reason why you might agree with Toste's mum.

c Suggest something else that Toste and his mum might consider.

a The mean amount is about $1.40 more than the pocket money that Toste is given.

b The mode is about $1 more than the pocket money that Toste is given.

c There are lots of things that may be considered, such as whether Toste is prepared to help with chores, whether Toste has to buy his own magazines. They may even think about the size of the area in which Toste got his data. They may consider the number of people he asked – a small sample would not lead to reliable averages.

Exercise 6.2

1 For each of the following decide which would be the best average to use.

  a The average shoe size of students in your class.

  b The average number of children in each family in a village.

  c Judges at a dance competition to get an average score.

  d The average height of the students in your class.

  e The average hair colour for people in a village.

2 Colin records the kilometres he drives in his car every day.

  12, 15, 8, 9, 14, 137, 8

  a Find the mode, median and mean for these distances.

  b Which average best represents these distances? Give a reason for your answer.

6 Processing, interpreting and discussing data

3 A box of paper clips says 'average contents 30 paper clips'.

Anton counts the paper clips inside 170 boxes.

Clips	26	27	28	29	30	31	32	33	34	35	36	37
Frequency	5	3	8	12	48	32	17	20	15	5	1	4

a How many packets have 30 or more paper clips?

b How many packets have fewer than 30 paper clips in?

c Calculate the mode, median and mean number of paper clips in the packets.

d Use your answers to part **c** to comment on the statement 'average contents 30 paper clips'.

4 Hattie records the time taken to download files from the internet in seconds, to the nearest second.

The modal time is 4 seconds. The mean time is 9 seconds. The median time is 6 seconds.

State whether each of the following statements is true, false or not possible to say.

a Most of the download times are 4 seconds.

b Over half the download times are less than 7 seconds.

c There are a small number of very large download times.

d A file takes 5 seconds to download. This is longer than most files take to download.

e A file takes 6 seconds to download. Exactly half the files take longer to download.

5 Greta is comparing the cost of looking after different pets.

She finds the following information from 10 gerbil owners.

Monthly cost

$11, $10, $9, $24, $6, $16, $8, $4, $25, $32

She finds the following information from 8 hamster owners.

Monthly cost

$8, $22, $18, $3, $5, $21, $4, $19

Greta says 'On average, it costs more to look after a gerbil than a hamster.'

Is Greta correct? Show how you decide.   *Choose suitable averages.*

6 The tables show information about the wages of employees at a factory.

Workers	
Pay, p (thousands)	Frequency
$10 < p \leqslant 20$	8
$20 < p \leqslant 30$	28
$30 < p \leqslant 40$	14
$40 < p \leqslant 50$	3
$50 < p \leqslant 60$	0

Managers	
Pay, p (thousands)	Frequency
$10 < p \leqslant 20$	0
$20 < p \leqslant 30$	0
$30 < p \leqslant 40$	5
$40 < p \leqslant 50$	12
$50 < p \leqslant 60$	8

a Calculate an estimate of the mean pay for the workers.

b Calculate an estimate of the mean pay for the managers.

c Combine the information given in the two tables. Calculate the mean pay for all the employees at the factory.

d The workers want a pay rise. Which value could they use to try to get the best pay rise?

e A newspaper article suggests the people working at the factory are already well paid. Which value are they likely to give in their article?

Review

1 A group of teachers is asked how many hours they work each week.

The table shows the results.

a Calculate an estimate for the mean time worked using mid-class values.

b Find the class that contains the median time worked.

Time worked, h (hours)	Frequency
$35 < h \leq 40$	1
$40 < h \leq 45$	4
$45 < h \leq 50$	7
$50 < h \leq 55$	15
$55 < h \leq 60$	9
$60 < h \leq 65$	8
$65 < h \leq 70$	3

2 Members of a running club complete a race.

The table shows the times.

a Calculate an estimate of the mean time.

b The race also has some runners who are not in a running club.

Time, t (minutes)	Frequency
$30 < t \leq 40$	15
$40 < t \leq 50$	17
$50 < t \leq 60$	11
$60 < t \leq 70$	7

They have a mean time of 49 minutes.

Compare the times of the member of the running club with the runners who are not in a running club.

3 Dakarai and Julia are comparing the service of RedBus and BlueBus.

They collect information about buses from both companies on the same route.

Number of minutes late

RedBus: 5, 10, 15, 30, 5, 20, 5, 10, 5, 10

BlueBus: 0, 0, 30, 40, 0, 10, 15, 25, 10, 0

a Calculate the mean, median and mode for the number of minutes late for each bus company.

b Dakarai says that RedBus is better. Give one reason why Dakarai could be correct.

c Julia says that BlueBus is better. Give one reason why Julia could be correct.

4 A company employs 50 people.

The table shows job titles, number of employees and annual wages.

Job title	Chairman	Director	Manager	Supervisor	Skilled worker	Manual worker	Cleaner
Number of employees	1	2	4	8	13	18	4
Annual wage	$200 000	$150 000	$80 000	$40 000	$30 000	$25 000	$20 000

a Work out the mode, mean and median for the employees at the company.

b Which average best represents the data? Give a reason for your answer.

5 Jenny drives to work every day.

She can choose from route A or route B.

The tables show her times for both routes over a two-month period.

Route A	
Time, t (minutes)	Frequency
$5 < t \leq 10$	5
$10 < t \leq 15$	6
$15 < t \leq 20$	12
$20 < t \leq 25$	4
$25 < t \leq 30$	3

Route B	
Time, t (minutes)	Frequency
$5 < t \leq 10$	0
$10 < t \leq 15$	6
$15 < t \leq 20$	21
$20 < t \leq 25$	3
$25 < t \leq 30$	0

a Calculate an estimate for the mean time using route A.

b Calculate an estimate for the mean time using route B.

c Which route is generally faster?

d One day Jenny sleeps in.

She needs to get to work in 25 minutes otherwise she will be late.

Which route should she choose? Give a reason for your answer.

Examination-style questions

1 Mercedes rolls a six-sided dice thirty times.

She records her results.

3 5 4 3 3 6 4 3 1 4

3 1 5 6 4 1 2 4 3 2

6 5 2 6 4 6 4 5 3 3

a Copy and complete the frequency column of her table.

Number	Frequency
1	
2	
3	
4	
5	
6	
Total	30

[2]

b i Write down the mode. [1]

ii Write down the median. [1]

iii Work out the mean. [3]

2 Henri has four number cards.

[8] [3] [?] [?]

a If the mode of the numbers is 8 and their median is 6, what are the two missing numbers? [1]

b If the range of the numbers is 5 and their mean is 5, what are the two missing numbers? [1]

3 Here are the heights, in cm, of a group of people.

162   183   175   190   176   169   181   177   171

Work out:

a the median height [1]

b the mean height. [2]

4 John types 132 words in 4 minutes. Mark types 224 words in 7 minutes.

Who types the fastest? [3]

5 The number of minutes that buses arrive late at a bus station is shown below.

Number of minutes late, m	Frequency	Mid-class value
$0 < m \leq 10$	16	
$10 < m \leq 20$	10	
$20 < m \leq 30$	11	
$30 < m \leq 40$	8	
$40 < m \leq 50$	5	

a Copy the table and complete the mid-class value column. Use the mid-class values to calculate an estimate of the mean number of minutes that buses arrive late. [3]

b Which class interval contains the median number of minutes that buses arrive late? [2]

7 Length, mass and capacity

> **Learning outcomes**
> - Solve problems involving measurements in a variety of contexts.

A revolutionary idea

The system of units known as the metric system dates back to 1792.

In that year the French Revolutionary Council introduced a decimal system of measurement.

It has been added to and adapted over the years and is now in use in most countries of the world.

However, one idea that did not catch on was decimal time. This had 10 hours in a day and 100 minutes in an hour.

This ten-hour clock face looks very strange.

7.1 Converting metric units

Metric prefixes

The **metric system** has units for many different quantities.

Examples include force (newton), electrical current (ampere) and frequency (hertz).

You should be familiar with the units for length (**metre**), mass (**gram**) and capacity (**litre**).

Bigger and smaller measures are made by adding a prefix to the unit.

This table shows some of the prefixes for creating measures bigger than the basic unit.

	deca-	hecto-	kilo-	mega-	giga-	tera-	peta-
10^0	10^1	10^2	10^3	10^6	10^9	10^{12}	10^{15}
one	ten	hundred	thousand	million	billion	trillion	quadrillion

For example, the unit of power is the watt.

A million watts is called a megawatt.

81

Mathematics for Cambridge Secondary 1

The next table shows some of the prefixes for creating measures smaller than the basic unit.

	deci-	centi-	milli-	micro-	nano-	pico-	femto-
10^0	10^{-1}	10^{-2}	10^{-3}	10^{-6}	10^{-9}	10^{-12}	10^{-15}
one	tenth	hundredth	thousandth	millionth	billionth	trillionth	quadrillionth

The unit of energy is the joule.

One billionth of a joule is the nanojoule.

You need to remember the commonly used prefixes **kilo-**, **centi-**, and **milli-**.

Worked example 1

Complete these statements.

a 2790 centimetres = ... metres

b 5.8 litres = ... millilitres

c 867 kilograms = ... tonnes

a 1 cm = $\frac{1}{100}$ m

2790 cm = $\frac{1}{100}$ × 2790 m

2790 cm = 27.9 m

> Multiplying by $\frac{1}{100}$ is the same as dividing by 100.
> 2790 ÷ 100 = 27.9

b 1 l = 1000 ml

5.8 l = 1000 × 5.8 ml

5.8 l = 5800 ml

c 1000 kg = 1 tonne   remember that 1 tonne is 1000 kilograms

1 kg = $\frac{1}{1000}$ tonne

867 kg = $\frac{1}{1000}$ × 867 tonnes

867 kg = 0.867 tonne

Worked example 2

Adil has a piece of timber 2.4 metres long.

He needs to cut the following pieces from it to make a small toy:

  i 4 pieces each 125 mm long

  ii 7 pieces each 9.5 cm long

  iii 2 pieces each 0.45 m long.

a Work out the total length he needs to cut.

  Give the answer in centimetres.

b How much wood is left after he finishes cutting the pieces?

a **i** 4 × 125 = 500 mm

500 mm = 50 cm

The four pieces have a total length of 50 cm.

ii 7 × 9.5 = 66.5 cm

The seven pieces have a total length of 66.5 cm.

iii 2 × 0.45 = 0.9 m

0.9 m = 90 cm

The two pieces have a total length of 90 cm.

50 + 66.5 + 90 = 206.5 cm

The total length that he cuts is 206.5 cm.

b 2.4 m = 240 cm

240 − 206.5 = 33.5 cm

There is 33.5 cm left after he has cut the wood.

Exercise 7.1

1 Copy and complete these statements.

    a 84.9 kilometres = ………………………….. metres

    b 129 centimetres = ………………………….. millimetres

    c 37.5 tonnes = ………………………….. kilograms

    d 8436 centimetres = ………………………….. metres

    e 5.9 metres = ………………………….. millimetres

    f 21 434 millilitres = ………………………….. litres

    g 125 kilograms = ………………………….. grams

    h 8030 metres = ………………………….. kilometres

2 Choose the correct metric unit to complete these statements.

    a A calculator is about 16 ………. long.

    b A coffee mug holds about 300 ………….

    c A bed is about 2 ………. long.

    d The mass of my laptop is about 2.25 ……….

    e The distance between Washington DC and New York is about 370 ……….

    f A paper clip is about 30 ………. long.

3 A building brick weighs 2.7 kilograms.

Work out the weight of 840 bricks.

Give the answer in tonnes.

4 Cassie has a jug with 3 litres of orange juice.

Work out how many 180 millilitre cups she can fill with the juice.

5 Kane has a piece of wood 2.8 metres long.

He wants to cut it into short pieces each 12 centimetres long.

Work out how many short pieces he can cut.

6 A $1 coin has a mass of 9.5 grams.

Samir has a bag containing $500 in $1 coins.

Work out the mass of the coins in the bag in kilograms.

7 This is a recipe for buns.

1 cup of flour has a mass of 125 grams.

1 ml of water has a mass of 1 gram.

Work out the total mass of all the ingredients.

Give the answer in kilograms.

Recipe

150 g butter
175 g sugar
4 cups of flour
25 g salt
14 g yeast
200 ml water

8 One coin has a mass of 2.268 grams.

Vicki finds a bag containing 960 coins.

a Work out the total mass in kilograms of the coins in the bag.

Round your answer to 2 decimal places.

> Work out the total mass in grams first.

b Each coin is 1.35 millimetres thick.

Vicki tries to stack them up on top of each other.

What height would the pile be if all the coins are stacked up?

Give the answer in metres.

> Work out the total height in millimetres first.

9 a Work out how many centimetres there are in 1 kilometre.

> Convert this to metres first.

b Which metric unit is equal to 1 000 000 000 milligrams?

7.2 Converting imperial units

Miles and kilometres

For long distances the older, **imperial unit** is the mile.

In some countries, including Great Britain and the US, miles are still used on road signs.

The relationship between miles and kilometres is important and should be remembered

   8 kilometres = 5 miles

To change kilometres to miles, divide by 8 and multiply by 5.

To change miles to kilometres, divide by 5 and multiply by 8.

Worked example 1

The Australian cities Adelaide and Melbourne are about 450 miles apart.

Convert 450 miles to kilometres.

To change miles to kilometres, divide by 5 and multiply by 8.

   $450 \div 5 = 90$

   $90 \times 8 = 720$

The distance between the cities is about 720 kilometres.

Worked example 2

Paolo's car can cover about 70 miles before it runs out of fuel.

He sees this sign on the motorway.

Will he be able to get to the service station without running out of fuel?

Method 1: Change miles to kilometres by dividing by 5 and then multiplying by 8.

   $70 \div 5 = 14$

   $14 \times 8 = 112$

He is likely to run out of fuel in about 112 kilometres.

This is less than 120 kilometres so he is unlikely to reach the service station.

Method 2: Change kilometres to miles by dividing by 8 and then multiplying by 5.

   $120 \div 8 = 15$

   $15 \times 5 = 75$

The service station is 75 miles away.

This is greater than 70 miles so he is unlikely to reach the service station.

Next fuel on the motorway is at the service station in 120 km

Other imperial units

There are imperial units for length, mass and capacity as well many other quantities.

For everything except miles and kilometres, the conversion factor will be given in the question.

Worked example 3

Deniz has a car with a petrol tank that holds 9 gallons.

What is the capacity of his petrol tank in litres?

1 gallon is about 4.5 litres.

1 gallon = 4.5 litres

9 gallons = 4.5 × 9 = 40.5 litres

His petrol tank holds about 40.5 litres.

Worked example 4

Shaliya knows that her garage is 5.2 metres long.

She sees this advertisement for a ladder.

a Work out the length of her garage in feet.

b Find out if she would be able to fit the ladder in the garage.

1 foot is about 0.305 metres.

a 0.305 metres = 1 foot

1 metre = 1 ÷ 0.305 = 3.279 feet

5.2 metres = 3.279 × 5.2 = 17.05 feet

> This is the unitary method for proportionality. Find the equivalence for 1 metre by division first.

Her garage is just over 17 feet long.

b The ladder is 16 feet long.

This is shorter than 17 feet.

The ladder will fit into her garage.

Exercise 7.2

1 Convert the following distances to kilometres:

    a 30 miles   **b** 125 miles   **c** 55 miles   **d** 240 miles   **e** 64 miles

2 Convert the following distances to miles:

    a 32 km   **b** 72 km   **c** 160 km   **d** 240 km   **e** 75 km

7 Length, mass and capacity

3 An athlete runs 60 kilometres each week.

Work out how far that is in miles.

4 During a Grand Prix race the cars do 60 laps of the circuit.

Each lap is 5.141 kilometres.

  a Find the total distance driven in kilometres.

  b Work out the length of the race in miles.

5 Jon went on a walking holiday.

Over five days he walked 21 miles, 18 miles, 17 miles, $15\frac{1}{2}$ miles and 13 miles.

  a Find the total number of miles that he walked.

  b Convert the distance to kilometres.

6 Azka found an old 6-inch bolt in the garden shed.

Convert the length to centimetres.

1 inch is about 2.54 centimetres.

7 A fuel tank holds 150 litres.

How many gallons is this?

1 gallon is about 4.5 litres.

8 Carla has a 5-litre container of apple juice.

It is to be served in glasses that hold $\frac{1}{2}$ pint each.

Work out how many glasses she can fill from the container.

1 litre is about 1.75 pints.

9 Cristiano has a mass of 198 pounds.

Work out his mass in kilograms.

1 kilogram is about 2.2 pounds.

10 The distance around a racecourse is 2.9 kilometres.

Kaz plans to do a sponsored run around the racecourse to raise money for charity.

> Convert 2.9 km to miles.
> Remember, 8 km = 5 miles.

She wants to run at least 12 miles.

  a Work out the smallest number of laps of the course that she can run.

On the day she runs 9 laps of the course.

  b Work out how many miles she runs altogether.

11 The Olympic marathon race was originally set at 26 miles and 385 yards.

Work out the distance in kilometres.

There are 1760 yards in a mile.

> 385 yards is $\frac{385}{1760}$ mile.
> Change this to a decimal before working out how many kilometres there are.

87

12 George has a mass of 93.6 kilograms.

Hugo has a mass of 15 stones and 5 pounds.

a 1 kilogram is about 2.2 pounds.

Use this to convert George's mass to pounds.

> Multiply by 2.2 to convert kilograms to pounds.

b There are 14 pounds in 1 stone.

Use this to work out Hugo's mass in pounds.

c Write down which of them has the greater mass and by how much. Give your answer correct to the nearest pound.

Review

1 Convert the following to metres:

   a 467 cm   **b** 1254 mm   **c** 21.65 km   **d** 540 mm   **e** 26 cm

2 Convert the following to kilograms:

   a 32 600 g   **b** 875 g   **c** 2 tonnes   **d** 17.5 tonnes   **e** 0.3 tonne

3 Convert the following to millilitres:

   a 6 l   **b** 2.8 l   **c** 0.43 l   **d** 3 cl

4 Choose the correct metric unit to complete these sentences.

   a The mass of a pair of sunglasses is about 35 ....

   b A garden shed is about 3.5 .... long.

   c A car petrol tank holds 45 .... of fuel.

   d The lens of a pair of sunglasses is about 60 .... across.

5 Rahni has a jug that will hold 750 millilitres of water.

She wants to use it to fill a fish tank.

The tank will hold 12 litres of water.

Work out how many full jugs of water she will need to fill the tank.

6 A sack of rice has a mass of 30 kilograms.

What is the total mass of 45 sacks?

Give the answer in tonnes.

7 Rafad's bathroom wall is 2.4 metres high and 3.2 metres long.

He plans to cover the wall with tiles.

Each tile is a square with side of length 15 centimetres.

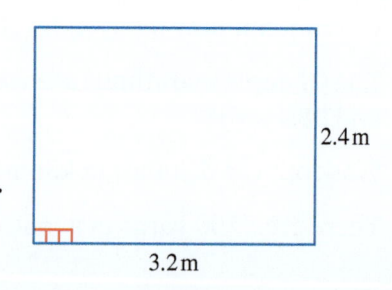

2.4 m

3.2 m

a Work out the number of rows of tiles that will fit up the wall.

b Work out how many tiles there are in each row.

c Find out how many tiles he needs to buy.

8 Convert the following to kilometres:

   a 10 miles   b 90 miles   c 250 miles   d 72 miles

9 Convert the following to miles:

   a 32 km   b 88 km   c 176 km   d 125 km

10 A salesperson drives 112 miles in the morning.

He then drives a further 97 miles in the afternoon.

In the evening he drives 76 miles back home.

a Work out the total number of miles that he drives.

He is paid to drive up to 500 kilometres in a day.

b Find out how many kilometres he drives.

Has he driven any unpaid kilometres that day?

11 The distance from London to Bristol is 190 kilometres.

The distance from London to Nottingham is 128 miles.

a Convert 128 miles to kilometres. Give the answer to the nearest whole number.

b Which is further from London, Nottingham or Bristol? How much further?

12 The Body Mass Index (BMI) is a measure of human body shape.

The formula for BMI is:

BMI $= \frac{m}{h^2}$, where m is the mass in kilograms and h is the height in metres

a Sarita is 1.48 m tall and has a mass of 50 kg.

   Work out her BMI. Give the answer to one decimal place.

b Marlon is 6 feet tall and has a mass of 200 pounds.

   i Convert his height to metres. (1 foot = 30 centimetres)

   ii Convert his mass to kilograms. (1 kilogram = 2.2 pounds)

   iii Work out his BMI. Give the answer to one decimal place.

Examination-style questions

1. Copy and complete the following:

   a 4520 millilitres = ... litres [1]

   b 432 millimetres = ... centimetres [1]

   c 7.4 tonnes = ... kilograms [1]

2. a A factory makes washing machines that weigh 62 kilograms each.

   Work out the total weight of 36 washing machines.

   Give your answer in tonnes. [2]

   b A lorry has a maximum load of 10 tonnes.

   How many washing machines can be loaded on to the lorry? [2]

3. The distance from Bristol to Birmingham is about 90 miles.

   Work out how many kilometres it is from Bristol to Birmingham. [2]

4. Jasper has a jug containing 2.4 litres of apple juice.

   He divides it equally into 6 glasses.

   Work out how many millilitres of apple juice there are in each glass. [2]

5. Renata bought a curtain rail 2.4 metres long.

   She needs a curtain rail 1.95 metres long for her window.

   Work out how many centimetres she needs to cut off the end to make it fit. [2]

6. Put these lengths in order starting with the smallest.

   0.28 m   29 mm   27 cm   2.8 m   285 mm [2]

7. Copy and complete the following:

   a 12.3 litres = ... millilitres [1]

   b 32.6 kilograms = ... grams [1]

   c 4235 centimetres = ... metres [1]

8. Ramesh weighed a book on a set of scales.

   The diagram to the right shows the scales.

   a Write down the weight of the book in grams. [1]

   Ramesh has 7 books to pack for posting.

   b Work out the total weight of all 7 books in kilograms. [2]

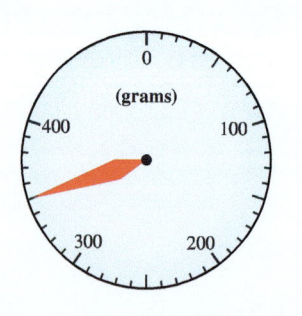

8 Equations and inequalities

> **Learning outcomes**
>
> - Construct and solve linear equations.
> - Understand and use inequality signs ($<$, $>$, $\leq$, $\geq$).
> - Construct and solve linear inequalities and represent solutions on a number line.
> - Use trial and improvement to find approximate solutions to equations.
> - Solve simultaneous equations.

Breaking records

In 2013 a Russian–American crew made a record-breaking journey from the Earth to the International Space Station.

The time taken from blast-off to docking was under 6 hours. This was a fraction of the usual journey time of over 50 hours.

This achievement could not have been possible without scientists and engineers solving some complicated equations.

8.1 Solving linear equations

The first two examples are a reminder of the work that you did in your Stage 8 Student Book on solving **equations**.

Worked example 1

Solve $\frac{x}{2} - 4 = 12$

$\frac{x}{2} - 4 = 12$ add 4 to both sides

$\frac{x}{2} = 16$ multiply both sides by 2

$x = 32$

Worked example 2

Solve $3(2x + 5) = 2(2x - 7) + 25$

$3(2x + 5) = 2(2x - 7) + 25$	expand the brackets
$6x + 15 = 4x - 14 + 25$	collect like terms
$6x + 15 = 4x + 11$	subtract $4x$ from both sides
$2x + 15 = 11$	subtract 15 from both sides
$2x = -4$	divide both sides by 2
$x = -2$	

To solve the equation $\frac{x+1}{4} = \frac{x+8}{5}$ you must multiply both sides of the equation by the lowest common multiple (LCM) of 4 and 5 which is 20.

$\cancel{20}^5 \times \frac{(x+1)}{\cancel{4}^1} = \cancel{20}^4 \times \frac{(x+8)}{\cancel{5}^1}$	cancel: $20 \div 4 = 5$ and $20 \div 5 = 4$
$5(x + 1) = 4(x + 8)$	expand brackets
$5x + 5 = 4x + 32$	subtract $4x$ from both sides
$x + 5 = 32$	subtract 5 from both sides
$x = 27$	

A quicker method can be used when there is a **single** fraction on both sides of the equals sign.

$\frac{x+1}{4} = \frac{x+8}{5}$	cross multiply
$5(x + 1) = 4(x + 8)$	expand brackets
$5x + 5 = 4x + 32$	subtract $4x$ from both sides
$x + 5 = 32$	subtract 5 from both sides
$x = 27$	

Exercise 8.1

1 Solve these equations.

- **a** $3x + 2 = 23$
- **b** $2x - 1 = -9$
- **c** $5x - 3 = 27$
- **d** $4x + 7 = 17$
- **e** $\frac{x+2}{5} = 4$
- **f** $\frac{2x-3}{4} = -2$
- **g** $\frac{x}{3} - 2 = 2$
- **h** $\frac{x}{5} + 6 = -2$
- **i** $15 - 2x = 9$
- **j** $30 - 4x = 24$
- **k** $3(2x - 7) = -27$
- **l** $2(5 - 2x) = 16$

2 Solve these equations.

- **a** $2(x + 3) + 4(2x - 1) = 22$
- **b** $3(2x - 2) + 5(x + 6) = 13$
- **c** $4(3x + 2) - 3(2x - 1) = 35$
- **d** $5(2x - 3) - 2(3x + 5) = -23$
- **e** $5x + 4 = 4x + 7$
- **f** $7x - 2 = 5x + 5$

g $8x + 1 = 5x - 8$ **h** $2x - 5 = 7 - x$

i $3(x + 1) = 2x - 8$ **j** $2 + 4(x - 3) = 2x + 7$

k $12 - 2x = 20 - 5x$ **l** $3(2x + 7) + 2 = 4(x - 8)$

3 Solve these equations.

  a $\dfrac{2x - 5}{2} = \dfrac{x + 1}{8}$ **b** $\dfrac{x - 1}{2} = \dfrac{2x + 10}{12}$ **c** $\dfrac{4x + 7}{3} = \dfrac{8x + 9}{5}$ **d** $\dfrac{2x + 2}{6} = \dfrac{x + 5}{5}$

  e $\dfrac{3x - 2}{7} = \dfrac{x - 2}{3}$ **f** $\dfrac{4x - 3}{9} = \dfrac{3x - 4}{5}$ **g** $\dfrac{2x + 1}{2} = \dfrac{5x + 7}{4}$ **h** $\dfrac{3x - 8}{6} = \dfrac{x - 2}{3}$

4 Rafiu thinks of a number.

  He adds 5 to his number and then divides by 7.

  The result is the same as when he subtracts 1 from his number and then divides by 4.

  Find the value of Rafiu's number.

5 The perimeter of this triangle is 15 cm.

  a Write down an equation in x.

  b Solve your equation in part **a** to find the value of x.

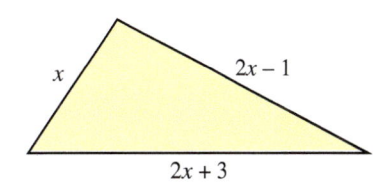

Triangle sides: x, $2x - 1$, $2x + 3$

6 Expression A is 7 less than expression B.

  Find the value of x.

  A $25 - 2x$ **B** $3(x + 4)$

7 Solve these equations.

  a $\dfrac{x}{3} + \dfrac{x}{4} = 14$ **b** $\dfrac{2x}{3} - \dfrac{x}{5} = 21$ **c** $\dfrac{6x}{7} - \dfrac{x}{3} = 22$ **d** $\dfrac{2x}{9} - \dfrac{x}{2} = 10$

8.2 Representing inequalities on a number line

There are four **inequality** signs that you need to know.

$a > 2$ means 'a is greater than 2'   or '2 is less than a'

$b \geqslant 5$ means 'b is greater than or equal to 5'   or '5 is less than or equal to b'

$c < 3$ means 'c is less than 3'   or '3 is greater than c'

$d \leqslant 4$ means 'd is less than or equal to 4'   or '4 is greater than or equal to d'

Inequalities are used in everyday life.

This speed limit sign says that the maximum speed, v km/h, of a car must be 100 km/h. This can be written as $v \leqslant 100$.

Inequalities are often represented on a number line.

$x < 2$ means that x is less than 2 and this can be shown as:

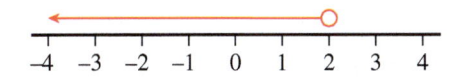

A hollow circle is used for $<$ or $>$.

$x \geqslant -3$ means that x is greater than or equal to -3 and this can be shown as:

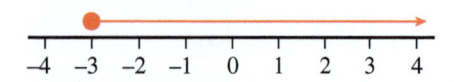

> A solid circle is used for $\leqslant$ or $\geqslant$.

$-2 < x \leqslant 3$ means that x is greater than -2 and it is also less than or equal to 3.

An easier way of describing this is to say:

x lies between -2 and 3. It cannot be equal to -2 but it can be equal to 3.

$-2 < x \leqslant 3$ can be shown as:

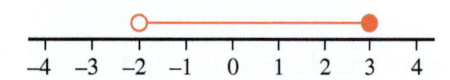

Worked example

Write down the inequalities shown on each of these number lines.

a   b   c

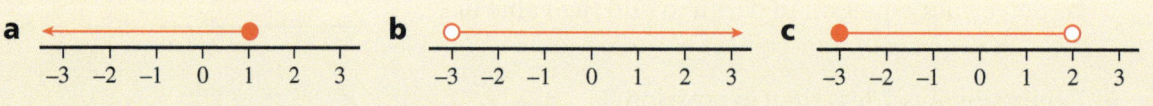

a $x \leqslant 1$ **b** $x > -3$ **c** $-3 \leqslant x < 2$

Exercise 8.2

1 Write down the inequalities shown on each of these number lines.

a   b   c

d   e   f

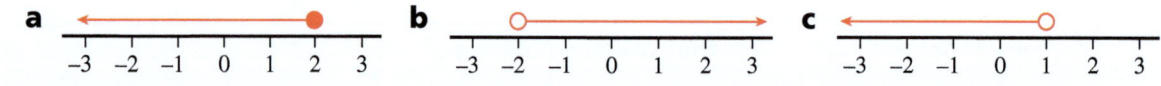

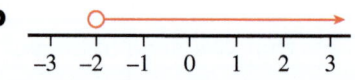

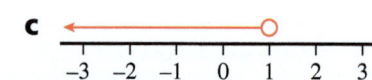

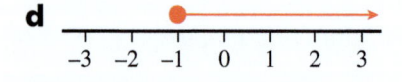

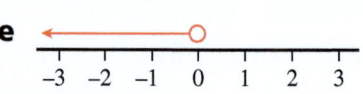

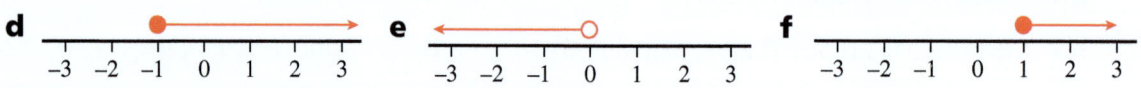

2 Show the following inequalities on number lines.

a $x > 2$ **b** $x < 4$ **c** $x \geqslant -5$ **d** $x \leqslant 1$

3 Write down the inequalities shown on each of these number lines.

a   b   c

d   e   f

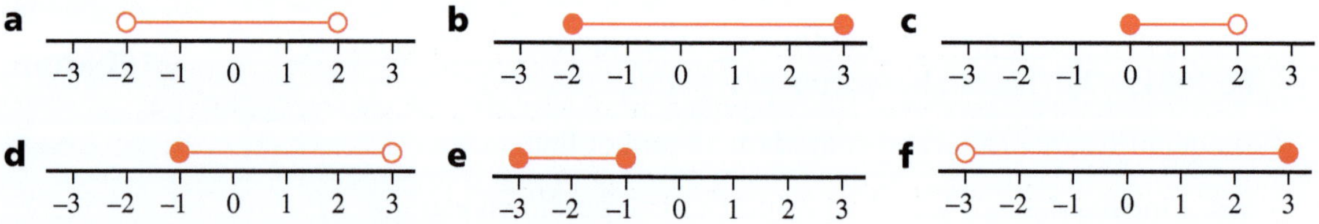

4 Show the following inequalities on number lines.

a $1 \leqslant x < 4$ **b** $-3 < x < 5$ **c** $-4 < x \leqslant 2$ **d** $3 \leqslant x \leqslant 5$

5 x is an integer and $-2 \leq x < 4$

List the possible values of x.

Integer means a whole number.

6 a Write down the smallest possible integer for which $-5 < x \leq 3$

  b Write down the greatest possible integer for which $-5 < x \leq 3$

8.3 Solving linear inequalities

The equation $x + 1 = 5$ can be shown on a balance diagram as:

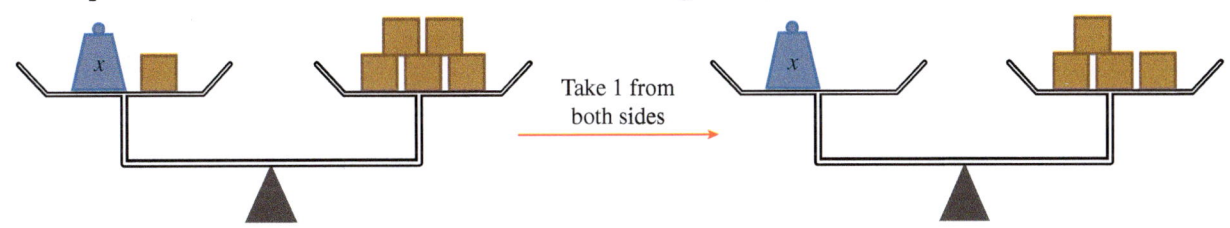

Take 1 from both sides

So the answer is $x = 4$.

There is only one value of x that satisfies the equation.

The inequality $x + 1 > 5$ can be shown on a balance diagram as:

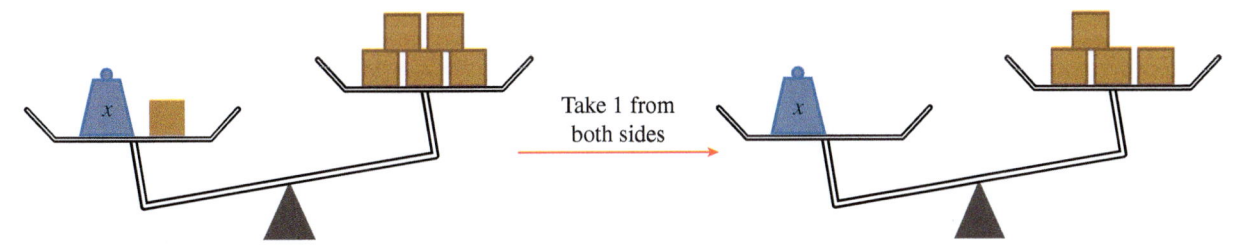

Take 1 from both sides

(The left-hand side of the inequality is heavier than the right-hand side of the inequality.)

So the answer is $x > 4$.

There is a range of values of x that satisfy the inequality.

x can be any number greater than 4.

Solving inequalities is very similar to solving equations.

When solving inequalities
- you can add the same thing to both sides
- you can subtract the same thing from both sides
- you can multiply or divide both sides by the same **positive** number
- if you multiply or divide by a **negative** number you must **reverse** the sign.

This last rule is illustrated by the following example:

Consider the inequality $\qquad -2 < 5$

If you multiply both sides by (-3) $\qquad 6 > -15$ — *The sign must be reversed to make the statement true.*

Mathematics for Cambridge Secondary 1

Worked example 1

Solve $2(3x - 4) < 10$ and represent your answer on a number line.

$2(3x - 4) < 10$	expand the brackets
$6x - 8 < 10$	add 8 to both sides
$6x < 18$	divide both sides by 6
$x < 3$	

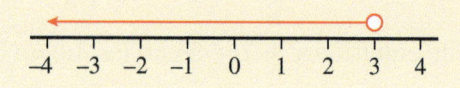

Worked example 2

Solve $9 - 3x \leqslant 3$

Method 1:

$9 - 3x \leqslant 3$	subtract 9 from both sides
$-3x \leqslant -6$	divide both sides by -3
$x \geqslant 2$	reverse the sign

Method 2:

$9 - 3x \leqslant 3$	add $3x$ to both sides
$9 \leqslant 3 + 3x$	subtract 3 from both sides
$6 \leqslant 3x$	divide both sides by 2
$2 \leqslant x$	

So, $x \geqslant 2$

Exercise 8.3

1 Solve these inequalities.

- **a** $x + 7 \leqslant 11$
- **b** $x - 2 > 3$
- **c** $x + 10 < 4$
- **d** $x - 5 > 4$
- **e** $x + 3 \geqslant -2$
- **f** $2x > 8$
- **g** $3x \geqslant 21$
- **h** $6x \leqslant 60$
- **i** $5x > -20$
- **j** $\frac{x}{3} < 5$
- **k** $\frac{x}{2} \geqslant -8$
- **l** $\frac{x}{7} < 15$
- **m** $2x + 1 < 9$
- **n** $4x + 3 \geqslant 15$
- **o** $3x - 2 > 1$
- **p** $5x + 2 \leqslant 12$
- **q** $2x + 3 > 8$
- **r** $5x - 2 < 3$
- **s** $4x + 1 \geqslant 7$
- **t** $2x + 9 > 6$

2 Solve these inequalities.

- **a** $7x + 2 > 3x + 22$
- **b** $6x - 3 \leqslant 5x + 4$
- **c** $12x + 5 < 8x + 13$
- **d** $3x - 5 \leqslant 2x + 8$
- **e** $4x - 5 \geqslant 2x + 7$
- **f** $3x + 8 > x - 2$
- **g** $6x + 7 < 3x - 5$
- **h** $9x + 10 < 7x - 3$
- **i** $3x + 2 \leqslant 12 - 2x$
- **j** $2x + 5 > 10 - 3x$
- **k** $3 - 5x \leqslant 6 - 7x$
- **l** $6 - 2x > 10 - 7x$

3 Solve these inequalities.

   a $-2x \geq 6$ **b** $-3x < 15$ **c** $-6x > -24$ **d** $-4x \leq 6$
   e $10 - 2x > 4$ **f** $15 - 4x \leq 3$ **g** $14 - 3x < 8$ **h** $2 - 5x \geq 7$
   i $6 - x \geq 10$ **j** $7 - 2x > 3$ **k** $60 - 6x \leq 12$ **l** $16 \geq 8 - 4x$

4 Solve these inequalities.

   a $x + 7 > 3x + 1$ **b** $2x + 14 \leq 6x - 2$ **c** $6 - 9x < 14 - 7x$
   d $36 - 27x > 15 - 20x$ **e** $40 - 16x \geq 14 + 2x$ **f** $24 - 9x \leq 12 - 6x$
   g $7 - 5x > 5 - 2x$ **h** $7 - 12x \geq 25 - 3x$ **i** $1 - x \leq 5x + 7$
   j $6 - 3x \leq 13 + x$ **k** $3 - 9x < 2(3x - 5)$ **l** $5 - 2(x + 3) > 8 + x$

5 The perimeter of the regular pentagon is less than 42 cm.

Write down and solve an inequality in x.

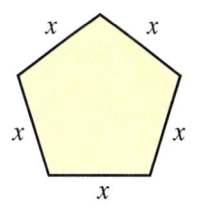

6 The perimeter of the rectangle is greater than 36 cm.

Write down and solve an inequality in x.

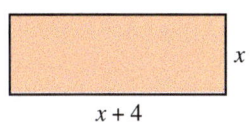

7 The perimeter of this triangle must not be more than 45 cm.

Find the maximum value that x can be.

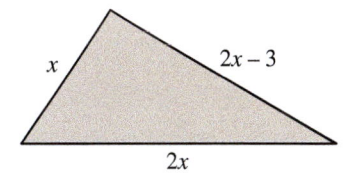

8 a Write down the smallest possible integer that satisfies the inequality $3x > 14$

   b Write down the smallest integer that satisfies the inequality $2x - 5 > 16$

   c Write down the largest integer that satisfies the inequality $4x - 15 < 6$

9 Solve the inequality $4 < 2x \leq 10$ and show your solution on a number line.

10 Solve these inequalities and then list the integer solutions.

   a $6 \leq 2x \leq 8$ **b** $-8 < 2x < 6$ **c** $3 \leq 3x \leq 18$ **d** $2 < \frac{x}{2} \leq 4$
   e $3 < x + 2 < 7$ **f** $2 \leq x - 4 < 5$ **g** $-3 \leq 2x + 1 \leq 7$ **h** $-5 < 3x - 2 \leq 7$

11 Solve these inequalities.

   a $x + 1 \leq \frac{x - 1}{4}$ **b** $\frac{x - 2}{2} > \frac{x + 3}{7}$ **c** $\frac{x - 5}{3} \leq \frac{5 - x}{2}$ **d** $\frac{2x - 3}{4} \geq \frac{3x - 2}{5}$

12 $x + 3 \geq 7$ $2x - 3 < 20$

   a Solve each of the inequalities.

   b Find the range of values of x which satisfy both inequalities.

13 $4x + 5y \leq 20$ and x and y are both positive integers (not including zero).

List all the possible pairs of values for x and y.

8.4 Trial and improvement

Some equations are too difficult to solve using algebra.

These equations are solved using a method called **trial and improvement**.

In trial and improvement you estimate the answer and then try to improve on your estimate.

Worked example 1

Use trial and improvement to solve the equation $x^3 - 3x = 322$

Try $x = 5$:	$5^3 - 3 \times 5 = 125 - 15 = 110$ →	too small
Try $x = 6$:	$6^3 - 3 \times 6 = 216 - 18 = 198$ →	too small
Try $x = 7$:	$7^3 - 3 \times 7 = 343 - 21 = 322$ →	correct

The solution is $x = 7$

In the last example it was possible to find the exact answer to the equation.

Sometimes it is not possible to find the exact answer so an approximate answer must be found.

It helps to use a table to show your working.

Worked example 2

The area of the rectangle is 72 cm².

Use trial and improvement to find the value of x correct to 1 decimal place.

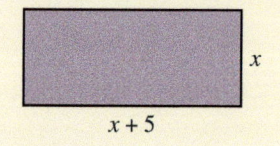

Area of rectangle = base × height

So $x(x + 5) = 72 \Rightarrow x^2 + 5x = 72$

Estimate	$x^2 + 5x$	Comment	
6	$6^2 + 5 \times 6 = 66$	too small	
7	$7^2 + 5 \times 7 = 84$	too big	So x lies between 6 and 7
6.3	$6.3^2 + 5 \times 6.3 = 71.19$	too small	
6.4	$6.4^2 + 5 \times 6.4 = 72.96$	too big	So x lies between 6.3 and 6.4
6.35	$6.35^2 + 5 \times 6.35 = 72.0725$	too big	

The solution is between $x = 6.3$ and $x = 6.35$

So, the solution is $x = 6.3$ cm to 1 d.p.

> The final trial of $x = 6.35$ is needed to find if the answer is closer to 6.3 than it is to 6.4. You must always work to one more decimal place than is required in your answer.

8 Equations and inequalities

Exercise 8.4

You must use trial and improvement for ALL questions in this exercise.

1. Theo is using trial and improvement to solve the equation $x^3 - 5x = 56$

   The table shows his first two estimates.

Estimate	$x^3 - 5x$	Comment
4	$4^3 - 5 \times 4 = 44$	too small
5	$5^3 - 5 \times 5 = 100$	too large

   Continue the table to find a solution to the equation.

   Give your answer correct to 1 decimal place.

2. In this question you must find the solution to the equation that is in the given interval.

   Give your answers correct to 1 decimal place.

   a $x^2 + 5x = 60$ if x lies between 5 and 6.

   b $x^2 - 2x = 58$ if x lies between 8 and 9.

   c $x + \dfrac{10}{x} = 8$ if x lies between 6 and 7.

3. Solve these equations correct to 1 decimal place.

   a $x^3 + x = 4$ **b** $x^3 - 2x = 10$ **c** $2x^3 - 7x = 9$ **d** $2^x = 20$

4. Write down an equation in x for each of these rectangles.

   Solve each of your equations to find the value of x correct to 1 decimal place.

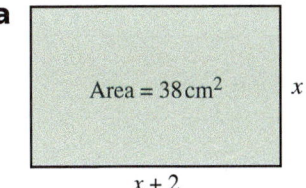

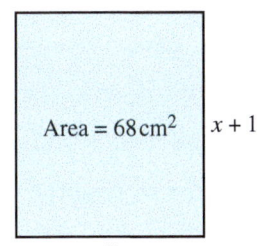

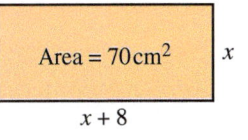

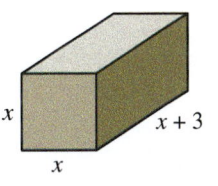

   a Area = 38 cm², sides x and $x + 2$

   b Area = 68 cm², sides x and $x + 1$

   c Area = 70 cm², sides x and $x + 8$

5. The cuboid has a volume of 250 cm³.

   a Show that $x^3 + 3x^2 = 250$

   b Find the value of x correct to 1 decimal place.

   (Cuboid with sides x, x, $x + 3$)

6. The area of the shape is 40 cm².

   a Show that $x^2 + 12x = 40$

   b Find the value of x correct to 1 decimal place.

   > Split the shape into two rectangles

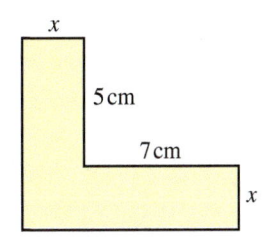

   (L-shape with measurements x, 5 cm, 7 cm, x)

7 Triangle A is an enlargement of triangle B.

  a Write down an equation in x and show that it simplifies to $x^2 - 3x = 35$

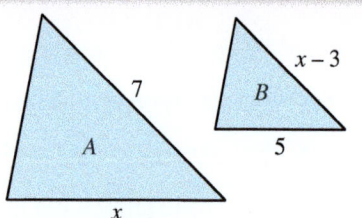

  b Find the value of x correct to 1 decimal place.

8 The surface area, A, of a solid cylinder with radius r and height h is $A = 2\pi r^2 + 2\pi rh$

The solid cylinder shown has a surface area of 300 cm²

  a Show that $\pi r^2 + 10\pi r = 150$

  b Find the value of r correct to 1 decimal place.

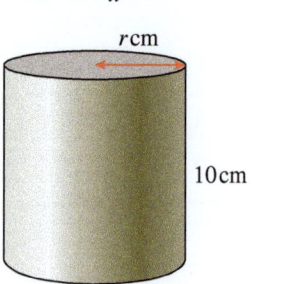

8.5 Simultaneous equations

The equation $2x + y = 6$ has two variables x and y.

There are many pairs of values of x and y which satisfy the equation $2x + y = 6$

Some possible pairs are: $x = 1, y = 4$ $x = 2, y = 2$ $x = 4, y = -2$

The equation $4x + y = 8$ also has two variables x and y.

Again, there are many pairs of values of x and y which satisfy the equation $4x + y = 8$

Some possible pairs are: $x = 0, y = 8$ $x = 1$ and $y = 4$ $x = 2$ and $y = 0$

When you solve a pair of **simultaneous equations** you need to find one pair of values for x and y that satisfy both equations at the same time.

The solution to the pair of simultaneous equations:

   $4x + y = 8$
   $2x + y = 6$

is $x = 1, y = 4$ because this pair of values satisfies both equations.

It is quicker to use an algebraic method for solving the simultaneous equations as shown below.

   $4x + y = 8$ [1]
   $2x + y = 6$ [2]

Equation [1] − equation [2] will eliminate the y terms.

   $2x = 2$
   $x = 1$

To find the value of y substitute $x = 1$ in equation [1] and then solve.

   $4 + y = 8$
   $y = 4$

The solution is $x = 1, y = 4$

> Always check your answer by substituting the x and y values into the original equation.

Worked example 1

Solve the simultaneous equations: $5x + 2y = 11$
$x + 2y = -1$

$5x + 2y = 11$ [1]

$x + 2y = -1$ [2]

Equation [1] − equation [2] will eliminate the y terms.

$4x = 12$

$x = 3$

To find the value of y substitute $x = 3$ in equation [1] and then solve.

$15 + 2y = 11$

$2y = -4$

$y = -2$

The solution is $x = 3, y = -2$

Check: $(5 \times 3) + (2 \times -2) = 15 + -4 = 11$ ✓ and $3 + (2 \times -2) = 3 + -4 = -1$ ✓

Worked example 2

Solve the simultaneous equations: $4x + 3y = 29$
$2x - 3y = -17$

$4x + 3y = 29$ [1]

$2x - 3y = -17$ [2]

Equation [1] + equation [2] will eliminate the y terms.

$6x = 12$

$x = 2$

To find the value of y substitute $x = 2$ in equation [1] and then solve.

$8 + 3y = 29$

$3y = 21$

$y = 7$

The solution is $x = 2, y = 7$

Check: $(4 \times 2) + (3 \times 7) = 8 + 21 = 29$ ✓ and $(2 \times 2) - (3 \times 7) = 4 - 21 = -17$ ✓

Exercise 8.5a

1 Solve the simultaneous equations.

  a $2x + 3y = 16$
     $x + 3y = 14$

  b $5x + y = 17$
     $2x + y = 8$

  c $4x + y = 9$
     $3x + y = 8$

  d $5x + 2y = 14$
     $x + 2y = 6$

  e $7x + 5y = 26$
     $3x + 5y = 14$

  f $x + 7y = 18$
     $x + 4y = 12$

  g $3x + 4y = 26$
     $x + 4y = 14$

  h $3x + 5y = 21$
     $3x - 2y = 0$

  i $4x + 3y = 31$
     $4x - 2y = 6$

  j $x - 4y = 12$
     $x - 7y = 6$

  k $7x + 3y = -5$
     $2x + 3y = 5$

  l $x + 2y = -2$
     $x - 5y = 19$

2 Solve the simultaneous equations.

  a $3x + y = 19$
     $2x - y = 6$

  b $3x + 2y = 22$
     $x - 2y = 2$

  c $x + 3y = 13$
     $2x - 3y = -10$

  d $5x - 3y = -21$
     $3x + 3y = -3$

  e $5x + 2y = 23$
     $3x - 2y = 17$

  f $6x - y = -16$
     $5x + y = -6$

  g $7x - 3y = 59$
     $2x + 3y = 13$

  h $8x - 5y = -14$
     $2x + 5y = -16$

3 Solve the simultaneous equations.

  a $3x + 2y = 12$
     $x + 2y = 0$

  b $6x + y = -8$
     $4x + y = -4$

  c $5x + 2y = 1$
     $5x - 7y = -26$

  d $3x - 5y = -11$
     $4x + 5y = -3$

  e $5x - 4y = -3$
     $3x - 4y = -13$

  f $4x - y = 14$
     $6x + y = 16$

  g $4x - 5y = 37$
     $4x + 2y = 2$

  h $3x - 2y = 4$
     $3x + 2y = -4$

  i $2x + 7y = 16$
     $2x - 3y = -24$

  j $4x - 3y = 21$
     $3x - 3y = 18$

  k $4x - 7y = -58$
     $3x + 7y = 30$

  l $7x - 3y = -4$
     $2x - 3y = 1$

4 The sum of two numbers is 22 and the difference is 9.

Let the two numbers be x and y.

  a Write down two equations in x and y.

  b Solve your two equations.

5 Find the values of x and y.

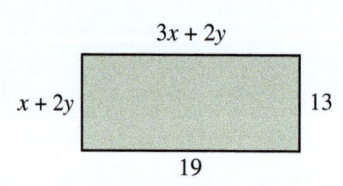

(Rectangle with sides labelled: $x + 2y$, $3x + 2y$, 13, 19)

6 The mean of two numbers is 44 and the difference is 17.

Find the values of the two numbers.

8 Equations and inequalities

To eliminate one of the variables you might first need to multiply one or both of the equations by a suitable number before adding and subtracting. This is shown in the next two examples.

Worked example 3

Solve the simultaneous equations: $x + 2y = 6$
$3x - 4y = 28$

$x + 2y = 6$ [1]
$3x - 4y = 28$ [2]

Multiply equation [1] by 3 to make the x terms match.

$3x + 6y = 18$
$3x - 4y = 28$

Subtracting these two equations will eliminate the x terms.

$10y = -10$ remember that $+6 - (-4) = 6 + 4$
$y = -1$

To find the value of x substitute $y = -1$ in equation [1] and then solve.

$x - 2 = 6$
$x = 8$

The solution is $x = 8, y = -1$

Check: $8 + (2 \times -1) = 8 + -2 = 6$ ✓ and $(3 \times 8) - (4 \times -1) = 24 - (-4) = 28$ ✓

Worked example 4

Solve the simultaneous equations: $5x + 3y = 26$
$8x + 5y = 43$

$5x + 3y = 26$ [1]
$8x + 5y = 43$ [2]

Multiply equation [1] by 5 and equation [2] by 3 to make the y terms match.

$25x + 15y = 130$
$24x + 15y = 129$

Subtracting these two equations will eliminate the y terms.

$x = 1$

To find the value of y substitute $x = 1$ in equation [1] and then solve.

$5 + 3y = 26$
$3y = 21$
$y = 7$

The solution is $x = 1, y = 7$

Check: $(5 \times 1) + (3 \times 7) = 5 + 21 = 26$ ✓ and $(8 \times 1) + (5 \times 7) = 8 + 35 = 43$ ✓

Exercise 8.5b

1 Solve the simultaneous equations.

   a $2x + y = 11$
      $5x + 3y = 30$

   b $8x - 3y = 26$
      $3x + y = 14$

   c $5x + 3y = 43$
      $4x + 6y = 38$

   d $3x + 2y = 27$
      $4x - y = 25$

   e $3x + y = 11$
      $x - 3y = -13$

   f $4x + y = 11$
      $5x + 2y = 13$

   g $4x - 5y = 31$
      $x - 3y = 13$

   h $5x - y = -16$
      $7x + 4y = 10$

   i $7x + 2y = 13$
      $3x + 4y = 15$

   j $x + 2y = 1$
      $3x - 4y = 23$

   k $4x + 3y = 13$
      $x + 5y = -1$

   l $2x + 3y = 5$
      $x + 5y = -8$

2 Solve the simultaneous equations.

   a $2x - 3y = 0$
      $5x + 2y = 38$

   b $3x + 2y = 19$
      $2x - 3y = -22$

   c $5x + 3y = 32$
      $2x + 4y = 24$

   d $4x + 3y = -2$
      $3x + 5y = 4$

   e $3x - 2y = 24$
      $5x - 7y = 51$

   f $2x + 3y = 7$
      $5x - 2y = 27$

   g $4x + 3y = 1$
      $5x + 2y = -4$

   h $2x + 5y = 8$
      $9x - 4y = -17$

   i $5x - 7y = 38$
      $3x - 4y = 23$

   j $2x + 2y = 10$
      $5x + 3y = 17$

   k $5x - 2y = 33$
      $3x + 7y = -13$

   l $2x + 3y = 5$
      $5x - 2y = 41$

3 Three times one number added to five times another number gives 2.

   The sum of the two numbers is 2.

   Find the two numbers.

4 Seven cans of cola and five cans of lemonade cost $15.55.

   Four cans of cola and three cans of lemonade cost $9.06.

   Find the cost of a can of cola.

5 Theatre tickets cost $27.50 per adult and $16 per child.

   One evening the theatre sells 342 tickets for a total of $7599.50.

   How many child tickets were sold?

6 Solve the simultaneous equations.

   a $3x + 2y = 11$
      $y - 2x = 3$

   b $y = x + 3$
      $3y + x = 5$

   c $3y - 5x = 21$
      $3x + y + 7 = 0$

   d $\dfrac{x}{2} + \dfrac{y}{5} = 9$
      $\dfrac{x}{4} - \dfrac{y}{3} = -2$

7 $\boxed{2x + y + 5}$ $\boxed{5x - 2y + 17}$ $\boxed{4x + 3y - 63}$

   Find the value of x and the value of y that makes each of the three expressions equal.

8 A, B and C are the centres of the three circles.

AB = 11, AC = 10 and BC = 8

Find the radii of the three circles.

9 Solve these simultaneous equations.

$2x + y + z = 11$

$2x - y + z = -1$

$x + 2y + 2z = 13$

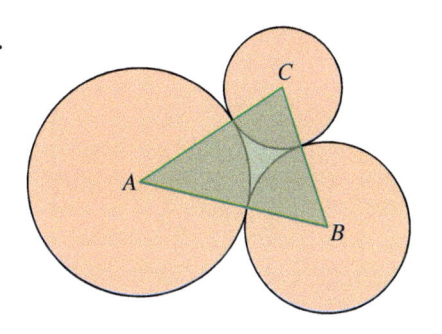

Review

1 Solve these equations.

   a $3x = 12$ **b** $x + 5 = 16$ **c** $3x - 4 = 11$

   d $5x + 4 = 6 + 3x$ **e** $\frac{x+1}{4} = 5$ **f** $\frac{7-2x}{3} = 5$

   g $\frac{x}{4} + 8 = 2$ **h** $3(2x - 7) = 4x + 3$ **i** $\frac{15+x}{3} = 7 - x$

   j $\frac{x+5}{2} = \frac{x-4}{3}$ **k** $5(3x - 2) - 2(3x - 4) = 13$

2 Solve these inequalities.

For each inequality show your solution on a number line.

   a $2x \geq 8$ **b** $x - 4 < 7$ **c** $3x + 1 \geq 8$ **d** $3x + 5 \leq x - 8$

3 Simone is using trial and improvement to find the answer to the equation $2x - \frac{1}{x} = 9$

She knows that the answer lies between 4 and 5.

The table shows her first two estimates.

Estimate	$2x - \frac{1}{x}$	Comment
4	$(2 \times 4) - (1 \div 4) = 7.75$	too small
5	$(2 \times 5) - (1 \div 5) = 9.8$	too large

Continue the table to find a solution to the equation.

Give your answer correct to 1 decimal place.

4 Solve these simultaneous equations.

   a $5x + 2y = 13$ **b** $3x + 2y = 18$ **c** $3x - 2y = 15$

       $x + 2y = 9$ $2x - y = 5$ $2x - 3y = 5$

Examination-style questions

1. Solve the equation: $2x + 7 = 19$ [2]

2. Solve the equation: $\dfrac{8-x}{3} = 6$ [2]

3. Solve the equation: $4(x - 1) + 3(x + 2) = 30$ [3]

4. The perimeter of the triangle is 50 cm.

    Write down and solve an equation in x. [3]

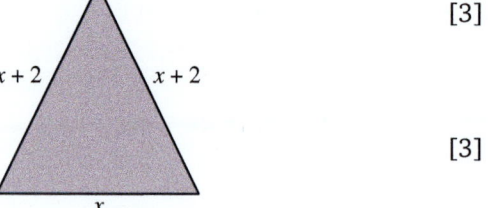

5. $7 < 2x \leqslant 14$

    If x is an integer write down all the possible values of x. [3]

6. a Solve $7x > 5x - 6$ [2]

    b Show your answer on a number line. [1]

7. Amelia is using trial and improvement to solve the equation $x^3 + 2x = 26$

    The table shows her first two estimates.

Estimate	$x^3 + 2x$	Comment
2	$2^3 + 2 \times 2 = 12$	too small
3	$3^3 + 2 \times 3 = 33$	too large

    Continue the table to find a solution to the equation.

    Give your answer correct to 1 decimal place. [3]

8. Solve the simultaneous equations.

    $3x + y = 10$

    $2x - y = 10$ [2]

9. Solve the simultaneous equations.

    $5x + 3y = 6$

    $2x + y = 1$ [3]

9 Shapes and geometric reasoning 2

> **Learning outcomes**
> - Analyse 3-D shapes through plans and elevations.
> - Draw 3-D shapes on isometric paper.
> - Identify reflection symmetry in 3-D shapes.
> - Use a straight edge and compasses to:
>   - construct the perpendicular from a point to a line and the perpendicular from a point on a line
>   - inscribe squares, equilateral triangles, and regular hexagons and octagons by constructing equal divisions of a circle.
> - Use bearings (angles measured clockwise from the north) to solve problems involving distance and direction.
> - Make and use scale drawings and interpret maps.

Building up a picture

Blueprints are accurate plans of buildings or other constructions.

They got their name because they were originally printed on blue coloured paper.

This example has been drawn by an architect to show exactly how a new house is to be built.

Computer drawing packages are now used to help with this process.

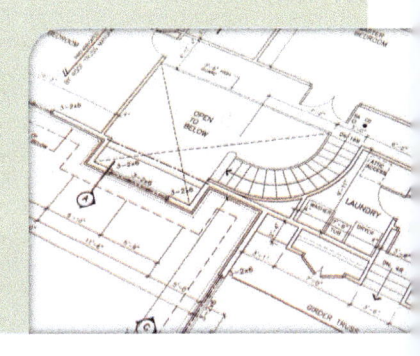

9.1 Three dimensions

Plans and elevations

Here is a picture of a **three-dimensional** solid made out of four cubes. The side of each cube is 2 cm long.

A drawing like this shows the shape but is not a scale drawing.

Using the **plan** and **elevations** it can be shown full size or to scale.

This diagram shows the direction of the plan and the elevations.

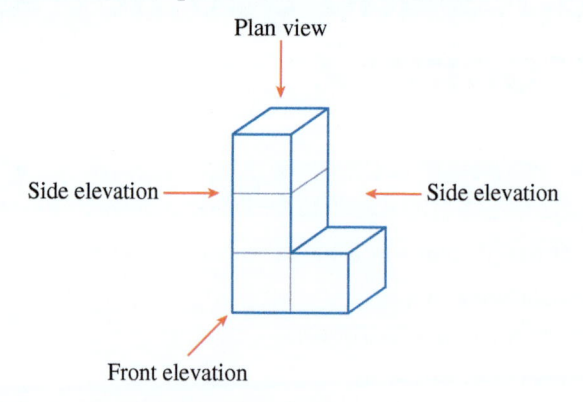

The plan view is looking vertically down on the shape.

The **front elevation** is looking horizontally from the front.

The **side elevations** are looking horizontally from each side.

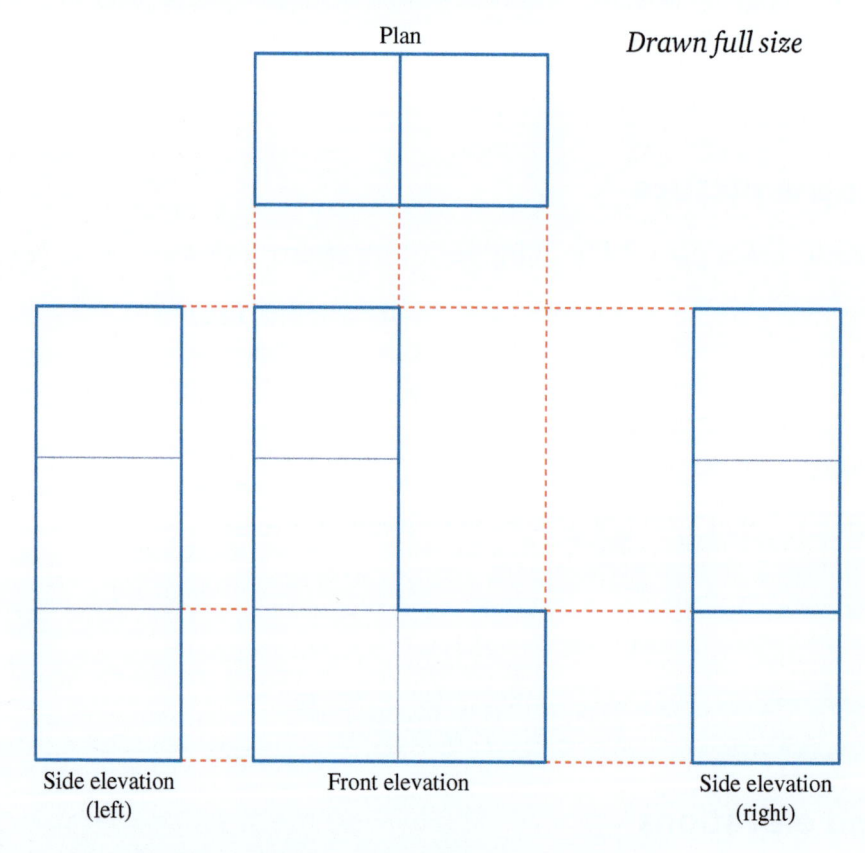

Drawn full size

This is drawn full size. For bigger shapes a scale drawing can be done.

The dotted red lines are not necessary but they help to show that the plan and the elevations have the same dimensions.

It is not usually necessary to draw both of the side elevations.

Note the darker line on the right elevation.

This shows that there is a change of level (or a step) facing the elevation.

9 Shapes and geometric reasoning 2

Isometric drawings

Another way to show three-dimensional objects in **two dimensions** is to use **isometric paper**.

Isometric paper has a triangular array of dots, usually 1 centimetre apart.

It is important that the paper is the right way up when drawing solids.

One of the rows of dots should be vertical on the paper.

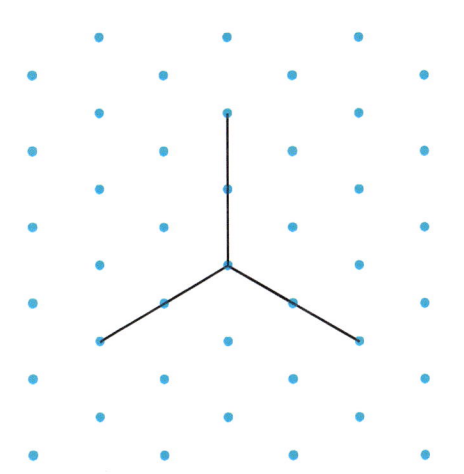

This is the correct way up.

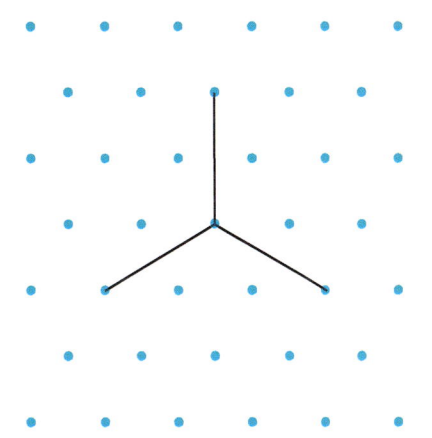

This is the wrong way up.

Here is an isometric drawing of the L-shaped solid.

Note that the vertical lines look vertical in the diagram.

All the lines follow the rows of dots.

The diagram is drawn full size.

Each line is the correct length.

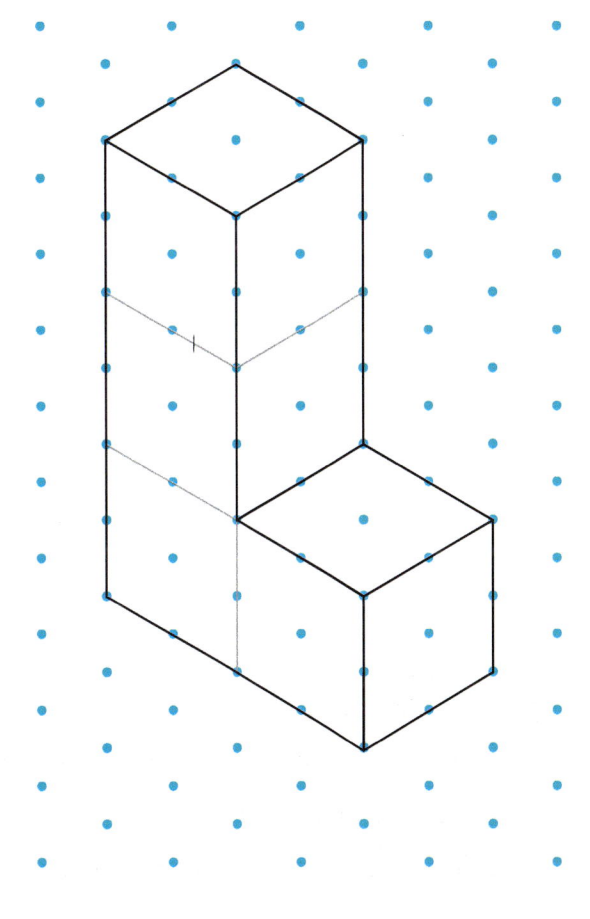

109

Worked example

These are the plan and elevations of a solid.

Plan

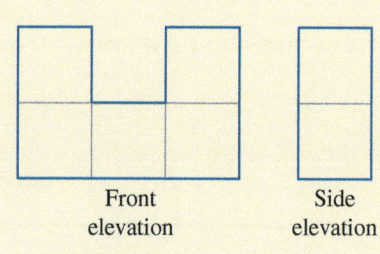

Front elevation

Side elevation

It often helps to sketch the solid first.

This gets easier with practice.

The sketch to the right shows the solid.

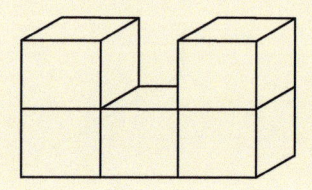

Start by drawing the top of the shape on isometric paper.

Make sure the paper is the right way round.

Each of the small squares appears as a rhombus.

The sides are each 1 centimetre long.

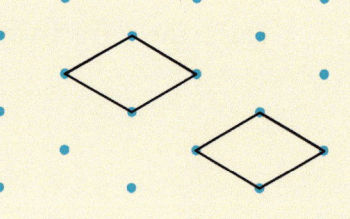

Then draw the vertical lines.

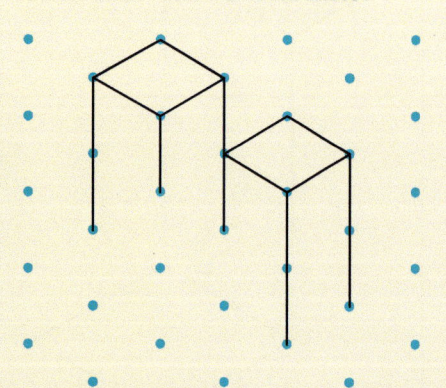

Now finish the diagram off with the rest of the lines.

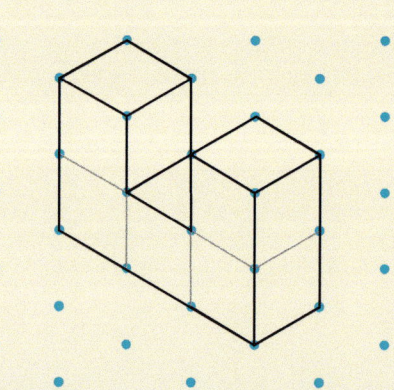

9 Shapes and geometric reasoning 2

Exercise 9.1

1 Draw the plan, the front elevation and one side elevation for each of these shapes.

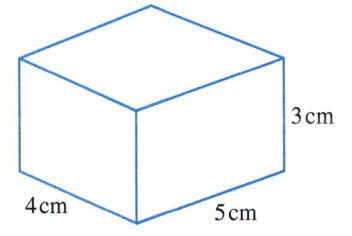

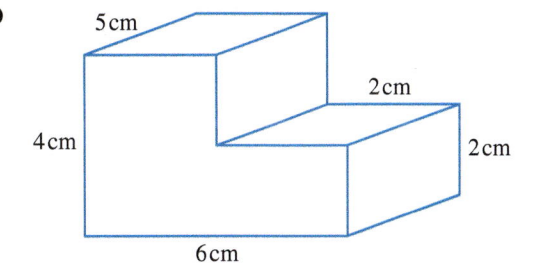

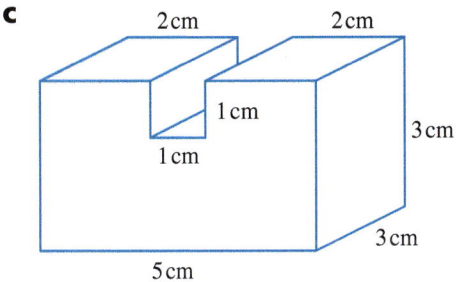

a (3 cm, 4 cm, 5 cm cuboid)

b (stepped shape: 5 cm, 4 cm, 2 cm, 2 cm, 6 cm)

c (shape with notch: 2 cm, 2 cm, 1 cm, 1 cm, 3 cm, 3 cm, 5 cm)

2 Use isometric paper to draw each of the shapes in question **1**.

3 The diagrams show the plan and elevations of a three-dimensional shape.

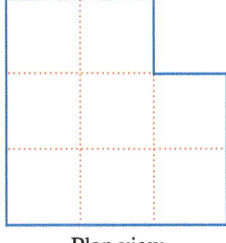

Plan view   Front elevation   Side elevation

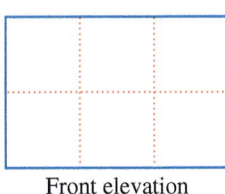

Draw the shape on to isometric paper.

4 In each part the diagram shows the plan, the front elevation and one side elevation of a three-dimensional shape.

Sketch each of the shapes and write down their mathematical name.

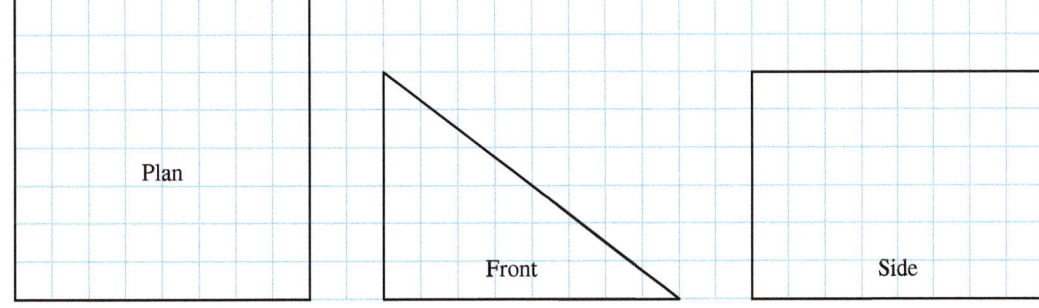

a   Plan   Front   Side

111

b

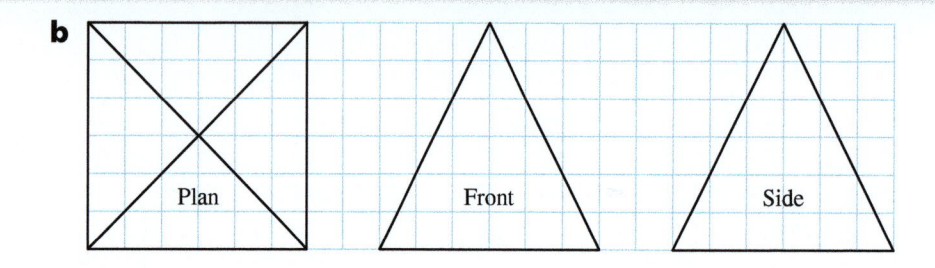

| Plan (square with diagonals) | Front (triangle) | Side (triangle) |

c

| Plan (circle) | Front (rectangle) | Side (rectangle) |

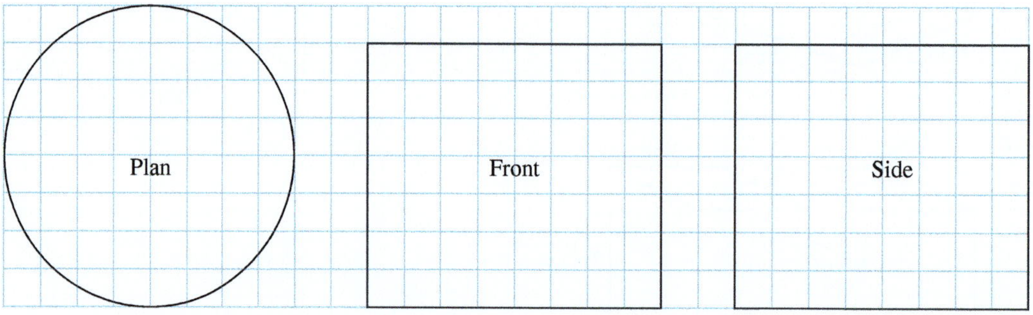

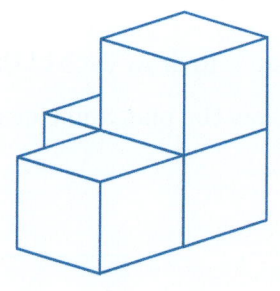

5 This solid object is made up of four cubes.

The side of each cube is 2 centimetres long.

Draw the plan and elevations of the object.

6 The three-dimensional shape in question **5** is made up of four cubes joined together.

Investigate other three-dimensional shapes that can be made by joining four cubes together. The cubes must join face to face with no overlaps.

> Use small cubes to make the shapes before drawing them.

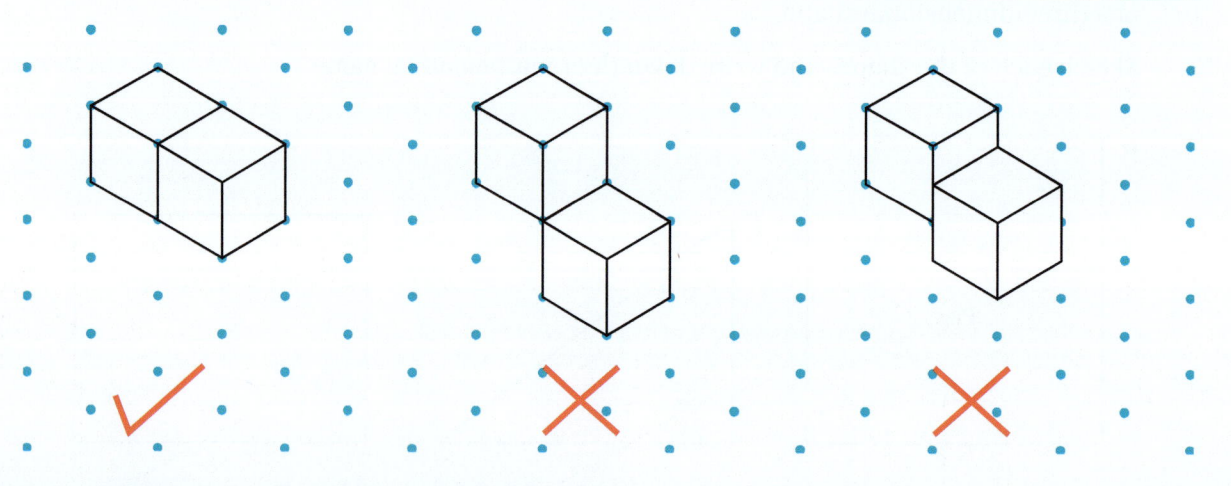

Draw each of the shapes on isometric paper.

112

9 Shapes and geometric reasoning 2

7 The diagrams show the plan and elevations of a three-dimensional shape.

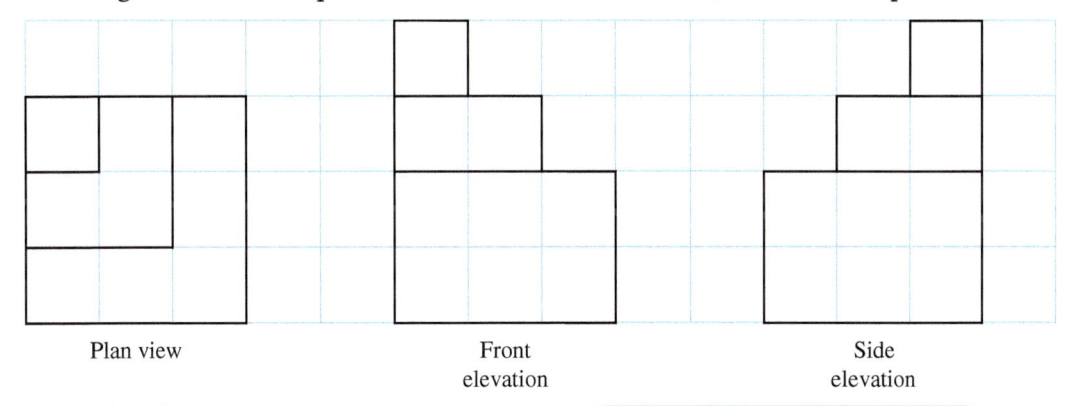

Plan view   Front elevation   Side elevation

Draw the shape on isometric paper.

> It may help to make the shape out of cubes before drawing it.

9.2 Reflection symmetry in three dimensions

In two dimensions a line of symmetry divides a shape into two **congruent** parts.

The two parts are mirror images of each other.

Here are some examples of **reflection symmetry** in two dimensions.

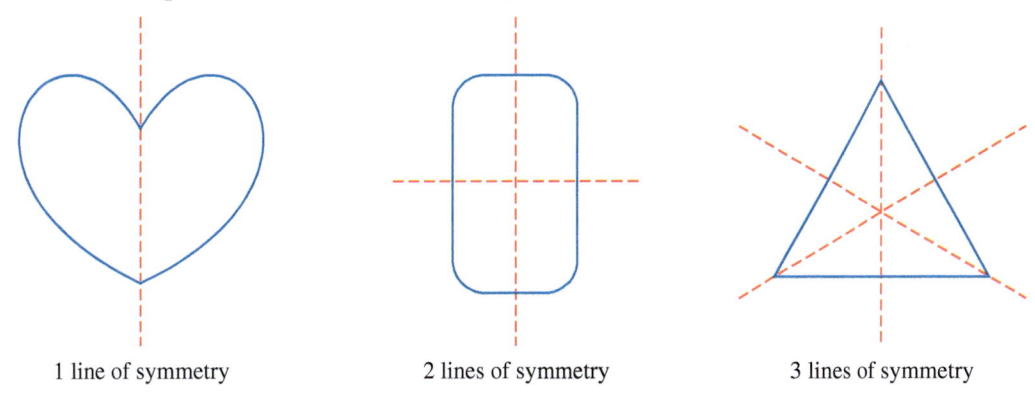

1 line of symmetry   2 lines of symmetry   3 lines of symmetry

For reflection symmetry in three dimensions a **plane of symmetry** divides a solid into two congruent parts.

> Plane is another word for a flat surface.

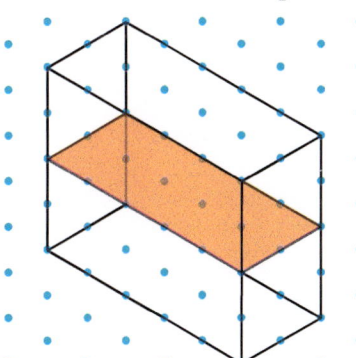

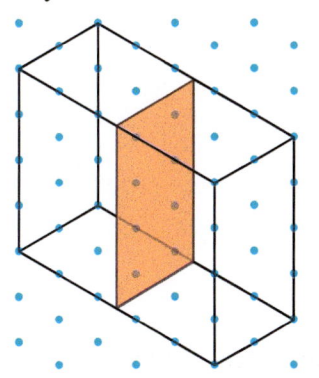

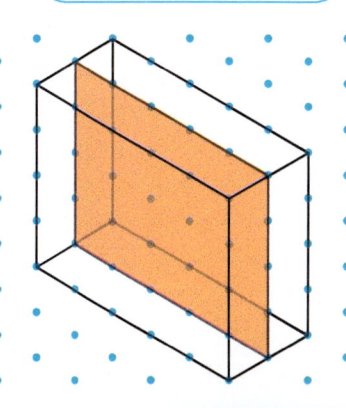

This cuboid has three planes of symmetry.

113

Mathematics for Cambridge Secondary 1

Worked example

Work out how many planes of symmetry there are in this solid.

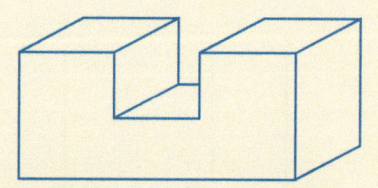

It is helpful to draw sketches to find the planes of symmetry.

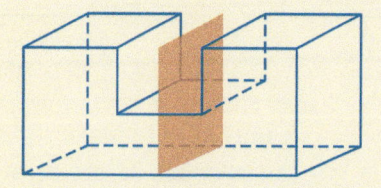

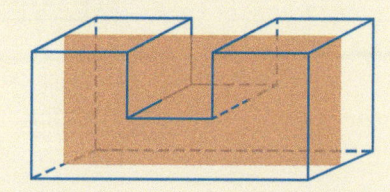

These show the two planes of symmetry for this shape.

Exercise 9.2

1 The diagram shows a square based pyramid.

One plane of symmetry is shown on the diagram.

The pyramid has three more planes of symmetry.

Draw diagrams to show the other three planes of symmetry.

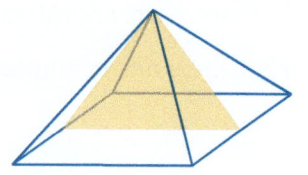

2 Find how many planes of symmetry there are for each of these solid shapes.

a A cube

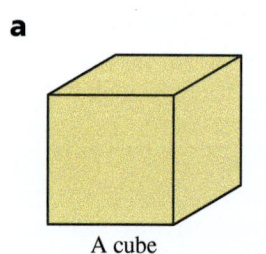

b An equilateral triangle prism

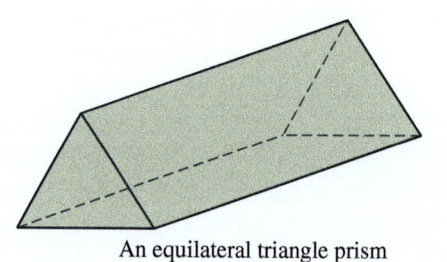

c An L-shaped prism

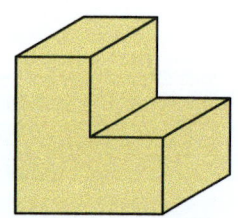

d A right-angled triangle prism

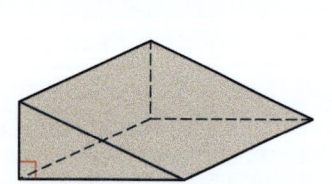

e A regular hexagonal prism

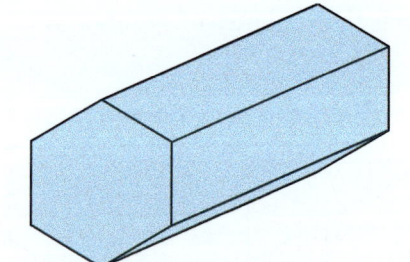

114

3 These three-dimensional shapes are each made from five cubes.

Find out how many planes of symmetry each one has.

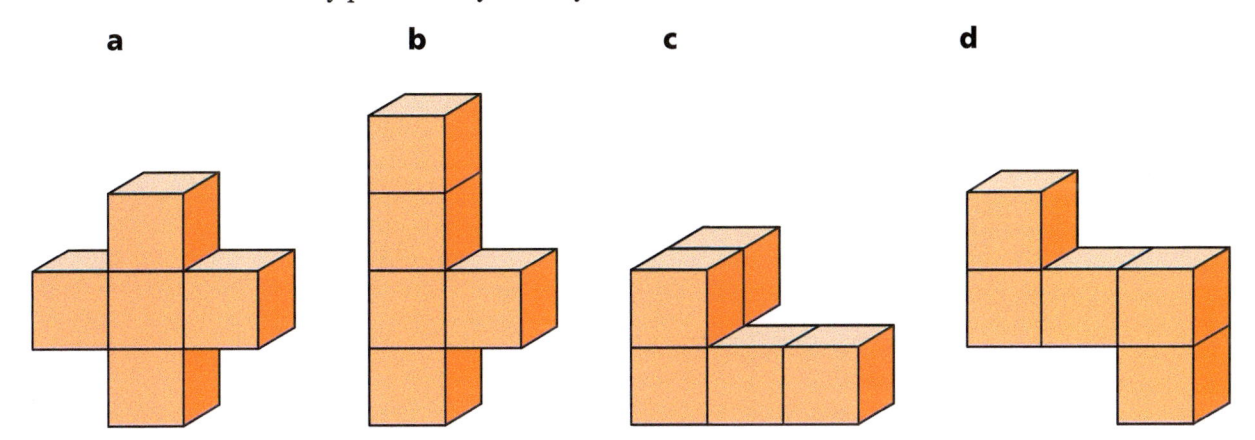

a b c d

4 The shapes in question **3** are made from five cubes.

Find another arrangement of five cubes that has two planes of symmetry.

> Use small cubes to make some shapes to see if you can find one with two planes of symmetry.

9.3 Construction of lines

Perpendicular line from a point to a line

The diagram shows a straight line AB with a point P.

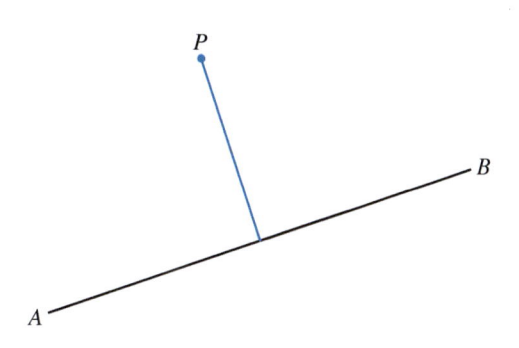

The shortest distance from P to the line is shown on the diagram.

The shortest distance is **perpendicular** to the line AB. This means that it is at right angles to the line AB.

This is how to construct the shortest distance using only a straight edge and compasses.

Mathematics for Cambridge Secondary 1

Worked example 1

Construct the line from P which is perpendicular to AB.

First use compasses centred at P to mark two points on the line AB.

Next draw the perpendicular bisector of the line between these two points.

Use compasses centred on one of the points to draw arcs either side of the line AB.

Repeat with the compasses centred on the other point.

Do not let the compasses change radius.

Place a straight edge between the pairs of crossed arcs and draw the lineas shown.

This is the perpendicular line from P to the line AB.

Perpendicular line from a point on a line

You also need to know how to draw a perpendicular line from a point on a line.

This example shows the steps to take.

Worked example 2

Draw a line perpendicular to the given line at the point *P*.

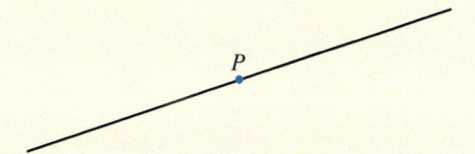

First use compasses centred at *P* to draw two arcs equidistant from *P* on the line.

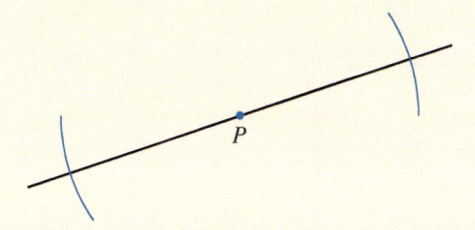

Now draw the perpendicular bisector of the line between these two arcs.

Use the same technique as in Worked example 1.

Draw arcs on one side of the line.

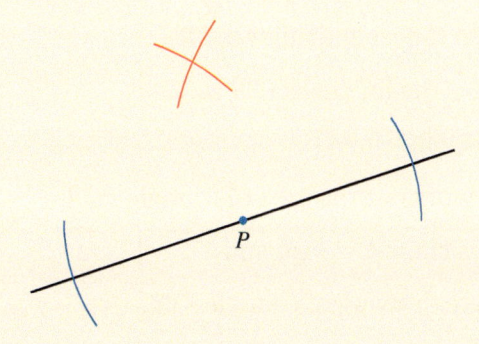

Join the crossed arcs to *P* to make the perpendicular line.

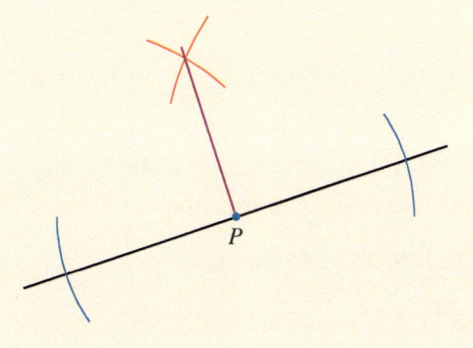

This is the required line.

Mathematics for Cambridge Secondary 1

Exercise 9.3

1 Make a copy of this diagram.

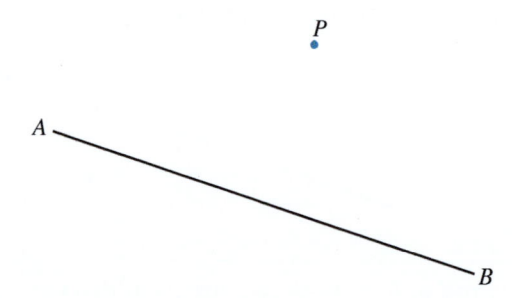

Construct a line through P which is perpendicular to the line AB.

2 Make a copy of this diagram.

Construct a line at M which is perpendicular to the line ST.

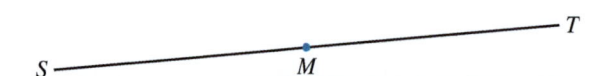

3 PQR is a triangle with PQ = 6 cm, QR = 7 cm and PR = 10 cm.

   a Construct an accurate drawing of the triangle.

   b Construct a line from Q perpendicular to PR.
   Label the point where the line meets PR as M.

   c Measure the length QM.

   d Use your answer to work out the area of triangle PQR.

4 XYZ is a triangle. XY = 5 cm, YZ = 7 cm and XZ = 11 cm.

   a Construct an accurate drawing of triangle XYZ.

   b Extend the line XY so that it is beneath point Z.

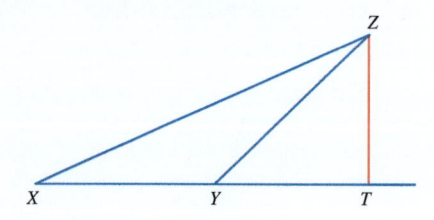

   Draw a perpendicular line from Z to this line.
   Label the point where it meets the line as T.

   c Measure the length ZT.

   d Work out the area of triangle XYZ.

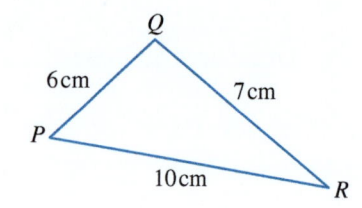

9 Shapes and geometric reasoning 2

5 a Draw a line about 10 cm long on plain paper.

Mark two points exactly 6 cm apart.

b By careful construction of perpendicular lines and accurate measurement, draw a rectangle 6 cm by 4 cm.

> A rectangle has 90° angles at each corner. Start by constructing right angles at the two points which are 6 cm apart.

c Measure the diagonal of the rectangle.

6 Construct an accurate drawing of quadrilateral *ABCD* as follows.

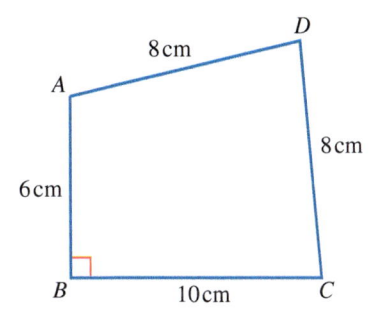

a Draw a line and mark points *B* and *C*, 10 cm apart.

b Construct a line at *B* which is perpendicular to *BC*.

Mark point *A*, 6 cm up that line.

> To draw the right angle, your line will need to extend to the left of point *B*.

c Set your compasses at 8 cm and draw an arc centred at *A*.

Repeat with the compasses centred at *C*.

Label the point where the arcs cross as *D*.

Join the lines *AD* and *CD*.

d Measure angle *ADC*.

9.4 Construction of polygons

This section shows how to **inscribe** a regular **polygon** inside a circle.

All of the vertices of the polygon will be on the circumference of the circle.

The examples show the method for a regular hexagon, an equilateral triangle, a square and a regular octagon.

Worked example 1

Construct a regular hexagon with all of its vertices on a circle of radius 5 cm.

First set the compasses to a radius of 5 cm and draw a circle.

Keep the compasses the same size; do not allow them to change.

Place your compass point on the circumference and make two arcs as shown in this diagram.

Now place the compass point on the point where one of the arcs crosses the circumference.

Make two more arcs either side of that point.

Repeat this step, putting the compass point at the point marked by the arcs.

Work around the circle making arcs until there are six points marked around the circumference.

Finally, join up the six points to create a regular hexagon.

Each side of the hexagon is equal to the radius of the circle.

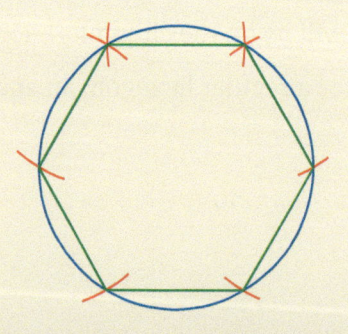

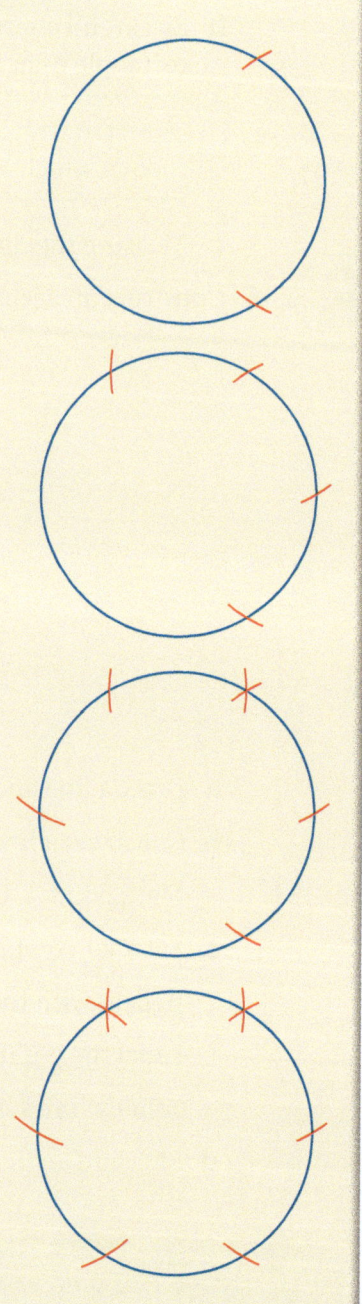

Worked example 2

Construct an equilateral triangle with all the vertices on the circumference of a circle.

Repeat the steps for drawing a hexagon to get six arcs around the circle.

Join up three of the arcs, leaving out alternate ones.

This gives an equilateral triangle inscribed in the circle.

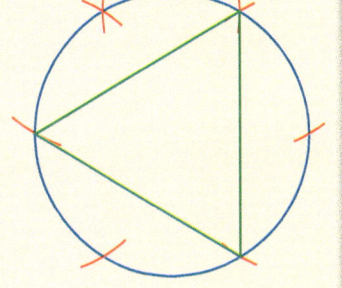

Worked example 3

Construct a square with all of the vertices on the circumference of a circle with diameter 10 cm.

First draw a circle with the correct radius.

Draw a diameter.

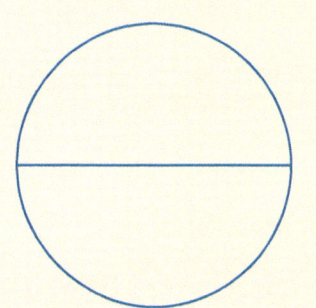

Construct the perpendicular bisector of the diameter using arcs above and below the line.

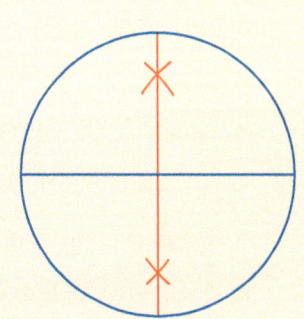

Join the four points where the lines meet the circumference.

This creates a square inscribed in the circle.

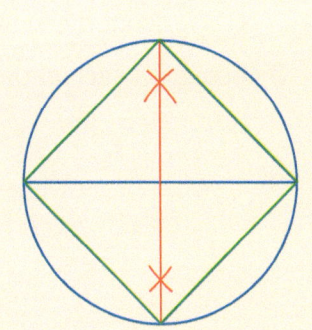

Worked example 4

Construct a regular octagon with all of the vertices on the circumference of a circle with diameter 10 cm.

Follow the first three steps for constructing a square.

You should have two lines at right angles across the circle.

Bisect one of the right angles. Mark arcs along each line then mark a pair of crossing arcs as shown.

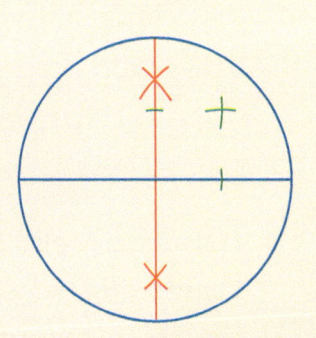

Draw a line through the point where the arcs cross and the centre of the circle.

Then bisect one of the remaining two right angles and draw a line across the circle.

Your diagram should now look like this.

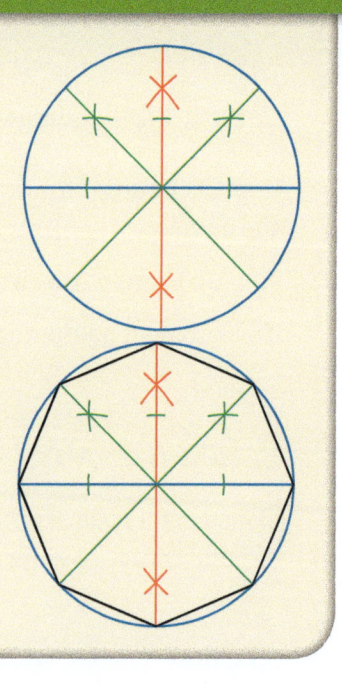

Join up the eight points around the circumference to create a regular octagon.

Exercise 9.4

1. Draw a circle with radius 6 cm.
   - **a** Construct a square with vertices on the circumference of the circle.
   - **b** Write down the length of the diagonal of the square.
   - **c** Measure the length of one side of the square.

2. Draw a circle with radius 6 cm.
   - **a** Construct an equilateral triangle with vertices on the circumference of the circle.
   - **b** Measure the length of one side of the triangle.
   - **c** Work out the area of the triangle.

3. Construct each of these patterns inside a circle with radius 5 cm.

   a　　　　　**b** 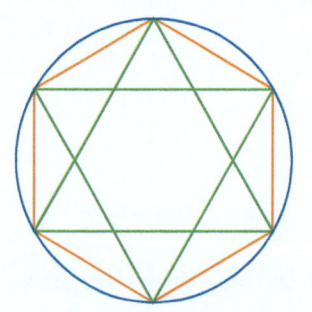　　　　　**c**

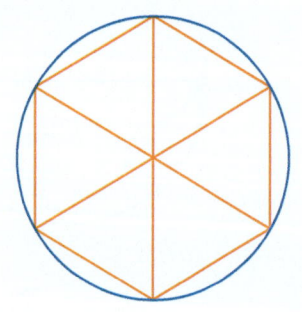

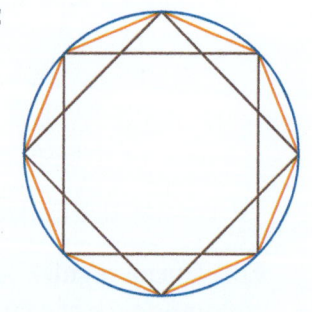

   In questions **4** and **5** draw your circles with a radius of at least 5 cm. This will give you enough space to draw the perpendicular bisectors.

4. Construct a regular hexagon inscribed in a circle.

   Construct the perpendicular bisector of each side.

   Use these to construct a regular twelve sided polygon with vertices on the circle.

9 Shapes and geometric reasoning 2

5 Construct a regular octagon in a circle.

Use this as a basis for constructing a regular sixteen sided polygon inscribed in the circle.

9.5 Maps and bearings

Bearings

Historically, travellers found directions by using the points of the **compass**.

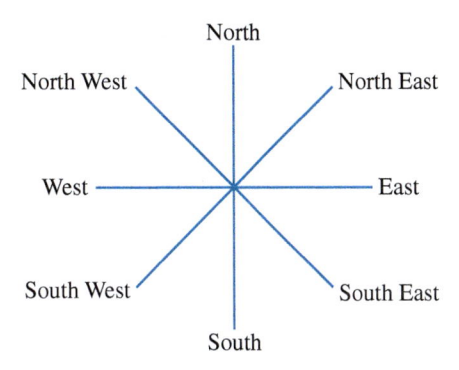

Here you can see the main eight points of the compass.

By combining these eight points it is possible to define 32 directions using a compass.

For modern travel this is not accurate enough.

Directions are now defined in terms of bearings.

A **bearing** is an angle, measured clockwise from north.

Bearings are usually stated as three figures. A bearing of 45° would be written as 045°.

Worked example 1 shows how to measure bearings using a protractor.

Worked example 1

A, B and *C* are three points on a map.

a Measure the bearing from *A* to *B*. **b** Measure the bearing from *A* to *C*.

123

a Draw a line between points *A* and *B*.

Set the protractor with the 0° mark lined up with the north line at *A*.

The cross at the centre of the protractor should be at point *A*.

Note that the inner scale is the one with zero on the north line.

Read the angle on the line to *B*.

The bearing is 128°.

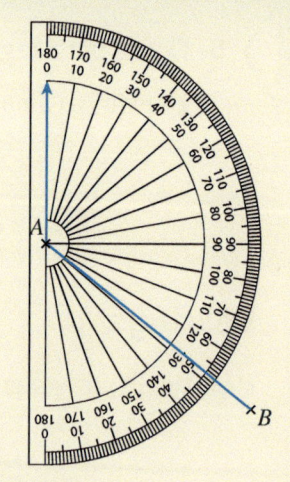

b Draw a line from *A* to *C*.

The protractor set as it was for *B* will not read far enough round.

To find the bearing of *C* set the protractor on the other side of the north line.

There are two ways of finding the bearing.

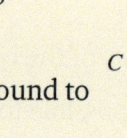

Method 1: Read clockwise from south round to the line *AC*.

The reading is 70°.

The angle from the north line to the south line is 180°.

The bearing from *A* to *C* is 180° + 70° = 250°

Method 2: Read the angle anticlockwise from north to the line *AC*.

The reading on the outer scale is 110°.

The bearing is 360° − 110° = 250°

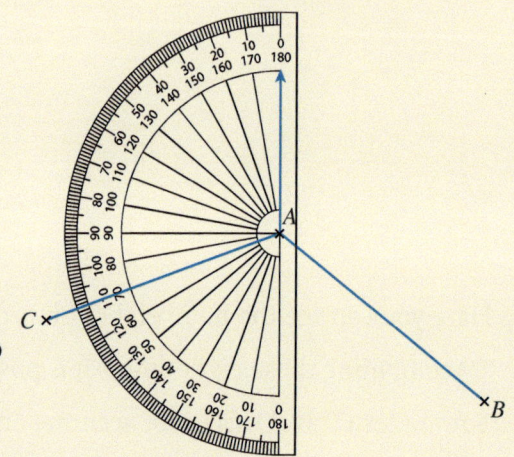

Map scales

Map scales are sometimes given as a ratio.

A scale of 1 : 200 000 on a map means that 1 centimetre on the map represents 200 000 centimetres on the ground.

    1 cm to 200 000 cm

    1 cm to 2000 m   Divide by 100 as there are 100 cm in 1 metre.

    1 cm to 2 km   Divide by 1000 as there are 1000 m in 1 km.

A scale of 1 : 200 000 is the same as a scale where 1 cm represents 2 km.

9 Shapes and geometric reasoning 2

Worked example 2

The scale of a map is 1 : 250 000

A road on the map is 6 cm long.

How long is the actual road?

1 cm to 250 000 cm

1 cm to 2500 m

1 cm to 2.5 km

Each centimetre on the map represents 2.5 kilometres.

A distance of 6 cm on the map represents 6 × 2.5 = 15 km

Worked example 3

This map shows an island with a town, a harbour and a lighthouse.

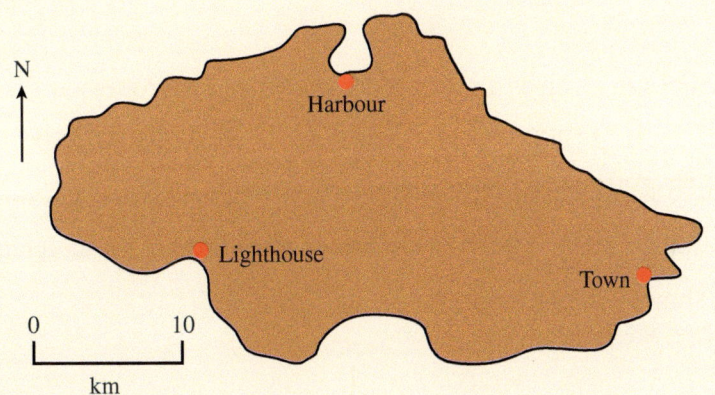

The scale of the map is 1 : 500 000

Find **a** the distance between the town and the harbour

b the bearing from the lighthouse to the harbour.

a The scale is 1 : 500 000 so

1 cm to 500 000 cm

1 cm to 5000 m

1 cm to 5 km

Each centimetre represents 5 kilometres.

The distance on the map between the town and the harbour is 4.8 cm.

The actual distance is 4.8 × 5 = 24 km

125

b Draw a north line at the lighthouse and draw the line between the lighthouse and the harbour.

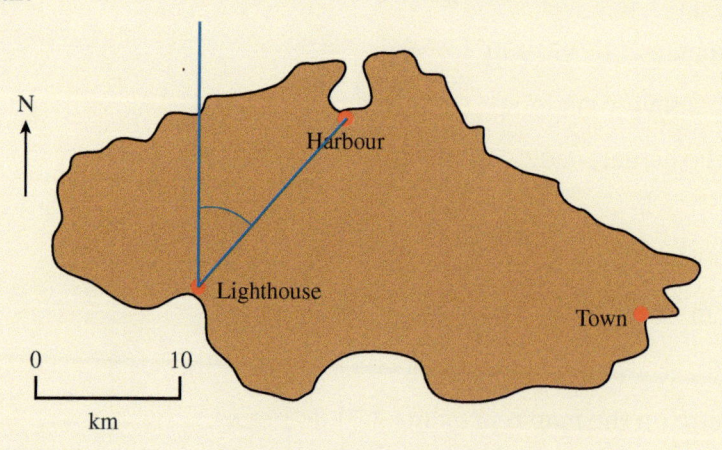

Measure the angle marked between the two lines.

The bearing is 042°.

Exercise 9.5

1 These map scales are given as a ratio. Work out what 1 cm represents in each case.

   a 1 : 500 000   **b** 1 : 2000   **c** 1 : 50 000   **d** 1 : 40 000 000   **e** 1 : 25 000 000

2 Write these map scales in ratio form.

   a 1 cm represents 1 km   **b** 1 cm represents 50 km   **c** 1 cm represents 250 m

3 The diagram shows five places labelled *A*, *B*, *C*, *D* and *E*.

Measure the following bearings.

   a From *A* to *B*.   **b** From *A* to *C*.   **c** From *D* to *A*.   **d** From *C* to *E*.

4 Use the map from question **3** to answer these questions.

   a Which point is north of *D*?

   b Which point is south west of *A*?

   c What direction is it from *C* to *A*?

5 The map shows a peninsula with four towns marked on it.

The scale of the map is 1 : 200 000

Scale 1 : 200 000

  a Work out how many kilometres are represented by 1 cm on the map.

  b Find out the distance between Southport and Northport in kilometres.

  c Find the bearing from Northport to Southport.

  A helicopter pilot needs to fly from Kipstown to Beach City.

  d Work out the bearing of Beach City from Kipstown.

  e Work out the distance from Kipstown to Beach City in kilometres.

6 This map shows part of a coastline.

Trace the map on to plain paper.

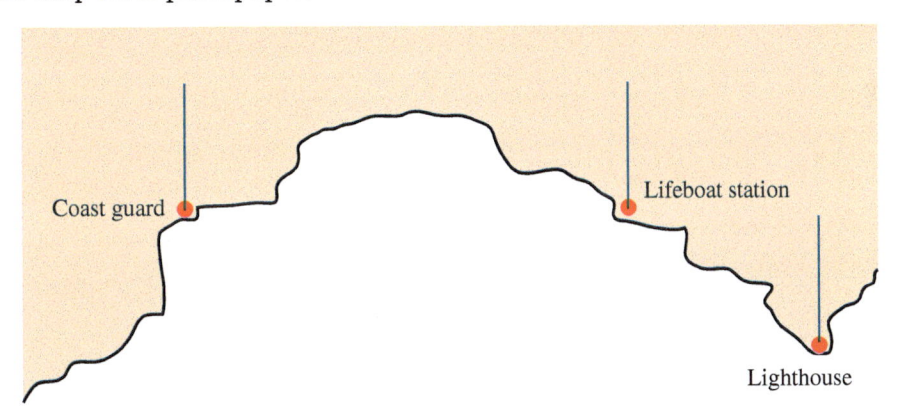

Scale 1 cm = 20 km

A ship in trouble sends out an emergency signal which is received by the lighthouse and the coast guard.

  a The ship is on a bearing of 135° from the coast guard station.

   Draw a line on the map to show the bearing of 135° from the coast guard station.

  b The ship is on a bearing of 260° from the lighthouse.

   Draw a line on the map to show the bearing of 260° from the lighthouse.

  c Where the lines cross shows the position of the ship.

   Measure the distance from the ship to the lifeboat station.

  d What is the bearing from the lifeboat station to the ship?

7 The diagram shows two points A and B.

The north lines are drawn in for you.

a Find the bearing of B from A.

b Without measuring the angle at B, work out the bearing of A from B.

This is known as the back bearing.

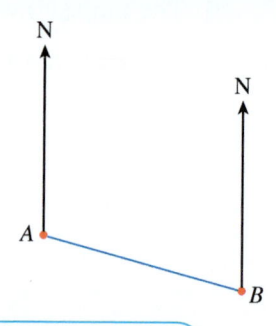

> The two north lines are parallel. Use your knowledge of parallel lines to work out the answer.

The bearing and the back bearing always differ by 180°.

If the bearing is less than 180° you add 180° to find the back bearing.

If the bearing is greater than 180° you subtract 180° to find the back bearing.

c Work out the back bearing for each of these bearings.

  i 065° **ii** 154° **iii** 217° **iv** 300° **v** 270°

Review

1 The diagram shows the plan and elevations of a three-dimensional object.

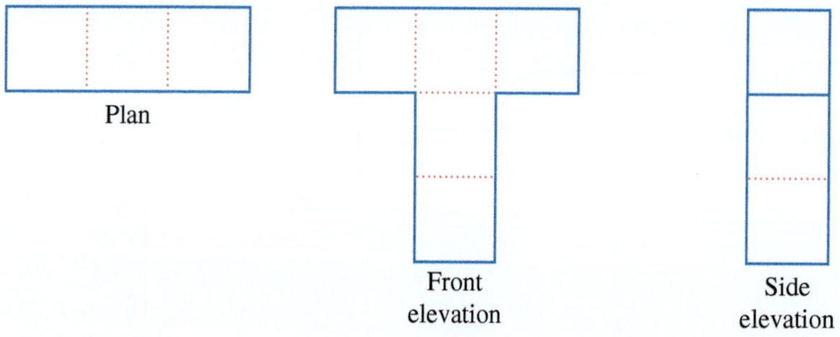

Plan Front elevation Side elevation

Draw the object on isometric paper.

2 Draw the plan and elevations for this three-dimensional solid.

9 Shapes and geometric reasoning 2

3 a The diagram shows an isosceles triangular prism.

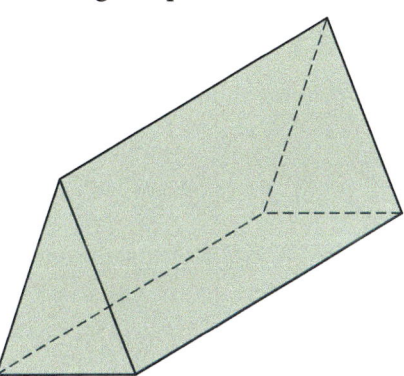

Copy the diagram and draw a plane of symmetry for the object.

b Work out how many planes of symmetry there are altogether.

4 Work out how many planes of symmetry these solids have.

   a A cuboid   **b** A semicircular prism   **c** A regular triangular pyramid

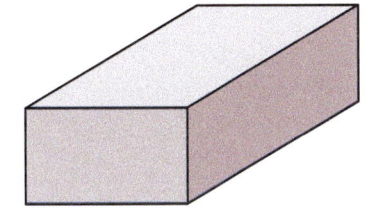

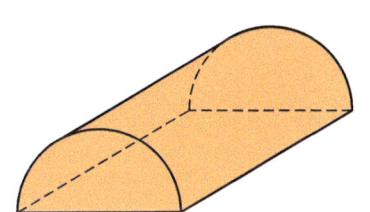

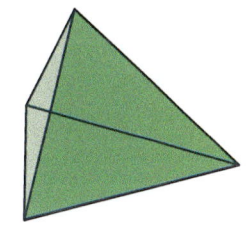

5 Copy this diagram on to plain paper.

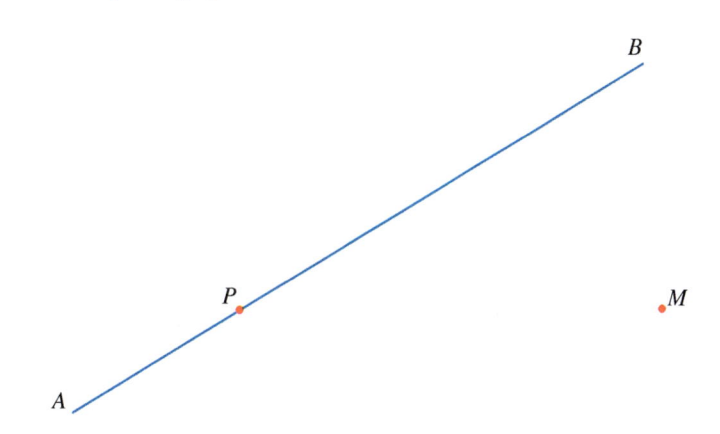

a Draw a line at *P* which is perpendicular to *AB*.

The line should be on the same side of *AB* as point *M*.

b Draw a line through *M* to meet your new line at 90°.

6 a Construct a triangle *ABC* as shown in this diagram.

b Draw a line from *B* to meet the line *AC* at right angles.

c Measure the perpendicular height of the triangle.

d Work out the area of the triangle.

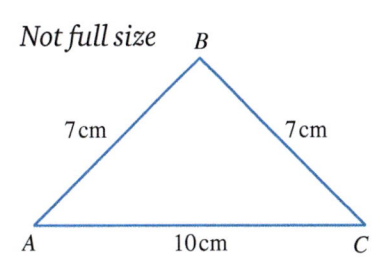

Not full size

7 cm, 7 cm, 10 cm

129

7 a Draw a circle with radius 5 cm.

  b Construct a regular octagon with the vertices on the circumference of the circle.

8 Use ruler and compasses to construct this pattern.

9 Write these in ratio form.

  a 1 cm = 10 km   **b** 1 cm = 250 m   **c** 2 cm = 1 km

10 A map scale is given as 1 : 400 000

  What distance is represented by 1 cm?

11 The map shows an island.

Scale 1 cm = 20 km

  a Find the distance and bearing from Cookstown to St George.

  b Find the distance and bearing from Peterport to Cookstown.

9 Shapes and geometric reasoning 2

Examination-style questions

1. Each interior angle of a regular pentagon is 108°.

   Use your ruler and protractor to draw a regular pentagon with sides 4 cm long. [2]

2. Copy the diagram on to plain paper.

   Use ruler and compasses to bisect the angle.

   Leave the construction lines on your answer. [2]

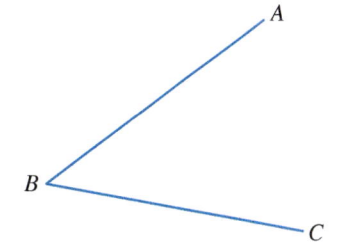

3. **a** Write down the mathematical name for this three-dimensional solid. [1]

   b Work out how many planes of symmetry the solid has. [1]

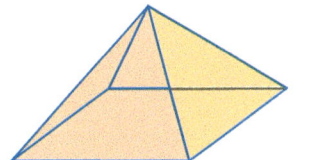

4. Use a ruler and compasses to construct a regular hexagon with sides of 5 cm.

   Leave any construction lines visible on your drawing. [2]

5. The diagram shows the plan and elevations of a three-dimensional solid.

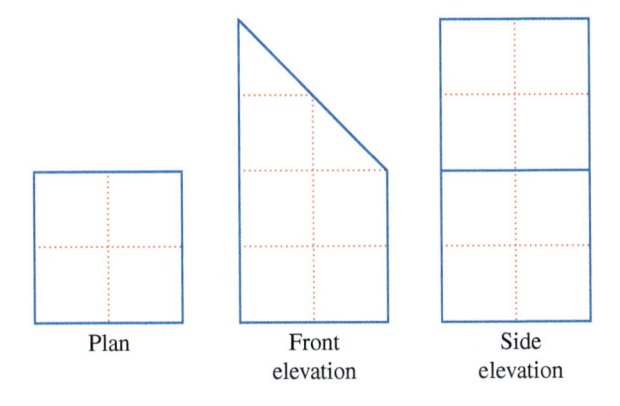

   Plan   Front elevation   Side elevation

   Draw the solid on isometric paper. [2]

131

Mathematics for Cambridge Secondary 1

6 The diagram shows the grounds of an old palace.

The grounds form a rectangle, 700 m by 400 m.

a Draw an accurate scale drawing of the palace grounds.

  Use a scale of 1 cm = 50 m

An old treasure map shows something buried in the grounds.

It is on a bearing of 115° from the well.

It is on a bearing of 340° from the castle.

b Draw a line from the well at a bearing of 115°.

c Draw a line from the castle at a bearing of 340°.

d Put a cross to mark the position of the buried treasure. [4]

7 Write down the name for a solid with six square faces. [1]

8 Copy this diagram on to plain paper. *XY* is 10 cm.

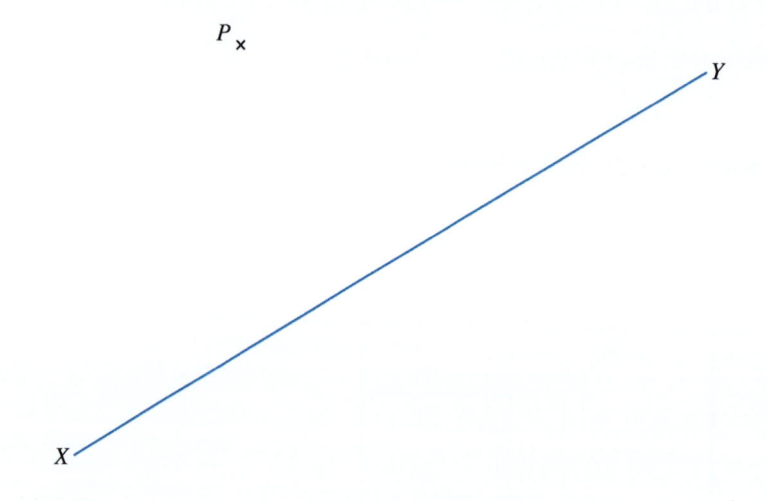

Use a ruler and compasses to draw a perpendicular line from point *P* to the line *XY*.

Leave all construction lines visible on the diagram. [2]

10 Presenting, interpreting and discussing data

Chapter 18 covers collecting data
Chapter 6 covers processing data

Learning outcomes

- Select, draw and interpret graphs and diagrams.
- Draw and interpret scatter diagrams.
- Have a basic understanding of correlation.
- Draw and interpret back to back stem-and-leaf diagrams.
- Draw and interpret frequency diagrams for discrete and continuous data.
- Draw and interpret line graphs for time series.
- Compare two or more distributions.
- Relate conclusions to the original question.

Camping and ice cream sales

Francis Galton was an English mathematician who lived in the 1800s.

He invented a number of statistical ideas including correlation.

Correlation involves looking for agreement.

For example, in warm weather less electricity is used.

The amount of electricity used is linked directly to the weather.

In warm weather more people go camping, and more ice creams are sold. There is agreement between ice cream sales and camping holidays, but they are not directly linked – they both depend on the weather.

10.1 Scatter diagrams and correlation

Scatter diagrams

Scatter diagrams are used to show paired data.

Points are plotted as crosses on two axes.

When you draw scatter diagrams you need to plot the points carefully.

133

Worked example 1

The table shows the height, h cm, and age, x years, of a number of children.

Age, x (years)	2	3	5	7	8	9	11	11	12	15
Height, h (cm)	90	95	115	130	128	140	150	160	150	170

Draw a scatter diagram to show the data.

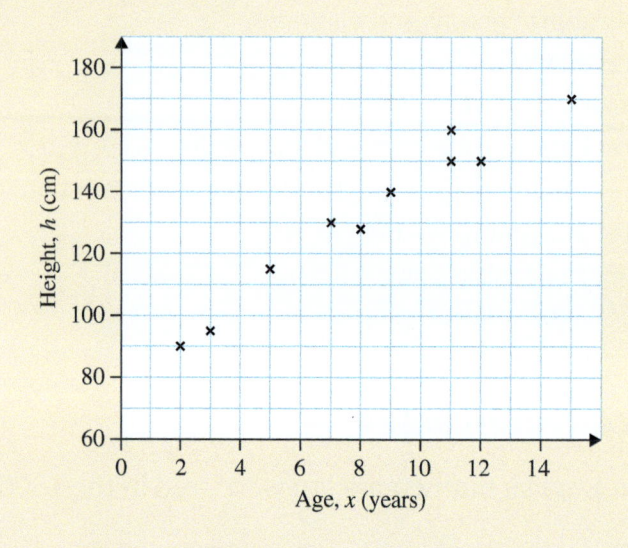

Correlation

Scatter diagrams are used to show **correlation**.

Correlation is the connection between two **variables**. When you interpret scatter diagrams you may find one or two points that do not fit the pattern.

This diagram shows strong **positive correlation**.

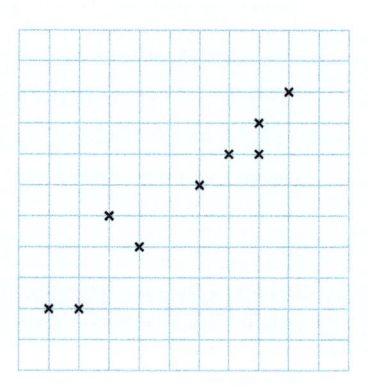

The points are close to a straight line going from 'bottom left to top right' on the scatter diagram.

If the points are on a straight line this is called perfect positive correlation.

This diagram shows weak positive correlation.

The points are not close to a straight line, but there is still a general pattern from 'bottom left to top right' in the scatter diagram.

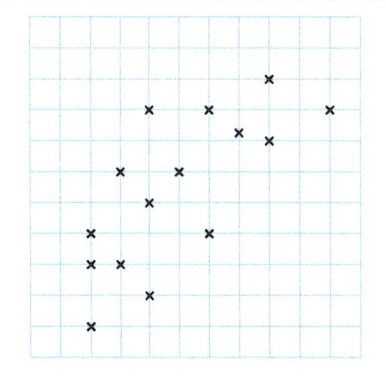

This diagram shows strong **negative correlation**.

The points are close to a straight line going from 'top left to bottom right' on the scatter diagram.

If the points are on a straight line this is called perfect negative correlation.

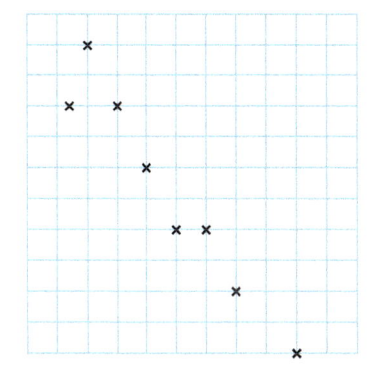

This diagram shows weak negative correlation.

The points are not close to a straight line, but there is still a general pattern from 'top left to bottom right' in the scatter diagram.

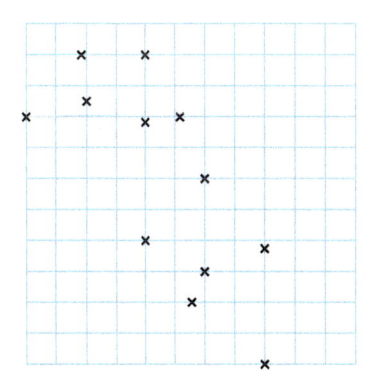

This diagram shows **no correlation**, or zero correlation.

There is no pattern shown in the scatter diagram.

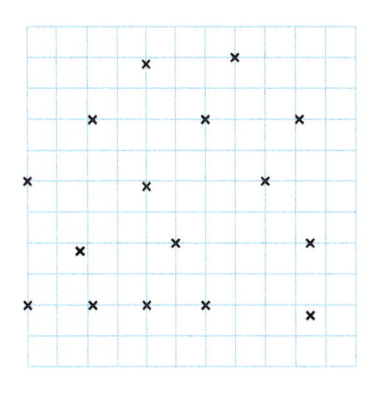

Mathematics for Cambridge Secondary 1

Worked example 2

The table shows the length, l cm, and width, w cm, of some pebbles from a beach.

Length, l (cm)	4	5	4	6	4.5	3	7	7	2	6
Width, w (cm)	2	3	3	4	4	2.5	4	5	1	5

a Plot a scatter diagram to show the lengths and widths of the pebbles.

b Comment on any correlation shown in your scatter diagram.

a

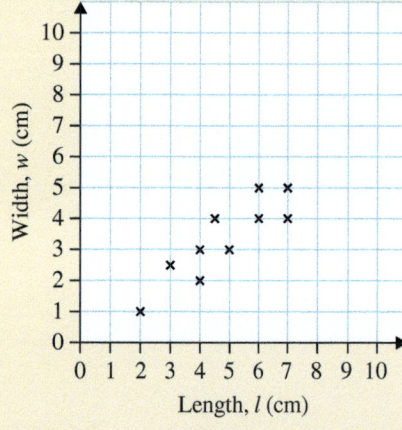

b The scatter diagram shows positive correlation.

This means longer pebbles have a greater width.

Exercise 10.1

1 As part of his homework Navarro draws these scatter diagrams.

Diagram 1 Diagram 2 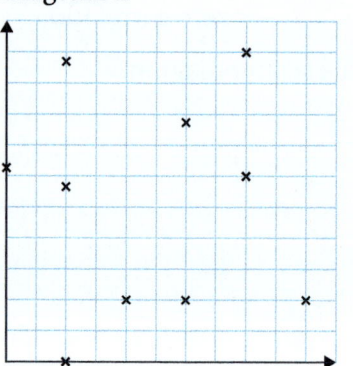 Diagram 3

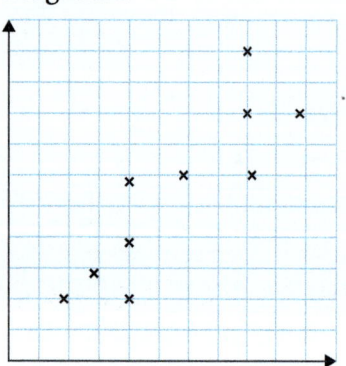

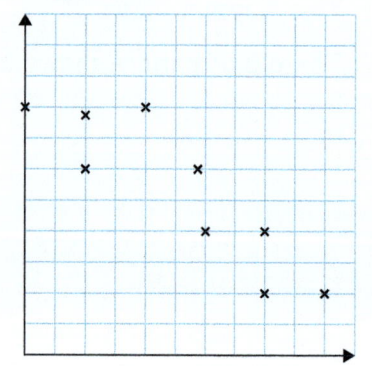

a Which diagram shows

  i negative correlation

  ii positive correlation?

b The horizontal axis on this scatter diagram is labelled 'Age of car'.

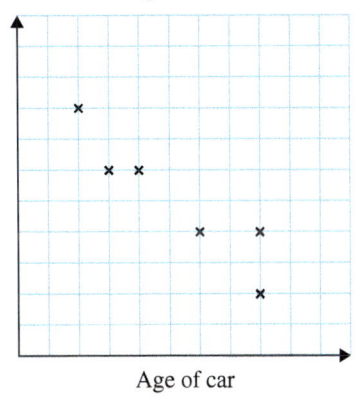

Age of car

Which of these could be the label for the vertical scale?

Engine size, Colour, Value, Mass, Length

2 The table shows the number of text messages sent and the number of calls made by a group of students.

Number of text messages	16	18	20	24	25	30	38	42
Number of calls	40	39	35	32	28	22	12	8

  a Plot the data on a scatter diagram.

  b Describe the correlation shown in the scatter diagram.

3 The scatter diagram shows the heights, h cm, and masses, m kg, of a group of footballers.

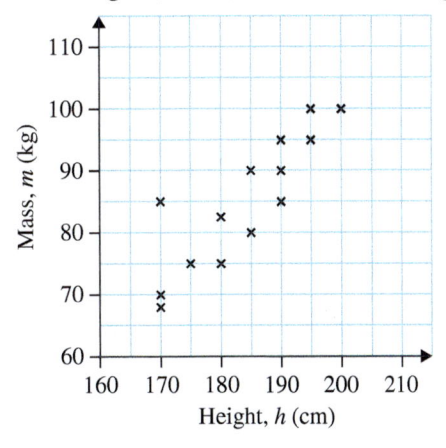

  a What is the mass of the lightest footballer?

  b What is the height of the tallest footballer?

  c One footballer is particularly heavy for their height. Write down the height and mass for this footballer.

  d Comment on any correlation shown in the scatter diagram.

4 Wendy records the time it takes to complete various walks.

Distance (km)	0.5	1	1.5	2	2.5	3	3	3.5	4	4.5
Time (minutes)	6	11	12	18	25	28	32	31	27	55

  a Plot the data on a scatter diagram.

Use the scales shown. Continue up to 5 on the horizontal axis and 60 on the vertical axis.

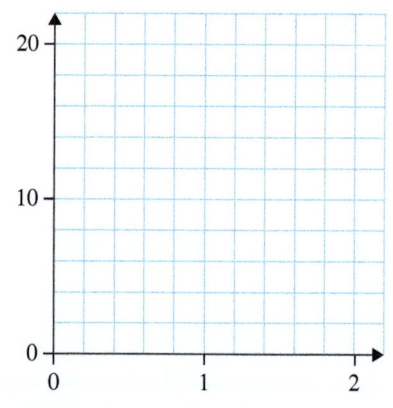

b One of the walks involved going up a lot of hills.

How many km long is this walk? How long did it take?

c Comment on the correlation shown in the scatter diagram.

5 The table shows the price and age of some second-hand cars.

Age (years)	1	1	2	2	3	4	5	6	7	8	9	10
Price ($, thousands)	15	12.5	10	9	10	7	12.5	6.5	4.5	2.5	3	2

a Plot the data on a scatter diagram.

b Comment on any correlation shown in the scatter diagram.

c One of the cars is a 'special edition'.

Which car do you think this is? Give a reason for your answer.

6 A driving school collects data about the number of lessons their students have, and the number of times their students take the test before passing.

Number of lessons	15	18	20	25	30	40	30	60	50	45	35	70	60
Number of tests taken	1	1	1	2	1	3	2	1	3	3	2	5	4

a Plot the information from the table on a scatter diagram.

b Comment on any correlation shown in your scatter diagram.

c Luiz wants to know how many driving lessons he can expect to need before passing his test.

> Use scales of 2 cm for 10 lessons on the vertical axis and 2 cm for 1 lesson on the horizontal axis.

Which pair of values would he be likely to ignore? Give a reason for your answer.

7 The table shows the scores in two papers of a maths exam.

Paper 1	25	50	35	59	45	56	28	23	32
Paper 2	30	52	40	64	48	59	30	30	83

a Plot the data on a scatter diagram.

b One of the test scores has been recorded incorrectly.

   i Which test score do you think this is?

   ii What do you think the test score may have been?

c Maisy was absent for Paper 2. She scored 40 in Paper 1. Use your graph to estimate her score in Paper 2.

10.2 Back to back stem-and-leaf diagrams

Stem-and-leaf diagrams

Stem-and-leaf diagrams are used to organise data in order of size.

The stem is to the left of the vertical line.

The leaves are always just one digit.

The leaves must be organised in order of size.

Keep the leaves in vertical lines.

You must always include a key.

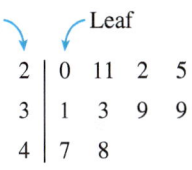

Stem | Leaf
2 | 0 1 1 2 5
3 | 1 3 9 9
4 | 7 8

Key: 4|7 represents 47

Worked example 1

The test scores for a group of 23 boys are:

41, 10, 23, 15, 52, 30, 26, 29, 58, 42, 19, 49, 38, 37, 25, 31, 55, 42, 24, 43, 35, 34, 25

a Draw a stem-and-leaf diagram to show the data.

b Write down the median for the test scores.

c Work out the range of the test scores.

a
1 | 0 5 9
2 | 3 4 5 5 6 9
3 | 0 1 4 5 7 8
4 | 1 2 2 3 9
5 | 2 5 8

Key: 5|2 represents 52

b The median is the middle value when they are in order of size.

There are 23 values. The middle value is the 12th.

The median is 34.

c The range = largest − smallest = 58 − 10 = 48

Mathematics for Cambridge Secondary 1

Back to back stem-and-leaf diagrams

Back to back stem-and-leaf diagrams are used when you need to work with two distributions, such as males and females.

They allow you to compare the distributions quickly and easily.

The stem is in the middle.

The leaves are at either side of the stem.

You must label the two sides of the stem-and-leaf diagram.

The leaves on the left-hand side of the stem are in reverse order, with the smallest nearest the stem.

The key should show the stem and two leaves.

Worked example 2

The test scores for a group of 25 girls are:

45, 20, 37, 57, 12, 43, 53, 57, 47, 57, 44, 11, 24, 36, 44, 31, 25, 50, 42, 40, 55, 25, 28, 58, 48

a Add the scores for the girls to the stem-and-leaf diagram in Worked example 1 to make a back to back stem-and-leaf diagram.

b Write down the median of the test scores for the girls.

c Work out the range of the test scores for the girls.

d Compare the test scores of the girls and boys.

a

```
            Girls              Boys
              2  1 | 1 | 0  5  9
        8  5  5  4  0 | 2 | 3  4  5  5  6  9
              7  6  1 | 3 | 0  1  4  5  7  8
  8  7  5  4  4  3  2  0 | 4 | 1  2  2  3  9
        8  7  7  5  3  0 | 5 | 2  5  8
```
Key: 0|5|2 means 50 for the girls and 52 for the boys

b The median is 43.

c The range is 58 − 11 = 47

d On average the girls do better in the test as they have a higher median.

The boys and girls test scores have almost the same range, so they are about as varied as each other.

Looking at the shape of the stem-and-leaf diagram the boys' test scores are mainly in the middle of the mark range. The girls' test scores are mainly in the 40s and 50s. This suggests that the girls do better overall.

10 Presenting, interpreting and discussing data

Exercise 10.2

1 The back to back stem-and-leaf diagram shows the number of runs scored by two cricketers in games last season.

```
         Isaac              Andrew
                      1 | 9
            9 8 7 3 | 2 | 1 7
    8 7 6 5 4 2 0 | 3 | 2 5 6 8 8
          8 5 4 2 1 | 4 | 0 2 4 5 5 6 8
              5 3 1 0 | 5 | 1 3 3 9
                      6 | 2 3
                      1 | 7
```
Key: 3|2|1 means Isaac scored 23 runs and Andrew scored 21 runs

  a What is the smallest number of runs scored by Andrew?
  b What is the largest number of runs scored by Andrew?
  c Write down the range for Andrew.
  d Find the range for Isaac.
  e Find the median number of runs for Andrew.
  f Find the median number of runs for Isaac.
  g Who has the largest median? What does this tell you about their number of runs?
  h Who has the largest range? What does this tell you about their number of runs?

2 The marks in a test by students in two classes are:

Class A:

71, 51, 25, 31, 38, 47, 53, 30, 34, 43, 59, 60, 35, 42, 45, 56, 63, 40, 47, 56, 64

Class B:

28, 46, 59, 66, 73, 19, 30, 46, 63, 64, 37, 43, 58, 62, 69, 74, 23, 34, 59, 60, 70, 41, 55, 68, 75

  a Draw a back to back stem-and-leaf diagram to show the test scores for the two classes.
  b For class A, find:
   i the median
   ii the range.
  c For class B, find:
   i the median
   ii the range.
  d Which class had the highest median? What does this tell you about their test scores?
  e Which class had the largest range? What does this tell you about their test scores?

141

Mathematics for Cambridge Secondary **1**

3 Kenny records the number of cars in a car park at 9 o'clock in the morning and at 5 o'clock in the afternoon for 27 days.

His results are:

9 o'clock:

05, 25, 57, 43, 30, 34, 34, 45, 53, 65,

38, 23, 76, 58, 40, 62, 18, 37, 45, 52,

16, 09, 34, 32, 50, 43, 22

5 o'clock:

09, 38, 43, 57, 54, 31, 77, 57, 68, 62,

58, 43, 57, 42, 37, 31, 30, 40, 60, 52,

58, 52, 39, 32, 30, 72, 62

a Draw a back to back stem-and-leaf diagram to show the data.

b Find the median and range for the number of cars in the car park at 9 o'clock.

c Find the median and range for the number of cars in the car park at 5 o'clock.

d Compare the number of cars in the car park at the two times.

4 A grower measures the masses, in grams, of two varieties of potato.

Variety A:

36, 44, 49, 54, 57, 68, 82, 34, 45, 57, 58, 63, 74, 79, 31, 48, 53, 67, 69, 81, 30, 48, 51, 66, 71

Variety B:

75, 76, 81, 39, 43, 56, 63, 75, 40, 49, 59, 66, 31, 47, 61, 65, 56, 59, 65, 73, 76, 80, 64, 67, 72

By drawing a back to back stem-and-leaf diagram compare the two distributions.

10.3 Comparing distributions

Frequency polygons

Frequency polygons are used to compare frequency distributions for grouped continuous data.

Points are plotted at the midpoints for each class, and at the height for each frequency.

These points are joined with straight lines.

Axes follow the same rules as for frequency diagrams.

The 'peak' of a frequency polygon shows you where data is concentrated.

Two distributions may be shown on the same axes.

When you do this you must draw the lines using different colours and give a key.

When you interpret you can comment on the positions of the peaks and the spread of the distributions.

10 Presenting, interpreting and discussing data

Worked example 1

For the heights, h cm, shown in the table, draw:

a a frequency diagram **b** a frequency polygon.

Height, h (cm)	Frequency
$0 < h \leq 10$	3
$10 < h \leq 20$	12
$20 < h \leq 30$	6
$30 < h \leq 40$	4

a

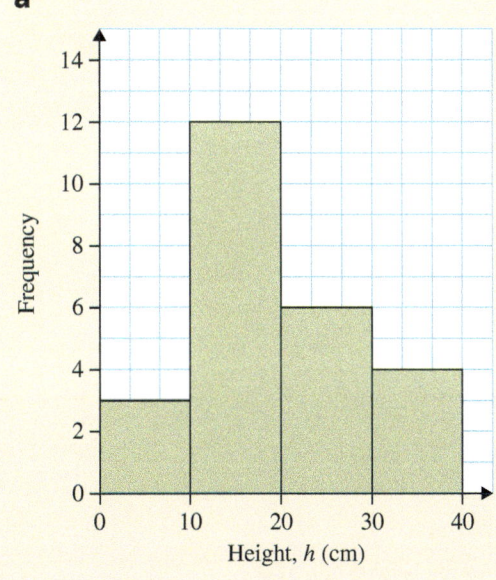

b

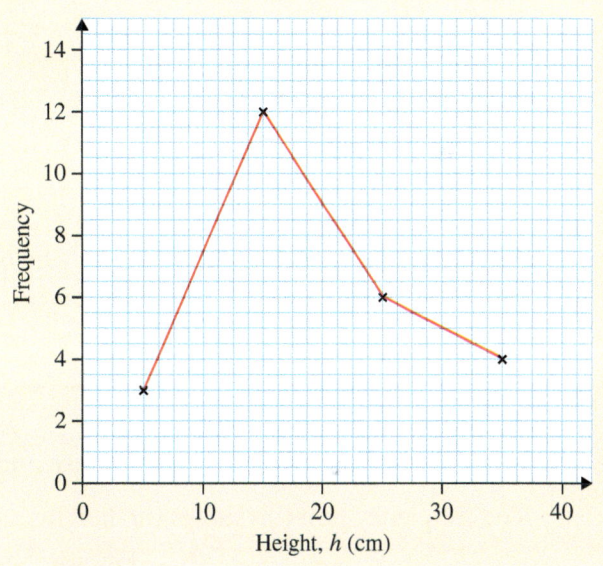

Bar charts

Bar charts can be used to show two, or more, sets of data at the same time.

These are called **multiple bar charts**.

Bar charts showing only two sets of data are called **dual bar charts**.

Multiple bar charts allow two, or more, distributions to be compared easily.

When you draw multiple bar charts you must include a key.

The rules for multiple bar charts are the same as for ordinary bar charts.

Worked example 2

The attendance of boys and girls at a club is recorded for one week.

	Monday	Tuesday	Wednesday	Thursday	Friday
Boys	3	7	5	3	1
Girls	5	6	6	4	6

a Show this information on a multiple bar chart.
b Compare the attendance of boys and girls.

a Bar chart showing attendance of boys and girls at a club

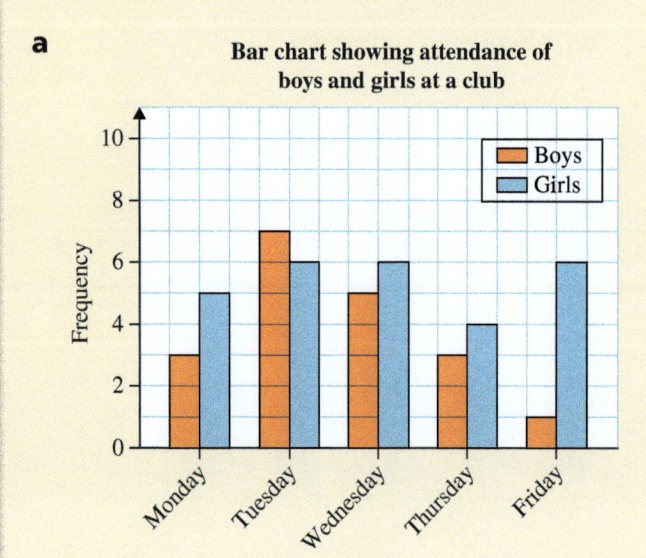

b The girls have more even attendance. The boys' attendance peaks on Tuesday, and gradually reduces over the rest of the week.

Exercise 10.3

1 Mei-Yin grows plants in her greenhouse and in her garden.

The frequency polygon shows the heights of her plants.

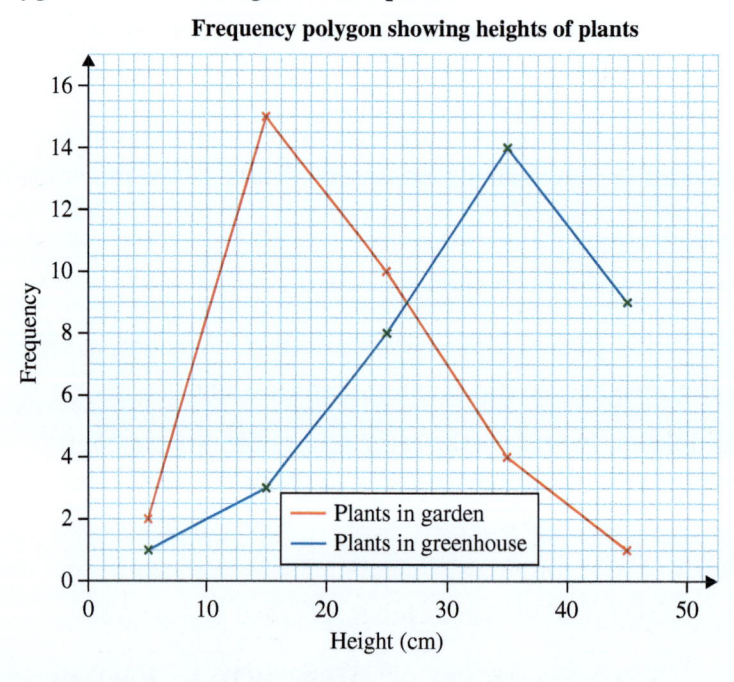

Frequency polygon showing heights of plants

a How many garden plants are between 10 cm and 20 cm in height?
b How many greenhouse plants are between 10 cm and 20 cm in height?

c How many garden plants grow to a height of 20 cm or greater?

d How many greenhouse plants grow to a height of 20 cm or greater?

e Which plants generally grow taller? How can you tell from the frequency polygon?

2 A survey of two groups of students records the time they spent using computers in one week.

The frequency polygon shows the results for group A.

Frequency polygon showing time spent on computers by group A

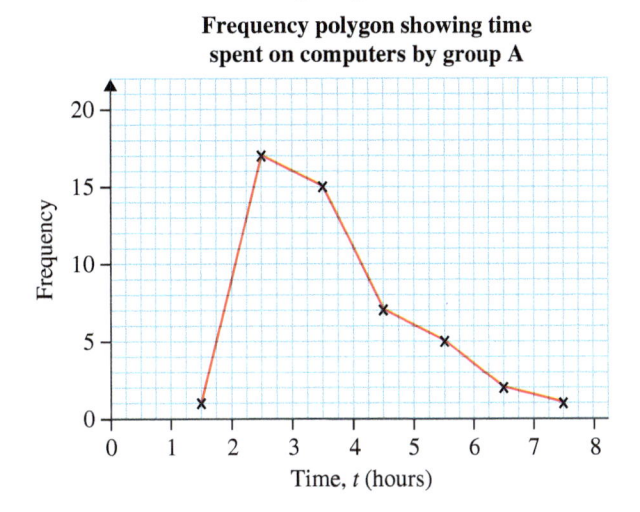

The table shows the time group B spent on computers in the same week.

Time, t (hours)	$1 < t \leq 2$	$2 < t \leq 3$	$3 < t \leq 4$	$4 < t \leq 5$	$5 < t \leq 6$	$6 < t \leq 7$	$7 < t \leq 8$
Frequency	0	3	5	8	14	13	3

a Draw a frequency polygon for group B using the same scales as used for group A.

b How many students in group B spent less than 4 hours using computers?

c How many students in group A spent less than 4 hours using computers?

d Use the frequency polygons to decide which group generally spent more time using the computer. How did you decide?

3 The dual bar chart shows the attendance of boys and girls at a swimming club one week.

Dual bar chart showing attendance at a swimming club

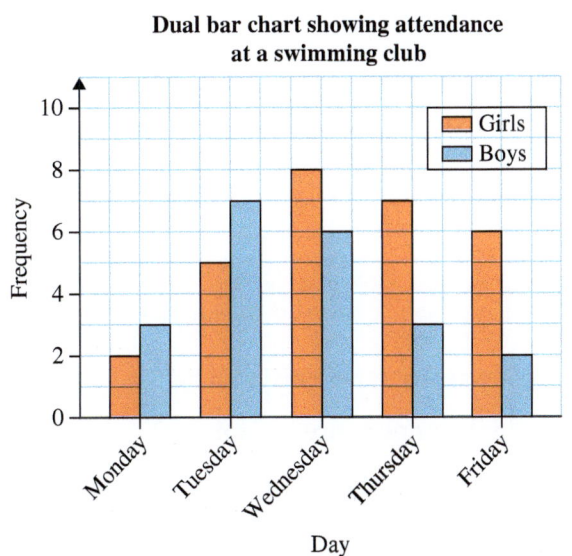

a How many girls attend on Tuesday?

b How many boys attend on Friday?

c On which day do the largest number of girls attend?

d On which day do the largest number of boys attend?

e Which day has the largest total attendance?

f Over the whole week, how many more girls attend than boys?

4 In a survey people were asked about their travel patterns.

The table shows the trips per person per year.

	2007	2008	2009
Shopping	186	198	193
Travel to work	162	156	147
Business	33	30	30

Source: Social trends: Transport ST41

a Draw a multiple bar chart to show this information.

b Comment on any patterns shown in your multiple bar chart.

5 The line graph shows the average high temperatures and the average low temperatures, by month, for Cyprus.

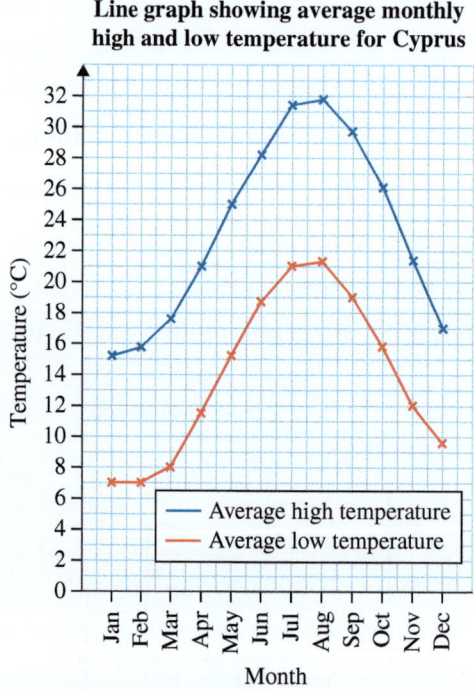

a What is the average low temperature in January?

b What is the average high temperature in January?

c What is the difference between the average high and average low temperature for January?

10 Presenting, interpreting and discussing data

d Which month has the largest difference between the average high and average low temperature?

e One day Jose checks the outside temperature. He says it is 28 °C. Which months are this temperature most likely to be in?

6 Katie says, 'Boys are taller than girls.'

Here is the data that she collects from some 13-year-old students.

Height, h (cm)	Boys' frequency	Girls' frequency
$140 < h \leq 150$	1	4
$150 < h \leq 160$	5	8
$160 < h \leq 170$	10	12
$170 < h \leq 180$	13	6
$180 < h \leq 190$	2	0

a Draw a frequency polygon to show the heights of the girls and the boys.

b Do you think that Katie is correct? Give reasons for your answer.

7 Two different makes of battery are tested.

The results are shown in the frequency tables.

Make X	
Time, t (hours)	Frequency
$0 < t \leq 5$	1
$5 < t \leq 10$	6
$10 < t \leq 15$	2
$15 < t \leq 20$	1

Make Y	
Time, t (hours)	Frequency
$0 < t \leq 5$	1
$5 < t \leq 10$	3
$10 < t \leq 15$	5
$15 < t \leq 20$	1

a Draw a frequency polygon to show the times.

b Greg wants to buy a battery. Which of the two makes would you recommend that he buys? Give a reason for your answer.

> Use the shape of the graph to help you decide.

Review

1 Isabella says, 'When there are fewer checkouts open at the supermarket I have to queue longer.'

To see if she is correct she collects the following data.

Number of checkouts open	3	4	8	6	12	5	10	9
Time waiting (minutes)	4	3	5	8	2	3	5	4

a Draw a scatter diagram to show the data.

b Do you think that Isabella is correct? Give a reason for your answer.

2 Frank stands in the centre of village A. Dermot stands in the centre of village B.

Frank says, 'village A is busier than village B.'

They count the number of people passing them in a half-hour period each day for 3 weeks.

Their results are:

Village A:

21, 34, 37, 45, 27, 36, 47, 51, 61, 19, 28, 46, 50, 23, 42, 38, 47, 20, 39, 48, 53

Village B:

55, 37, 24, 16, 24, 33, 39, 48, 13, 28, 37, 51, 62, 10, 25, 42, 44, 21, 35, 38, 56

  a Draw a back to back stem-and-leaf diagram to show this information.

  b Find the median and range for village A.

  c Find the median and range for village B.

  d Do you think Frank is correct? Give reasons for your answer.

3 Barbara and Marlon do jobs around the house to earn pocket money.

Barbara says, 'I work for longer than Marlon.'

They keep a record of the time they work each week.

Time, t (minutes)	Barbara's frequency	Marlon's frequency
$30 < t \leq 60$	2	3
$60 < t \leq 90$	3	10
$90 < t \leq 120$	5	9
$120 < t \leq 150$	14	4
$150 < t \leq 180$	4	2

  a Draw a frequency polygon to show this information.

  b Is Barbara correct? Give a reason for your answer.

4 The temperature in a greenhouse is measured every 30 minutes.

The table shows the temperature in the greenhouse one afternoon.

Time	1200	1230	1300	1330	1400	1430	1500	1530	1600	1630	1700	1730
Temperature (°C)	23	25	27	29	30	30	31	32	31	30	29	26

  a Show this information on a line graph.

  b At what time did the temperature first reach 30 °C?

  c For how long was the temperature 30 °C or above?

  d Estimate the length of time for which the temperature was above 26 °C.

10 Presenting, interpreting and discussing data

Examination-style questions

1 Here are the heights, given to the nearest centimetre, of a group of 13-year-old girls.

   154   162   148   158   163   168   155   160
   152   164   162   166   163   167   170   161

   a Copy and complete the frequency table to summarise the heights of the girls.

Height, h (cm)	Tally	Frequency
$145 < h \leq 150$		
$150 < h \leq 155$		
$155 < h \leq 160$		
$160 < h \leq 165$		
$165 < h \leq 170$		

   [2]

   b Copy and complete the frequency polygon to summarise the heights of the girls.

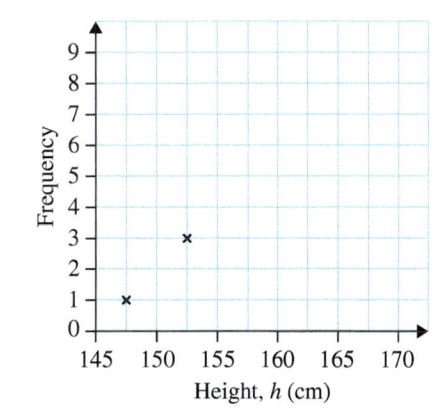

   [1]

2 A scientist is testing a new fertiliser. She grows some plants in two plots, A and B.

   Plot A has the new fertiliser. Plot B has no fertiliser.

   The scientist measures the heights of the plants after a month of growth. The median for plot A is 76.5 cm and the range is 46.

   The back to back stem-and-leaf diagram shows her results.

   ```
              Plot A              Plot B
                         4 | 0  1  1  7
            9 7 6 4 4 1  5 | 1  2  2  3  3  3  4  6  8
                5 5 5 3 2  6 | 0  0  3  4  5  5  6  7  9
        9 7 7 6 6 4 2 1  7 | 0  1  2  2  3  7  8
            9 8 5 3 2 2 0  8 | 0  1  1  4  6
                  7 7 4 1 1 0  9 |
   ```

   Key: 1|7|0 means plot A height 71 cm and plot B height 70 cm

   a Find the median height for the plants in plot B. [1]

   b Find the range of heights for the plants in plot B. [1]

   c Compare the plants grown in plots A and B.

      Make two comparisons. [2]

11 Area, perimeter and volume

Learning outcomes

- Convert between metric units of area, e.g. mm² and cm², cm² and m² and volume, e.g. mm³ and cm³, cm³ and m³; know and use the relationship 1 cm³ = 1 ml.
- Know that land area is measured in hectares (ha), and that 1 hectare = 10 000 m²; convert between hectares and square metres.
- Solve problems involving the circumference and area of circles, including using the π key of a calculator.
- Calculate lengths, surface areas and volumes in right-angled prisms and cylinders.

Measuring the land

Hectares are the metric measure of land area.

Most countries use hectares but some still use the acre.

An acre is thought to be the area that could be ploughed in one day by a team of oxen.

An acre is about 0.4 hectare.

Acres are still used today in several countries including the United States, Pakistan, India and Australia.

Anglo-Saxon Plough.

11.1 Area

Units of area

Think of a 1 metre by 1 metre square.

The area is $1 \times 1 = 1\,\text{m}^2$

It is also 100 cm by 100 cm.

The area is $100 \times 100 = 10\,000\,\text{cm}^2$

$$1\,\text{m}^2 = 10\,000\,\text{cm}^2$$

If you measure the square in millimetres it is 1000 mm by 1000 mm.

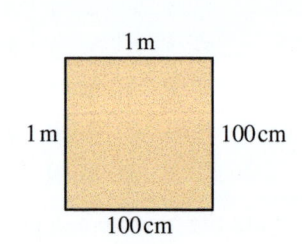

The area is 1000 × 1000 = 1 000 000 mm²

$1\,m^2 = 1\,000\,000\,mm^2$

To convert units of area you need to square the length conversion factor.

For example, 1 km = 1000 m

So 1 km² = 1000 × 1000 = 1 000 000 m²

Worked example 1

A sheet of paper has an area of 220 cm².

Convert this to mm².

1 cm = 10 mm   this is the length conversion factor

1 cm² = 10 × 10 = 100 mm²   this is the area conversion factor

220 cm² = 220 × 100 = 22 000 mm²

> To change area units you square the length conversion factor.

Worked example 2

An industrial site has an area of 0.3 km².

A builder wants to put houses on the site.

Each house needs 600 m² of land.

a Convert 0.3 km² to m².

b Work out how many houses can be built.

a   1 km = 1000 m   this is the length conversion factor

1 km² = 1000 × 1000 = 1 000 000 m²   this is the area conversion factor

0.3 km² = 0.3 × 1 000 000 = 300 000 m²

b 300 000 ÷ 600 = 500

500 houses can be built on the site.

> To change area units you square the length conversion factor.

Hectares

The area of 0.3 km² in Worked example 2 was enough for 500 houses.

This shows that 1 km² is a very large area.

There is a need for a unit of area between 1 m² and 1 km².

The unit that is used is the **hectare (ha)**.

$1\,ha = 10\,000\,m^2$

Hectares are used for areas of farms, fields, parks, etc.

Mathematics for Cambridge Secondary 1

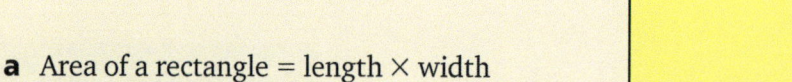

> **Worked example 3**
>
> Convert the following:
>
> **a** 4.5 ha to m²   **b** 150 000 m² to ha
>
> **a** 1 ha = 10 000 m²
>
> 4.5 ha = 4.5 × 10 000 = 45 000 m²
>
> *To change from a larger unit to a smaller one you multiply.*
>
> **b** 10 000 m² = 1 ha
>
> 150 000 m² = $\frac{150\,000}{10\,000}$ = 15 ha
>
> *To change from a smaller unit to a larger one you divide.*
>
> **Worked example 4**
>
> The diagram shows the plan of a farmer's field.
>
> **a** Work out the area in m².
>
> **b** Work out the area in hectares. 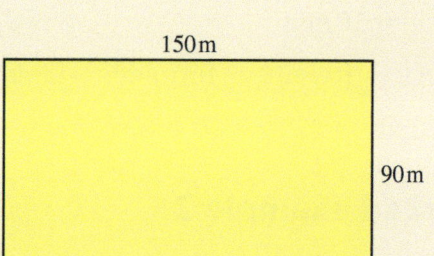
>
> 150 m
>
> 90 m
>
> **a** Area of a rectangle = length × width
>
> Area = 150 × 90
>
> Area = 13 500 m²
>
> **b** 1 ha = 10 000 m²
>
> 13 500 m² = $\frac{13\,500}{10\,000}$ = 1.35 ha

Exercise 11.1

1 Copy and complete these statements.

    a 1 m² = ... cm²   **b** 1 cm² = ... mm²

    c 1 km² = ... m²   **d** 1 ha = ... m²

2 Convert to cm²

    a 3 m²   **b** 2.8 m²   **c** 400 mm²   **d** 1000 mm²

    e 0.5 m²   **f** 20 mm²   **g** 100 m²   **h** 32 mm²

3 Convert to mm²

    a 4 cm²   **b** 7.5 cm²   **c** 2 m²   **d** 6.8 m²

4 Convert to m²

    a 3 km²   **b** 8000 cm²   **c** 9 000 000 mm²   **d** 7.2 km²

5 Convert to m².

   a 4 ha   **b** 12 ha   **c** 7.5 ha   **d** 2.35 ha

6 Convert to hectares.

   a 60 000 m²   **b** 150 000 m²   **c** 85 000 m²   **d** 5000 m²

7 Find the area of these rectangles.

   a 8 cm × 3 cm   **b** 2.6 cm × 1.5 cm   **c** 12 mm × 15 mm

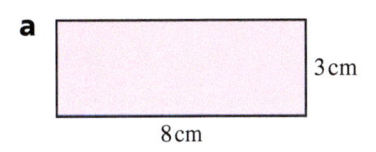

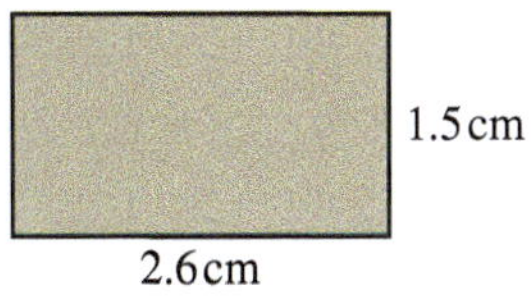

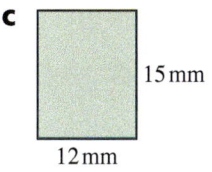

   Give the answer in: **i** cm²   **ii** mm²

8 A playing field measures 165 metres by 88 metres.

   Work out the area of the field in hectares.

9 A horse breeder expects to need 4000 m² of field space per animal.

   Work out how many horses could be kept on a field with an area of 2 hectares.

10 The diagram shows a large field.

   It is made up of two rectangles.

   a Work out the area of the field in m².

      Divide the field into two rectangles to work out the area.

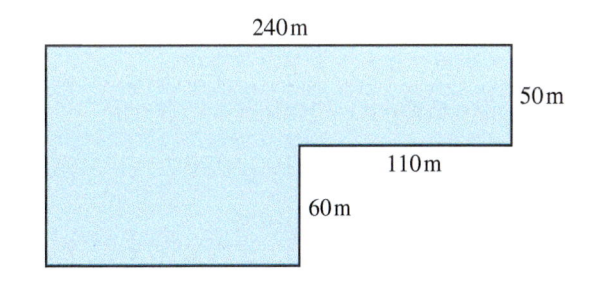

   Dimensions: 240 m, 50 m, 110 m, 60 m.

   b Convert the area to hectares.

11 A small tin of paint will cover 1.5 m².

   Hassan has 12 pieces of wood each measuring 35 cm by 40 cm.

   Be careful when the units are mixed. The lengths are in cm and the area is in m².

   a Work out the area of one piece of wood.

   b Work out whether one tin of paint is enough to cover all 12 pieces of wood.

12 A sheet of A4 paper has half the area of a sheet of A3.

   A sheet of A3 paper has half the area of a sheet of A2.

   A sheet of A2 paper has half the area of a sheet of A1.

   A sheet of A1 paper has half the area of a sheet of A0.

   A sheet of A0 has an area of 1 m².

   a What fraction of the area of an A0 sheet is a sheet of A4?

   b Work out the area of a sheet of A4 in cm².   *Start by converting the area into cm².*

Mathematics for Cambridge Secondary 1

13 A large industrial site is 0.3 km².

   a Convert this to hectares.

                                         Convert it to m² first.

   b How many hectares are there in 1 km²?

14 A forest is shown on a map as 12 cm by 8 cm.

   The scale of the map is 1 cm = 200 m

   a Work out the area of the forest on the map in cm².

   b Work out the actual area of the forest in m².

      Use the scale to change the lengths into metres before working out the area.

     12 cm

     Scale 1 cm = 200 m

   c Convert the answer to hectares.

11.2 Volume

Units of volume

This cube is 1 metre by 1 metre by 1 metre.

The volume is $1 \times 1 \times 1 = 1\,m^3$

It is also 100 cm by 100 cm by 100 cm.

The volume is $100 \times 100 \times 100 = 1\,000\,000\,cm^3$

$$1\,m^3 = 1\,000\,000\,cm^3$$

If you measure the cube in millimetres the volume is $1000 \times 1000 \times 1000 = 1\,000\,000\,000\,mm^3$

So 1 billion cubic millimetres is 1 cubic metre.

$$1\,m^3 = 1\,000\,000\,000\,mm^3$$

To convert units of volume you need to cube the length conversion factor.

1 km = 1000 m

So $1\,km^3 = 1000 \times 1000 \times 1000 = 1\,000\,000\,000\,m^3$

Worked example 1

Convert to the given units.

a 5.3 m³ to cm³   **b** 12 cm³ to mm³   **c** 4 500 000 cm³ to m³

a 1 m = 100 cm   *This is the length conversion factor.*

    $1\,m^3 = 100 \times 100 \times 100 = 1\,000\,000\,cm^3$ *This is the volume conversion factor.*

    $5.3\,m^3 = 5.3 \times 1\,000\,000 = 5\,300\,000\,cm^3$

154

b 1 cm = 10 mm

1 cm³ = 10 × 10 × 10 = 1000 mm³

12 cm³ = 12 × 1000 = 12 000 mm³

c 1 m = 100 cm

1 m³ = 100 × 100 × 100 = 1 000 000 cm³

4 500 000 cm³ = 4 500 000 ÷ 1 000 000 = 4.5 m³

Capacity and volume

The capacity of a container is the amount that it can hold.

The volume of a container is the amount of space it takes up.

A container with a volume of 1000 cm³ has a capacity of 1 litre.

1000 cm³ = 1000 ml

1 cm³ = 1 ml

Worked example 2

The diagram shows a fish tank.

The tank is 45 cm long, 24 cm wide and 30 cm deep.

a Work out the volume in cm³.

b Find the capacity of the tank in litres.

a Volume = length × width × depth

= 45 × 24 × 30

= 32 400 cm³

b Capacity = 32 400 ml

= 32 400 ÷ 1000

= 32.4 l

The fish tank holds 32.4 litres when it is full.

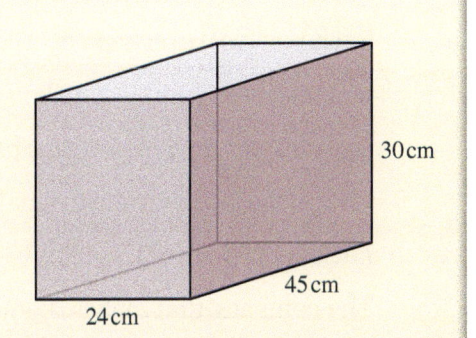

Exercise 11.2

1 Convert to cm³

   a 2 m³   **b** 5.6 m³   **c** 6000 mm³   **d** 9600 mm³

2 Convert to mm³

   a 3 m³   **b** 0.75 m³   **c** 9 cm³   **d** 1.4 cm³

Mathematics for Cambridge Secondary 1

3 Convert to m³

    a 7 000 000 cm³   **b** 12 500 000 cm³

4 **a** Work out the volume of this cuboid in cm³.

    b Work out the volume of the cuboid in mm³.

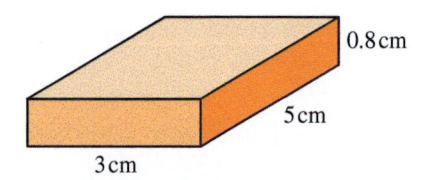

(Cuboid: 5 cm by 3 cm by 0.8 cm)

5 Work out how many millilitres of liquid the cuboid in question **4** would hold.

6 Work out the capacity in millilitres of containers with volume:

    a 25 cm³   **b** 125 cm³   **c** 2500 cm³   **d** 8000 mm³

7 Suzette has a bag containing 1500 cm³ of bird food.

She plans to put it into small boxes measuring 6 cm by 4 cm by 15 mm.

    a Work out the volume of one box.   *Be careful when there is a mixture of units.*

    b How many boxes can she fill from the bag of bird food?

    c How much seed will be left over?

8 Eric has 40 boxes of grass seed.

Each box measures 4 cm by 5 cm by 5 cm.

    a Find the volume of one box in cm³.

    b Work out how many litres of grass seed he has altogether.

9 Choco chocolates come in small packs.

Each small pack is 15 cm by 10 cm by 4 cm.

The company that makes them wants to pack them into big boxes.

Each big box is 80 cm by 30 cm by 36 cm.

    a Work out how many packs of Choco will fit into each big box.   *You may need to turn them round to make them fit with no gaps.*

    b Draw a diagram to show how they must be packed.

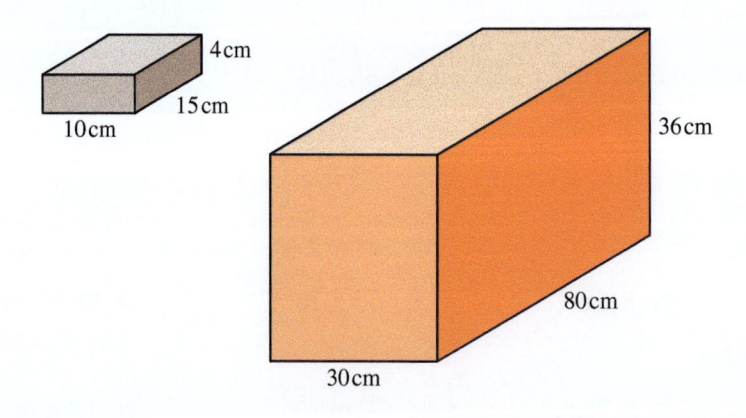

(Small pack: 10 cm by 15 cm by 4 cm; Big box: 30 cm by 80 cm by 36 cm)

11 Area, perimeter and volume

10 The diagram shows a swimming pool that is 25 metres long.

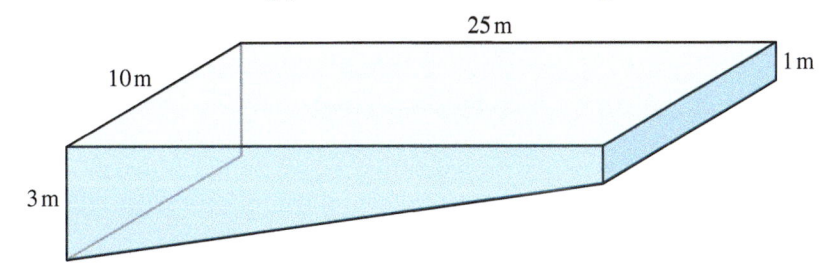

It is 1 metre deep at the shallow end and 3 metres at the deep end.

The side wall is a trapezium.

a Work out the area of the side wall in m².

The swimming pool is 10 metres wide.

b Work out the volume of the pool in m³.

c Find the capacity of the pool in litres.

Remember that 1 litre = 1000 cm³.

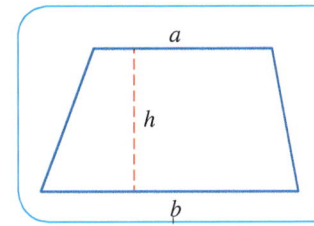

> The formula for the area of a trapezium is $A = \frac{1}{2}(a + b)h$

11.3 Circles

Circumference of circles

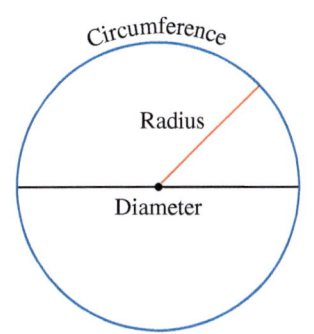

In your Stage 8 Student Book you learned how to find the **circumference** of a circle, C, if you know the diameter, d, or the radius, r.

$$C = \pi d \text{ or } C = 2\pi r$$

π, pronounced **pi**, is about 3.142

Your calculator has a π button which will give the value more accurately.

The calculator probably says 3.141592654

Some calculators will give a more accurate value than others.

Mathematics for Cambridge Secondary **1**

Worked example 1

Find the perimeter of these shapes. Give your answers to 3 significant figures.

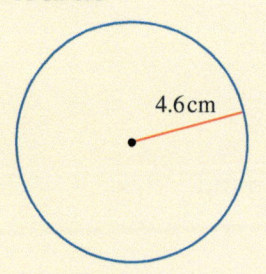

a A circle   **b** A semicircle   **c** A quarter circle

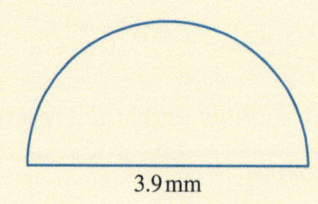

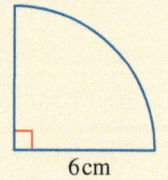

a The radius is 4.6 cm.

$C = 2 \times \pi \times 4.6$ this is the formula $C = 2\pi r$

$ = 28.90265241$ using the calculator value for π

$ = 28.9$ cm   to 3 significant figures

b The perimeter is made up of a curved part and a straight line.

The curved part of the perimeter, x, is half of the circumference of the full circle.

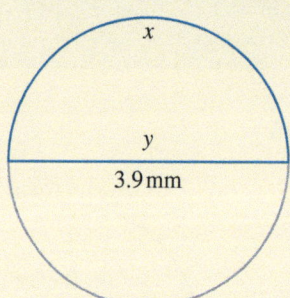

$x = \dfrac{\pi \times 3.9}{2}$ as $\dfrac{\pi d}{2}$ gives half the circumference

$x = 6.126105675$ using the calculator value for π

The straight part of the perimeter is the diameter.

$y = 3.9$

Total perimeter $= x + y$

$\phantom{\text{Total perimeter }} = 6.126105675 + 3.9$

$\phantom{\text{Total perimeter }} = 10.02610567$

$\phantom{\text{Total perimeter }} = 10.0$ mm   to 3 significant figures

c This is quarter of a circle.

The perimeter is made up of a curved part and two straight lines.

The curved part, a, is quarter of the circumference of the circle.

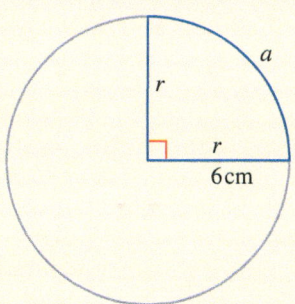

$a = \dfrac{2 \times \pi \times 6}{4}$ as $\dfrac{2 \times \pi \times r}{4}$ gives quarter of the circumference

  $= 9.424777961$ using the calculator value for π

The two straight lines are both equal to the radius, $r = 6$ cm.

Total perimeter $= a + 2r$

                $= 9.424777961 + 2 \times 6$

                $= 21.424777961$

                $= 21.4$ cm   to 3 significant figures

Area of circles

The formula for the area, A, of a circle with radius r is

$A = \pi r^2$

Worked example 2

Find the area of the shapes in Worked example 1. Give your answers to 2 decimal places.

a The radius of the circle is 4.6 cm.

    $A = \pi \times 4.6^2$ from $A = \pi r^2$

       $= 66.47610055$ using π from the calculator

       $= 66.48$ cm²   to 2 decimal places

The units are cm² as this is an area.

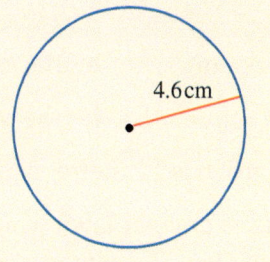

b The area of a semicircle is half the area of the circle.

The diameter is 3.9 mm so the radius is $3.9 \div 2 = 1.95$ mm

    $A = \dfrac{\pi \times 1.95^2}{2}$ half the area $= \dfrac{\pi \times r^2}{2}$

       $= 5.972953033$ using π from the calculator

       $= 5.97$ mm²   to 2 decimal places

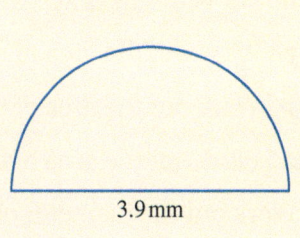

c The area of quarter of a circle is the area of the circle divided by 4.

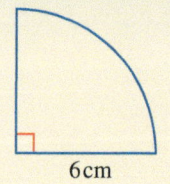

The radius is 6 cm.

$A = \dfrac{\pi \times 6^2}{4}$ quarter of the area $= \dfrac{\pi \times r^2}{4}$

$= 28.27433388$ using π from the calculator

$= 28.27\,\text{cm}^2$ to 2 decimal places

6 cm

Exercise 11.3

Give your answers to 3 significant figures in this exercise.

1 Find the perimeter of these shapes.

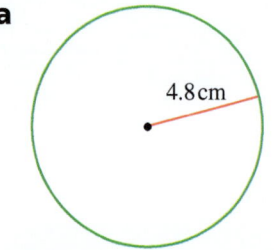

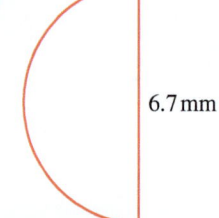

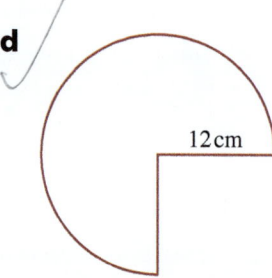

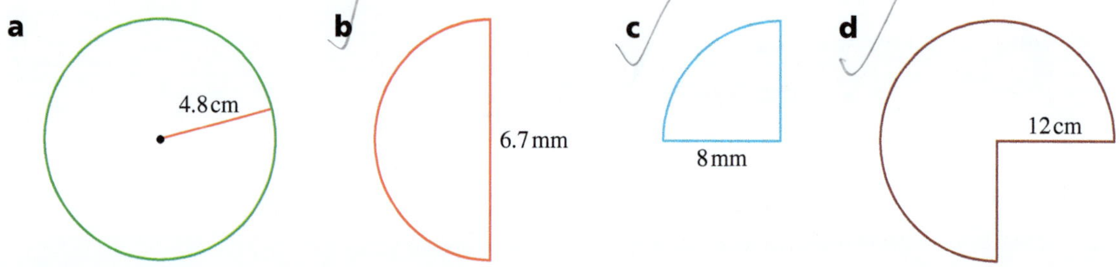

a 4.8 cm

b 6.7 mm

c 8 mm

d 12 cm

2 Find the area of the shapes in question **1**.

3 This shape is a rectangle and a semicircle.

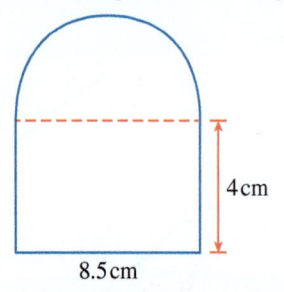

4 cm

8.5 cm

Work out: **a** the area of the shape

b the perimeter of the shape.

4 A garden is 12 metres square.

It has a circular lawn which touches all the edges. Each corner is a flower bed. The diagram shows a plan of the garden.

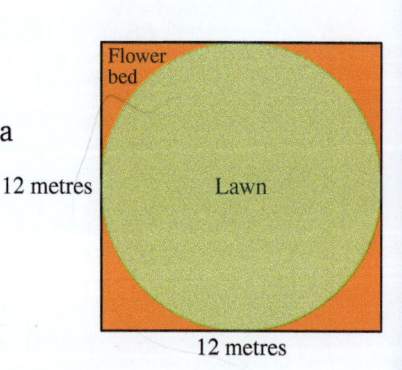

a Work out the area of the whole garden.

b Work out the area of the lawn.

c Work out the area of one flower bed.

Flower bed

12 metres

Lawn

12 metres

5 Work out the diameter of circles with circumference:

a 25 cm **b** 76 mm **c** 5 m

6 Work out the radius of circles with circumference:

a 65 cm **b** 5.9 mm **c** 2.5 m

11 Area, perimeter and volume

7 This company logo is made from two semicircles.

   a Work out the area of the logo.

   b Work out the perimeter of the logo.

> The shape is made up of two semicircles.

> Don't forget to add the straight lines on to the length of the arcs.

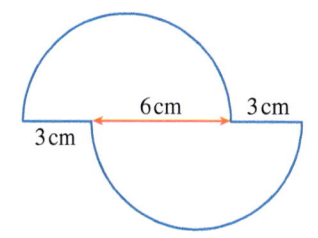

8 These three shapes each have the same perimeter.

Work out the lengths of the sides marked with letters.

> Work out the circumference of the circle first. Then use the other shapes to make equations to solve.

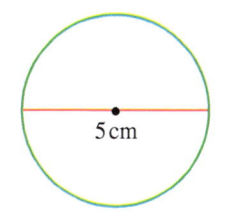

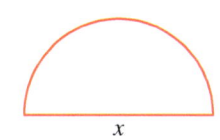

5 cm

x

y

Not drawn to scale

9 A circle has an area of 14.8 cm².

Work out the radius of the circle.

> Solve the equation $\pi r^2 = 14.8$

10 This picture shows a circular crater on the surface of the Moon.

The circumference is estimated to be 120 kilometres.

Work out the area covered by the crater.

> You will need to work out the radius first.

11.4 Prisms and cylinders

Volume of prisms

A **prism** is a three-dimensional solid.

The **cross section** is the same along the whole length of the prism.

This diagram shows a triangular prism.

All of the triangular cross sections shown are the same.

Here are some other prisms.

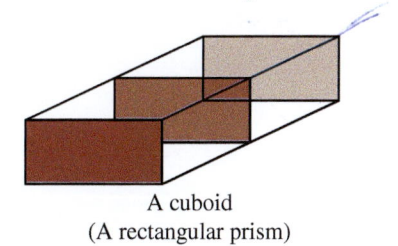

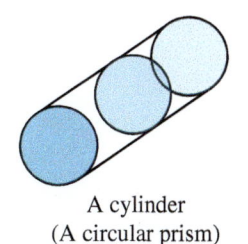

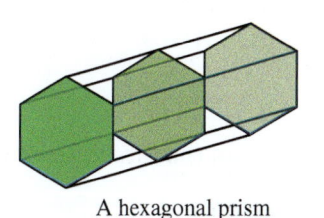

A cuboid
(A rectangular prism)

A cylinder
(A circular prism)

A hexagonal prism

Mathematics for Cambridge Secondary 1

The formula for finding the volume of a prism is

volume of prism = area of cross section × length

Worked example 1

Find the volume of this triangular prism.

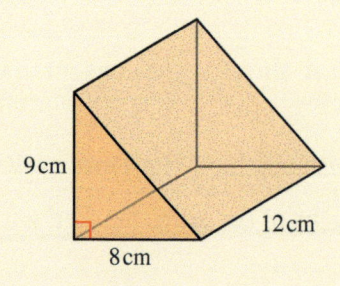

First find the area of the cross section.

Area of a triangle = $\frac{1}{2}$ × base × perpendicular height

$= \frac{1}{2} \times 8 \times 9 = 36 \, cm^2$

Volume of prism = cross-sectional area × length

$= 36 \times 12 = 432 \, cm^3$

> The units are cm^3 as this is a volume.

Worked example 2

Find the volume of this cylinder. Give the answer to 2 significant figures.

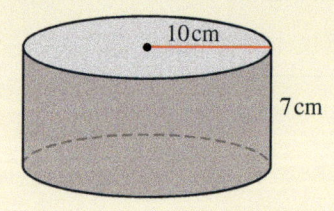

A cylinder has a circular cross section.

The area of a circle is πr^2

The volume of a cylinder, $V = \pi r^2 h$, where r is the radius and h is the height of the cylinder.

$V = \pi r^2 h$

$= \pi \times 10^2 \times 7$

$= 2199.114858$ using the calculator value for π

$= 2200 \, cm^3$ to 2 significant figures

Surface area

The surface area of a solid is found by adding the area of each face.

For a prism the faces are all flat.

Find the area of each face and add them together.

This is how to find the surface area of a cylinder.

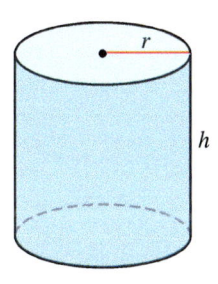

This cylinder has radius r and height h.

The surface area is made up of two flat ends and a curved side.

Each end is a circle for which the area $= \pi r^2$

To find the area of the curved face, imagine peeling it off, like the label from a tin.

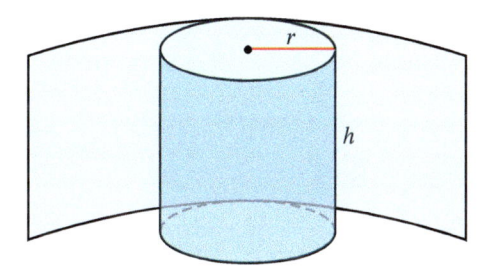

The curved face is the same as a rectangle.

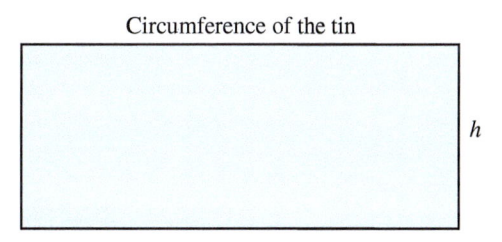

The height of the rectangle is h, and the length is the same as the circumference of the tin.

The circumference is $2\pi r$ so the area of the rectangle is $2\pi rh$.

The surface area of a cylinder = area of the curved face + area of the two circles

Surface area of a cylinder $= 2\pi rh + 2\pi r^2$

Worked example 3

Find the surface area of this cylinder. Give the answer to 3 significant figures.

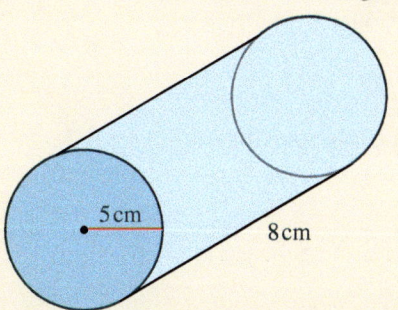

5 cm
8 cm

Mathematics for Cambridge Secondary 1

Area of one end	$= \pi r^2$
	$= \pi \times 5^2$
	$= 78.5398$ do not round to 3 significant figures until the end
Area of other end	$= 78.5398$
Curved area	$= 2\pi rh$
	$= 2 \times \pi \times 5 \times 8$
	$= 251.3274$
Total area	$= 78.5398 + 78.5398 + 251.3274$
	$= 408 \text{ cm}^2$ to 3 significant figures

The units are cm² as this is an area.

Exercise 11.4

Give your answers in this exercise to 3 significant figures where necessary.

1 Work out the volume of these prisms.

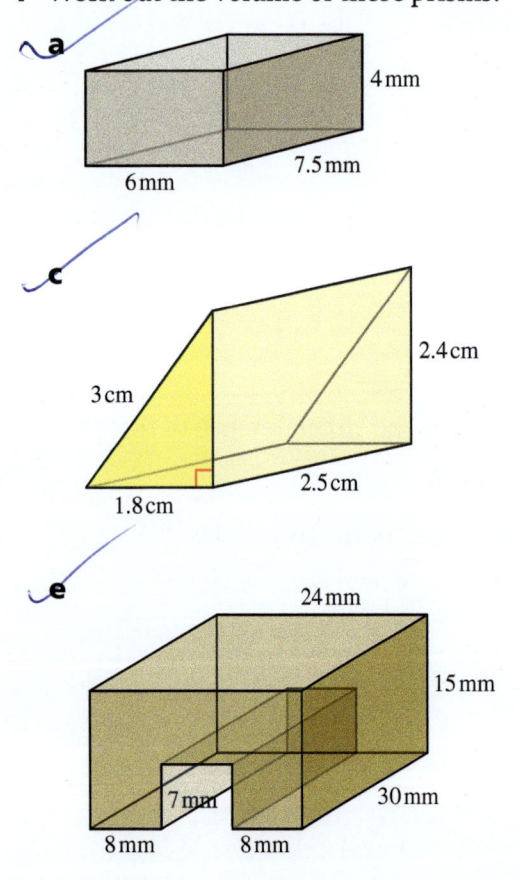

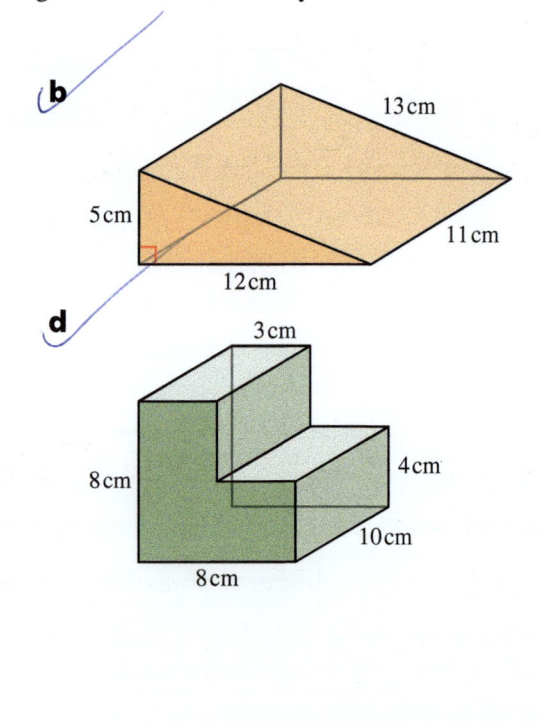

a) 6 mm, 7.5 mm, 4 mm

b) 5 cm, 12 cm, 13 cm, 11 cm

c) 3 cm, 1.8 cm, 2.5 cm, 2.4 cm

d) 3 cm, 8 cm, 4 cm, 10 cm, 8 cm

e) 24 mm, 15 mm, 30 mm, 7 mm, 8 mm, 8 mm

2 Work out the surface area of the prisms in question **1**.

3 Work out the volume of these cylinders.

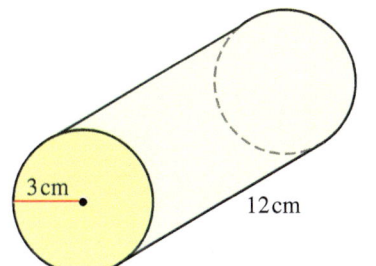

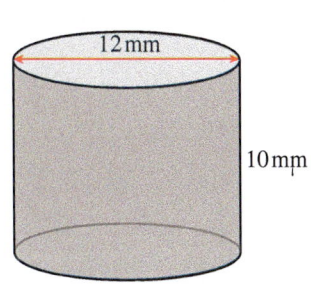

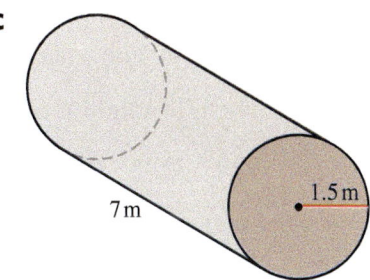

a 3 cm, 12 cm

b 12 mm, 10 mm

c 7 m, 1.5 m

4 Work out the surface area of the cylinders in question **3**.

5 Find the volume of these solids.

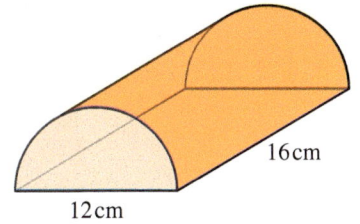

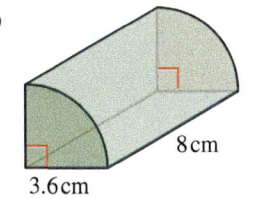

a 16 cm, 12 cm

b 8 cm, 3.6 cm

6 Work out the surface area of the solids in question **5**.

7 A water barrel is a cylinder with radius 30 cm.

The water in the barrel is 80 cm deep.

a Work out the volume of water in the barrel.

b How many litres of water are in the barrel?

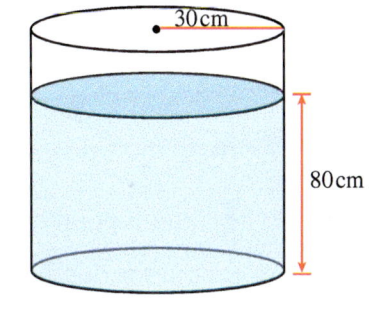

> 1 cm³ = 1 millilitre

8 a Work out the volume of this solid.

b Work out its surface area.

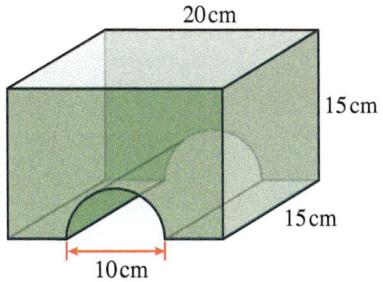

(20 cm, 15 cm, 15 cm, 10 cm)

9 Terry's polythene greenhouse is built around three hoops.

Each hoop is a semicircle.

The width of the greenhouse is 4 metres.

a Work out the length of each hoop.

Terry sees a greenhouse heater for sale.

It is big enough to heat up to 40 m³ of space.

b Work out the volume of the greenhouse.

c Is the heater big enough for Terry's greenhouse?

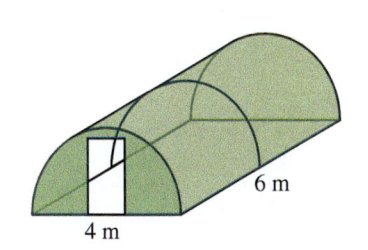

(6 m, 4 m)

> Work out the area of the semicircular end.
> Then multiply by the length to find the volume.

165

Mathematics for Cambridge Secondary 1

10 A coin is 3.15 mm thick and has diameter 22.5 mm.

A charity collects these coins and stacks them in a pile.

The pile contains 56 identical coins.

  a Work out the total height of the pile of coins.

  b Work out the volume of the pile of coins in cm³.

> Take care with mixed units. Change the lengths to centimetres before working out the volume.

Each coin has a mass of 9.5 grams.

  c Work out the total mass of the pile.

  Give your answer in kilograms.

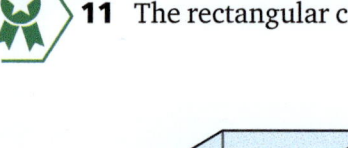

11 The rectangular container on the left is full of water.

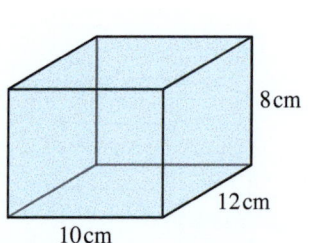

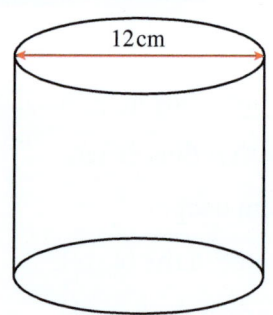

> Work out the volume of the water in the rectangular container first.

The water is poured into the cylindrical container shown on the right.

Work out the depth of the water in the cylinder.

Give your answer to the nearest mm.

Review

1 a Work out the area of this rectangle in cm².

  b Find the area in mm².

2 Convert these areas to cm²

  a 2 m²   **b** 200 mm²   **c** 6.5 m²   **d** 5000 mm²

3 Convert to m²

  a 4 km²   **b** 80 000 cm²   **c** 9.6 km²   **d** 5 000 000 mm²

4 A football field measures 195 metres by 96 metres.

  Work out the area in hectares.

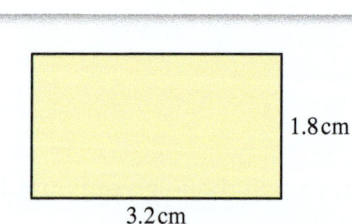

11 Area, perimeter and volume

5 The diagram shows a plan of a housing estate.

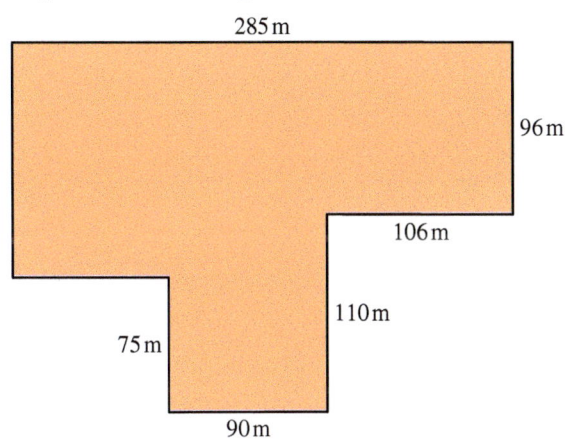

285 m
96 m
106 m
110 m
75 m
90 m

   a Work out the area of the estate in m².

   b Convert your answer to hectares.

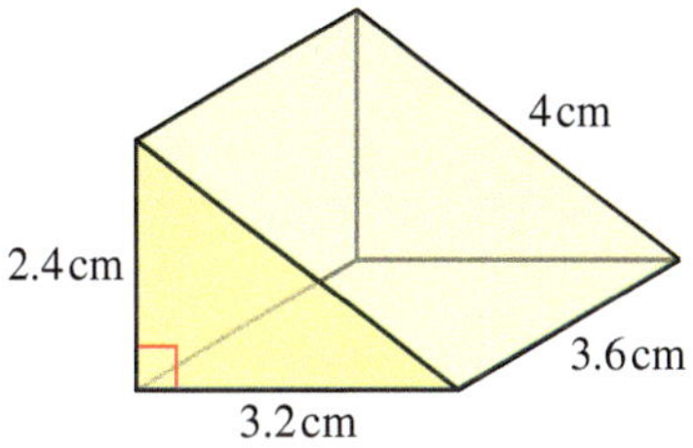

6 a Work out the volume of this prism in cm³.

   b Work out the volume of the prism in mm³.

4 cm
2.4 cm
3.6 cm
3.2 cm

7 Work out the surface area of the prism in question **6**.

   a Give your answer in cm².

   b Give your answer in mm².

8 Convert these volumes to the given units.

   a 3 m³ to cm³   **b** 8000 mm³ to cm³   **c** 4 cm³ to mm³   **d** 3 500 000 cm³ to m³

9 The diagram shows a cylindrical water tank.

   a Work out the volume of the tank in cm³.

   b Write down the capacity of the tank in litres.

   Give your answer to the nearest litre.

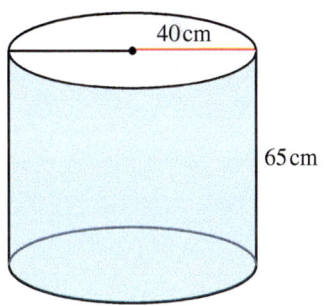

40 cm
65 cm

10 Write down the capacity in millilitres of containers with volume:

   a 450 cm³   **b** 1000 cm³   **c** 5000 mm³

11 The volume of a container is 2 m³.

   Work out how many litres it holds.

12 Work out the perimeter of these shapes. Give the answers to 3 significant figures.

   a 7 cm

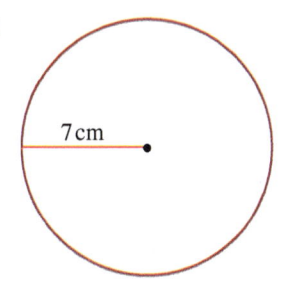

   b 12 mm

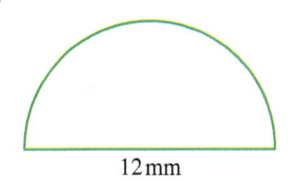

   c 5.6 cm

167

13 Work out the area of the shapes in question **12**.

14 This design is made up of two circles, one with radius 20 cm and the other with radius 10 cm.

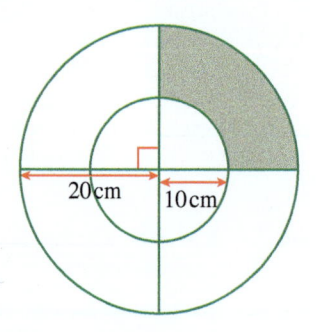

It is divided into 4 quarters.

Work out the area of the shaded part.

> Subtract $\frac{1}{4}$ of the area of the small circle from $\frac{1}{4}$ of the area of the large circle.

15 Zakariah has a container holding 25 litres of liquid.

He wants to fill some small cylindrical tins.

> Remember that 1 litre = 1000 cm^3.

Each tin is 10 cm tall and has radius 3 cm.

He has 100 tins to fill.

Work out whether he has enough liquid to fill them all.

You must show all of your working.

Examination-style questions

1 This cuboid is 4 cm by 3 cm by 2.5 cm.

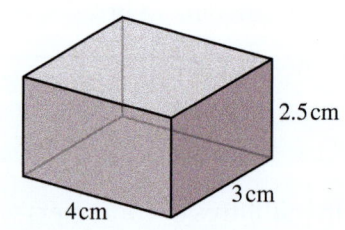

    a Work out the volume of the cuboid. [1]

    b Work out the surface area of the cuboid. [2]

2 This semicircle has diameter 6.8 cm.

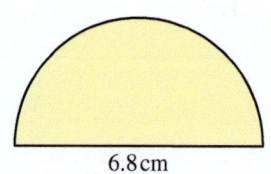

Work out the area of the semicircle. Give the answer correct to 3 significant figures. [2]

11 Area, perimeter and volume

3 This diagram shows the plan of a field.

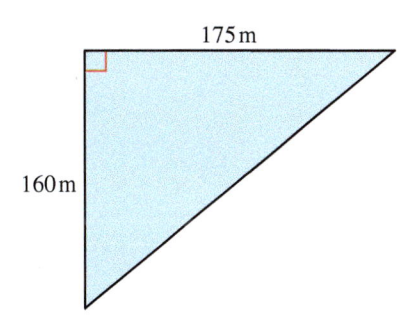

- 175 m
- 160 m

a Work out the area of the field in square metres. [2]

b Convert the answer to hectares. [1]

4 This diagram shows the net of a three-dimensional solid.

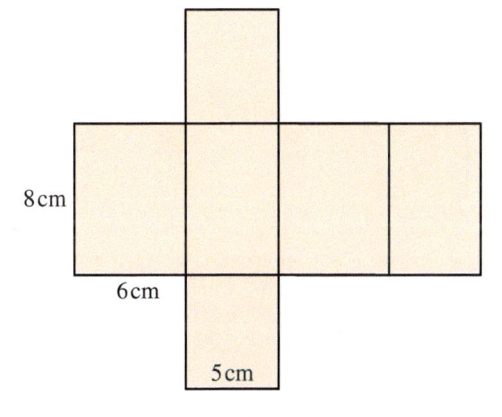

- 8 cm
- 6 cm
- 5 cm

a Work out the surface area of the solid. [2]

b Work out the volume of the solid. [2]

5 How many mm³ are there in 1 cm³? [1]

6 Work out which of these two shapes has the larger perimeter.

You must show your working.

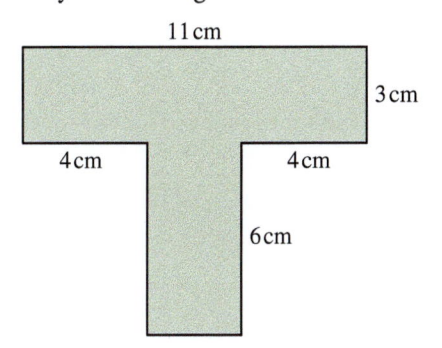

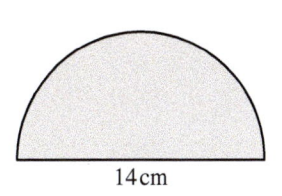

- 11 cm
- 3 cm
- 4 cm
- 4 cm
- 6 cm
- 14 cm

[2]

169

7   How many cm² are there in 1 m²? [1]

8   Work out the volume of this prism.

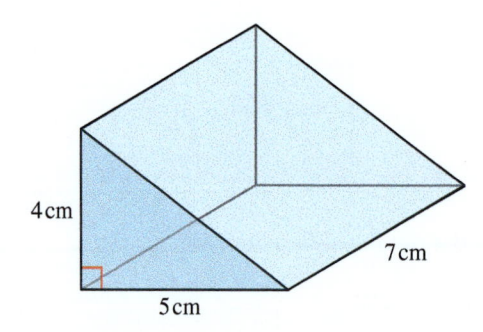

4 cm
5 cm
7 cm

[2]

9   The diagram shows a cylinder with diameter 6 cm and height 3 cm.

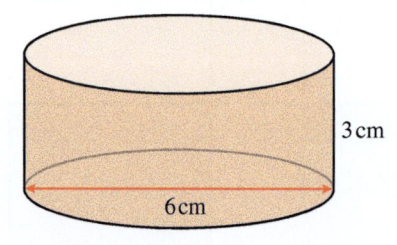

3 cm
6 cm

a   Work out the surface area of the cylinder. [2]

b   Work out the volume of the cylinder [2]

12 Formulae

> **Learning outcomes**
> - Substitute positive and negative numbers into expressions and formulae.
> - Derive formulae.
> - Change the subject of a formula.

Energy and mass

The most famous formula used in science is $E = mc^2$

It says that energy, E, and mass, m, are the same thing and that you can convert from one to the other using the constant c^2, where c is the speed of light.

A scientist called Einstein discovered this relationship at the start of the 20th century.

In this chapter you will use various formulae that are used in mathematics and science.

12.1 Deriving formulae and substituting numbers into formulae

This section revises the work that you did in your Stages 7 and 8 Student Books on **substituting** numbers into **formulae** and on deriving formulae.

Worked example 1

$$P = \frac{4h^2 + 5f}{2h}$$

a Find the value of P when $f = 2$ and $h = -5$
b Find the value of f when $P = 8$ and $h = 3$

a $P = \dfrac{4h^2 + 5f}{2h}$ replace the letters with the numbers given

$= \dfrac{4 \times (-5)^2 + 5 \times 2}{2 \times (-5)}$ $4 \times (-5)^2 = 4 \times -5 \times -5 = 4 \times 25 = 100$

$= \dfrac{100 + 10}{-10}$

$= \dfrac{110}{-10}$ remember that $+ \div - = -$

$= -11$

171

b $P = \dfrac{4h^2 + 5f}{2h}$ replace the letters with the numbers given

$8 = \dfrac{4 \times (3)^2 + 5f}{2 \times 3}$ $4 \times (3)^2 = 4 \times 3 \times 3 = 4 \times 9 = 36$

$8 = \dfrac{36 + 5f}{6}$ solve the equation

$48 = 36 + 5f$ multiply both sides by 6

$12 = 5f$ subtract 36 from both sides

$f = 2.4$ divide both sides by 5

Worked example 2

A chocolate has a mass of 23 grams and a sweet has a mass of 18 grams.

Ali has x chocolates and $(x - 2)$ sweets.

Write a formula for the total mass, T grams, of the chocolates and sweets.

Mass of x chocolates $= 23x$

Mass of $(x - 2)$ sweets $= 18(x - 2)$

$T = 23x + 18(x - 2)$ expand the brackets

$T = 23x + 18x - 36$ collect like terms

$T = 41x - 36$

Exercise 12.1

1 Find the value of y in each of the formulae when: **i** $x = 10$ **ii** $x = -5$ **iii** $x = -10$

  a $y = 5x - 12$ **b** $y = \dfrac{x}{5} + 7$ **c** $y = \dfrac{5x - 1}{2}$ **d** $y = 50 - 4x$

  e $y = 2x - x^2$ **f** $y = 5x^2 - 1$ **g** $y = x^2 + 5x + 4$ **h** $y = 3x^2 - 5x - 2$

2 Find the value of p in each of the formulae when:

  i $a = 2, b = 5$ **ii** $a = 7, b = -3$ **iii** $a = -1, b = -4$

  a $p = 2ab$ **b** $p = 5 - ab$ **c** $p = a - 2b$ **d** $p = 7a - 3b + 6$

  e $p = a^2 + 2b$ **f** $p = a(2b - 5)$ **g** $p = a^2 + 2ab$ **h** $p = 3a^2 - 2ab - b^2$

3 Find the value of T in each of the formulae when:

  i $a = 3, b = 2, c = 5$ **ii** $a = -1, b = 4, c = 3$ **iii** $a = -2, b = -3, c = -4$

  a $T = ab + c$ **b** $T = 3a + 2b - 4c$ **c** $T = (2a + b - c)^2$ **d** $T = 2a^2 + 5b - c$

4 Find the value of y in each of the formulae when: **i** $x = 2$ **ii** $x = -2$

  a $y = \dfrac{x - 4}{4}$ **b** $y = \dfrac{x + 3}{x - 1}$ **c** $y = 8 - \dfrac{6}{x}$ **d** $y = \dfrac{x^2 + 1}{1 - x}$

5 The formula for the perimeter, P, of the rectangle is $P = 2l + 2b$

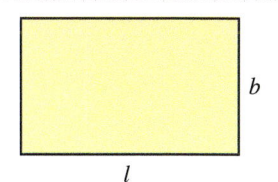

   a Find the value of P when $l = 7.5$ and $b = 3.6$

   b Find the value of l when $P = 50$ and $b = 4.7$

6 The formula for the area, A, of the trapezium is $A = \frac{1}{2}(a + b)h$

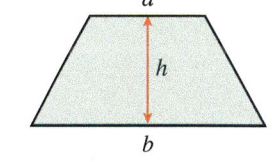

   a Find the value of A when $a = 7$, $b = 6$ and $h = 8$

   b Find the value of h when $A = 40$, $a = 3$ and $b = 5$

   c Find the value of b when $A = 60$, $a = 4$ and $h = 12$

7 $F = \dfrac{mv - mu}{t}$ is a formula used in physics.

   a Find the value of F when $m = 5$, $v = 8$, $u = 3$ and $t = 2$

   b Find the value of F when $m = 7$, $v = 10$, $u = -2$ and $t = 5$

   c Find the value of m when $F = 20$, $v = 9$, $u = 4$ and $t = 2$

8 A concert sells x tickets at $35 each and $(x + 18)$ tickets at $48 each.

   Write a formula for the total cost, C dollars, of the tickets.

9 Magazines cost $$b$ each and newspapers cost $(b - 3)$ each.

   Write a formula for the total cost, T dollars, of 4 magazines and 8 newspapers.

10 Helen buys x candles.

   Each candle costs $3. She pays with a $50 note.

   Write a formula for the amount of change, C dollars, that Helen receives.

11 Hero's formula can be used to find the area, A, of any triangle when all three sides are known.

You can find out more about Hero on the internet.

$A = \sqrt{s(s-a)(s-b)(s-c)}$

where $s = \frac{1}{2}(a + b + c)$

Use the formula to find the area of a triangle with sides 5 cm, 7 cm and 10 cm.

12 The diagram shows a torus.

A torus is formed by revolving a small circle, radius r, along a line made by a second circle of radius R.

The formulae for the volume, V, and the surface area, S, of a torus are:

$$V = 2 \times \pi^2 \times R \times r^2 \quad \text{and} \quad S = 4 \times \pi^2 \times R \times r$$

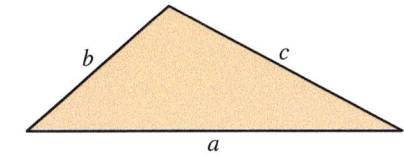

   a Find the values of V and S when $R = 7$ cm and $r = 4$ cm.

   b Find a formula connecting V, S and r.

Mathematics for Cambridge Secondary 1

12.2 Changing the subject of a formula

In the formula $P = ax + b$, P is called the **subject** of the formula.

The formula can be rearranged to make x the subject using the following two methods.

Method 1: Reverse function machine method

The function machine is:

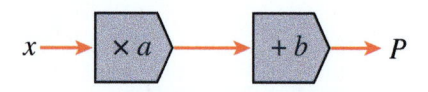

$x \rightarrow \boxed{\times a} \rightarrow \boxed{+b} \rightarrow P$

The reverse function machine is: $\frac{P-b}{a} \leftarrow \boxed{\div a} \leftarrow P-b \leftarrow \boxed{-b} \leftarrow P$

So $x = \frac{P-b}{a}$

Method 2: Balance method

$$P = ax + b$$

$-b$ $P - b = ax$ $-b$

$\div a$ $\frac{P-b}{a} = x$ $\div a$

> The methods for rearranging formulae are the same as the methods for solving equations.
>
> It is usually quicker to use the balance method.

Worked example

Make x the subject of the formula $a(bx - c) = d$

$a(bx - c) = d$	multiply out the brackets
$abx - ac = d$	add ac to both sides
$abx = d + ac$	divide both sides by ab
$x = \frac{d + ac}{ab}$	

Exercise 12.2

1 Make x the subject of these formulae.

   a $x + 2 = y$ **b** $x + a = b$ **c** $x - 5 = pq$ **d** $x - p = 3q$

   e $t = x - 2u$ **f** $v^2 = x + y$ **g** $a^2 + x = b$ **h** $x - a^2 = 2a$

   i $x - p = q + r$ **j** $b - c = x + a$ **k** $ab + x = c$ **l** $r = x - pq$

2 Make y the subject of these formulae.

   a $5y = a$ **b** $ay = b$ **c** $ay = 2e$ **d** $3by = c$

   e $pqy = r$ **f** $ay = \pi r^2$ **g** $3y = 2a + b$ **h** $cy = a - b$

   i $5y = 9x - 15$ **j** $xy = x + 2$ **k** $xy = fg - 5$ **l** $7y = 5x - 1$

   m $y(a + b) = c$ **n** $y(p - 3q) = 2r$ **o** $y(2p - q) = 5p + 2q$ **p** $y(3 - a) = a^2 + b^2$

3 Make x the subject of these formulae.

- **a** $2x + a = b$
- **b** $5x - q = r$
- **c** $y = 2x - 5$
- **d** $p = 3x + 2q$
- **e** $c = 7x + 4y$
- **f** $y = mx + c$
- **g** $2x + y = z$
- **h** $4x - y = 8$
- **i** $ax + b = c$
- **j** $2ax - b = 10$
- **k** $3x - 4 = y$
- **l** $a^2 + b^2x = c^2$
- **m** $abx + c = d$
- **n** $3fx - g^2 = h^2$
- **o** $8x - 2f = 5g + h$
- **p** $ab + c = 5d + 2x$

4 Kiran and Angelo are asked to make x the subject of the formula $w = 2(x + b)$

Kiran writes:

$$w = 2(x + b)$$
$$w = 2x + 2b$$
$$w - 2b = 2x$$
$$\frac{w - 2b}{2} = x$$

Angelo writes:

$$w = 2(x + b)$$
$$\frac{w}{2} = x + b$$
$$\frac{w}{2} - b = x$$

Who is correct? Explain your answer.

5 Make x the subject of these formulae.

- **a** $3(x + 2) = y$
- **b** $a(x + b) = c$
- **c** $a = 8(x - b)$
- **d** $3p = 5(x + 2q)$
- **e** $2a = c(x - b)$
- **f** $4g(h + x) = h$
- **g** $p = q(2 + x)$
- **h** $a^2(x - b) = c^2$
- **i** $p(x + 5q) = 8$
- **j** $ab(a + x) = c$
- **k** $a(2x + 3b) = c$
- **l** $a^2(3 + 4x) = b$

6 $v = u + at$ is a formula used in physics.

- **a** Make u the subject.
- **b** Make a the subject.
- **c** Make t the subject.

7 The formula for finding the surface area, A, of a solid cylinder is $A = 2\pi rh + 2\pi r^2$, where r is the radius and h is the height.

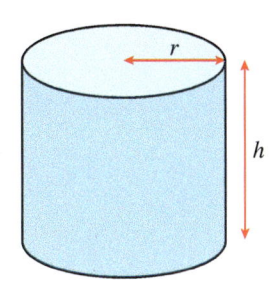

- **a** Find the surface area when $r = 5$ cm and $h = 8$ cm.

  Write your answer as a multiple of π.
- **b** Make h the subject of the formula.
- **c** Find the value of h when $A = 600$ cm² and $r = 8$ cm.

  Write your answer correct to 3 significant figures.

Mathematics for Cambridge Secondary 1

8

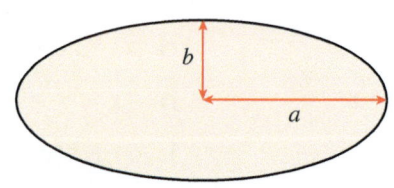

The diagram shows an ellipse.

The formula for finding the area, A, of an ellipse is $A = \pi ab$

a Show that the shaded area, S, in the diagram below can be found using the formula $S = \pi a + 6\pi$

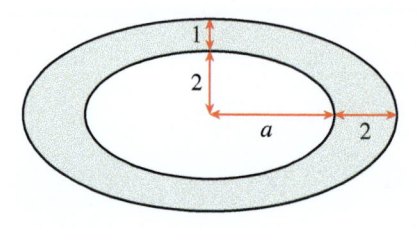

b Make a the subject of the formula $S = \pi a + 6\pi$

c Use your answer to part **b** to find the value of a when the shaded area is $30 \, \text{cm}^2$.

Write your answer correct to 3 significant figures.

12.3 Formulae involving negative x terms

The formula $a = b - 3x$ has a negative x term.

It is easier to work with positive x terms so you must take the x term to the other side of the equation to make it positive.

Worked example

Make x the subject of the formula $a = b - 3x$

$a = b - 3x$	add $3x$ to both sides
$a + 3x = b$	subtract a from both sides
$3x = b - a$	divide both sides by 3
$x = \dfrac{b - a}{3}$	

Exercise 12.3

1 Make x the subject of these formulae.

a $y = 2 - x$
b $a = b - x$
c $p = 2q - x$
d $3pq - x = 7$
e $a^2 - x = b^2$
f $p + q = r - x$
g $2a - 3b = 5c - x$
h $a = b + c - x$

176

i $2f = g - 2x$ **j** $f - gx = h$ **k** $mn = 8 - 5x$ **l** $p^2 = 2a - 3x$

m $a^2 + ab = b - cx$ **n** $5b^2 - a^2x = c$ **o** $4y - 3x = 5z$ **p** $m^2 - 2mnx = n^2$

2 Make y the subject of these formulae.

    a $9 = 2(x - y)$ **b** $d = 5(2 - y)$ **c** $a = 3(b - y)$ **d** $h(g - y) = a$

    e $a(b - y) = c$ **f** $h(h - y) = c$ **g** $a^2 = b(c - y)$ **h** $p = q(r + t - y)$

    i $5(a^2 - y) = b$ **j** $8(pq - 2y) = r$ **k** $7p = 2(q - 3y)$ **l** $2fg(3 - y) = 4h$

3 $Ft = mv - mu$ is a formula used in physics.

    a Make t the subject.   **b** Make v the subject.   **c** Make u the subject.

12.4 Formulae involving fractions

To rearrange formulae involving fractions you need to remember the rules that you used in Chapter 8 when solving algebraic fractions.

Worked example 1

Make x the subject of the formula $\dfrac{x - a}{2} = y$

$\dfrac{x - a}{2} = y$ multiply both sides by 2

$x - a = 2y$ add a to both sides

$x = 2y + a$

Worked example 2

Make x the subject of the formula $\dfrac{b}{4} = \dfrac{5a}{x}$

$\dfrac{b}{4} = \dfrac{5a}{x}$ there is a single fraction on each side so 'cross multiply'

$b \times x = 4 \times 5a$

$bx = 20a$

$x = \dfrac{20a}{b}$

Exercise 12.4

1 Make x the subject of these formulae.

    a $\dfrac{x}{3} = y$ **b** $\dfrac{x}{a} = p$ **c** $\dfrac{x}{7} = ab$ **d** $h = \dfrac{x}{g}$

    e $p^2 = \dfrac{x}{q}$ **f** $a + 2 = \dfrac{x}{c}$ **g** $\dfrac{x}{4} = p + q$ **h** $\dfrac{x}{5} = 2a - b$

    i $\dfrac{x}{2 + y} = 4$ **j** $\dfrac{x}{a - b} = c$ **k** $\dfrac{x}{a^2 - c^2} = b^2$ **l** $\dfrac{x}{2g + 3h} = 5f$

177

2 Make x the subject of these formulae.

a $p = \dfrac{x+7}{3}$ **b** $\dfrac{x-8}{5} = y$ **c** $\dfrac{x-a}{b} = c$ **d** $\dfrac{x-2m}{5} = n$

e $a = \dfrac{5-x}{4}$ **f** $\dfrac{6-x}{a} = b$ **g** $\dfrac{m-x}{2} = n$ **h** $ab = \dfrac{a-3x}{b}$

3 Make x the subject of these formulae.

a $\dfrac{3}{x} = y$ **b** $\dfrac{a}{x} = t$ **c** $\dfrac{2p}{x} = q$ **d** $\dfrac{y}{x} = 2$

e $\dfrac{x}{4} = \dfrac{a}{b}$ **f** $\dfrac{x}{p} = \dfrac{q}{r}$ **g** $\dfrac{5x}{a} = \dfrac{2b}{c}$ **h** $\dfrac{px}{q} = \dfrac{q}{p}$

i $y = \dfrac{a+b}{x}$ **j** $a^2 = \dfrac{b^2 - c^2}{x}$ **k** $2ab = \dfrac{c+d}{x}$ **l** $\dfrac{a+b}{x} = c+d$

4 Make x the subject of these formulae.

a $\dfrac{5}{x-3} = y$ **b** $\dfrac{7}{x+2} = p$ **c** $\dfrac{8}{3-x} = a$ **d** $\dfrac{a}{x+b} = c$

e $\dfrac{f}{g-x} = h$ **f** $\dfrac{a^2}{a-x} = 5b$ **g** $\dfrac{m}{8-3x} = 2n$ **h** $\dfrac{ab}{c-bx} = c$

5 Make x the subject of these formulae.

a $\dfrac{1}{x} + 2 = y$ **b** $\dfrac{1}{x} - a = b$ **c** $\dfrac{1}{x} - pq = p$ **d** $\dfrac{1}{x} - y^2 = 5y$

6 Make x the subject of these formulae.

a $\dfrac{1}{2}x + 5 = y$ **b** $\dfrac{1}{3}x - a = b$ **c** $y = \dfrac{1}{2}x + c$ **d** $p = q - \dfrac{x}{4}$

e $m = n - \dfrac{x}{5}$ **f** $y = 2 - \dfrac{x}{5}$ **g** $\dfrac{x}{2} + \dfrac{y}{4} = 1$ **h** $\dfrac{x}{a} + \dfrac{y}{b} = 12$

7 The formula for calculating simple interest is $I = \dfrac{PRT}{100}$

  a Make P the subject. **b** Make R the subject. **c** Make T the subject.

8 $s = \dfrac{1}{2}(u+v)t$ is a formula used in physics.

  a Make t the subject. **b** Make u the subject. **c** Make v the subject.

9 $s = \dfrac{d}{t}$ is a formula connecting speed, distance and time.

  a Make d the subject. **b** Make t the subject.

10 $A = \dfrac{1}{2}(a+b)h$ is the formula for finding the area, A, of a trapezium.

  a Make h the subject of the formula.

  b Find h when $A = 75\,\text{cm}^2$, $a = 8\,\text{cm}$ and $b = 12\,\text{cm}$.

  c Make b the subject of the formula.

  d Find b when $A = 58\,\text{cm}^2$, $a = 10\,\text{cm}$ and $h = 8\,\text{cm}$.

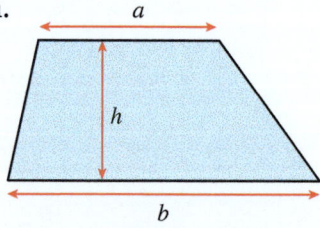

11 The formula $F = \dfrac{9C}{5} + 32$ is used to change degrees Celsius (°C) to degrees Fahrenheit (°F).

  a Show that the formula can be rearranged to $C = \dfrac{5(F-32)}{9}$

  b Use the formula in part **a** to find the value of C when $F = -31$

  c Show that when C is equal to -40, F is also equal to -40.

12.5 Formulae involving squares and square roots

It is important to remember that:

if $x^2 = 25$

The reverse of 'square' is 'square root'.

then $x = \pm\sqrt{25}$

so $x = \pm 5$ check: $5^2 = 25$ ✓ and $(-5)^2 = -5 \times -5 = 25$ ✓

To make x the subject of the formula $x^2 - b = a$

$$x^2 - b = a$$
$$x^2 = a + b$$
$$x = \pm\sqrt{a+b}$$

(+b, +b, √, √)

Worked example 1

Make x the subject of the formula $ax^2 - 5 = b$

$ax^2 - 5 = b$ add 5 to both sides

$ax^2 = b + 5$ divide both sides by a

$x^2 = \dfrac{b+5}{a}$ take the square root of both sides

$x = \pm\sqrt{\dfrac{b+5}{a}}$

Don't forget the ± symbol in front of the square root.

Worked example 2

Make x the subject of the formula $\sqrt{2x-3} = y$

$\sqrt{2x-3} = y$ square both sides

$2x - 3 = y^2$ add 3 to both sides

$2x = y^2 + 3$ divide both sides by 2

$x = \dfrac{y^2 + 3}{2}$

Worked example 3

The formula for the area of a circle is $A = \pi r^2$

Make r the subject of the formula.

$A = \pi r^2$ divide both sides by π.

$\dfrac{A}{\pi} = r^2$ take the square root of both sides.

$r = \sqrt{\dfrac{A}{\pi}}$

Since r is a length it must be a positive number, so you do not include the ± symbol.

Exercise 12.5

1 Make x the subject of these formulae.

   a $x^2 = a$ **b** $x^2 = pq$ **c** $2x^2 = y$ **d** $x^2 + 4 = m$

   e $x^2 - 5 = 4b$ **f** $x^2 + a = b$ **g** $x^2 - 2f = g$ **h** $2x^2 + 3 = y$

   i $ax^2 + b = c$ **j** $px^2 + 3q = 5r$ **k** $7 - x^2 = 2y$ **l** $8 - 3x^2 = 2a$

2 Make x the subject of these formulae.

   a $\dfrac{x^2}{2} - 4 = y$ **b** $\dfrac{x^2}{a} + b = c$ **c** $5 - \dfrac{x^2}{4} = a$ **d** $pq - \dfrac{x^2}{3} = r$

3 Make x the subject of these formulae.

   a $\sqrt{x} = c$ **b** $\sqrt{x} = 5p$ **c** $\sqrt{x+2} = c$ **d** $\sqrt{x-4} = 3a$

   e $\sqrt{x+a} = b$ **f** $\sqrt{x-f} = 5$ **g** $\sqrt{x-2m} = 3n$ **h** $2\sqrt{x-4} = y$

   i $a\sqrt{x} = b$ **j** $\sqrt{ax} = b$ **k** $\sqrt{\dfrac{x}{4}} = d$ **l** $n = \sqrt{\dfrac{x}{m}}$

   m $\sqrt{\dfrac{x+2}{5}} = q$ **n** $\sqrt{\dfrac{a}{x}} = \dfrac{b}{c}$ **o** $\sqrt{\dfrac{2}{3x}} = y$ **p** $\sqrt{\dfrac{1}{x-4}} = a$

4 The formula for the surface area of a sphere is $A = 4\pi r^2$

   a Make r the subject of the formula.

   b Use your answer to part **a** to find the value of r when $A = 20\,\text{cm}^2$.

      Write your answer correct to 3 significant figures.

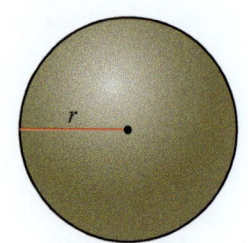

5 The formula for the volume of a cone is $V = \dfrac{1}{3}\pi r^2 h$

   a Make r the subject of the formula.

   b Find the value of r when $V = 50\,\text{cm}^3$ and $h = 12\,\text{cm}$.

      Write your answer correct to 3 significant figures.

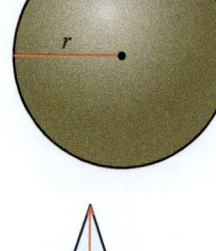

6 $T = 2\pi\sqrt{\dfrac{L}{10}}$

   The formula gives the time, T seconds, for one complete swing of a pendulum of length L metres.

   a Copy and complete the function machine for finding T.

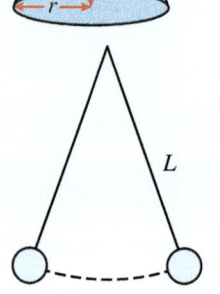

      $L \rightarrow \boxed{\div \ldots} \rightarrow \boxed{\sqrt{\ }} \rightarrow \boxed{\times \ldots} \rightarrow T$

   b Use your function machine from part **a** to find the value of T when $L = 0.2\,\text{m}$.

      Write your answer correct to 3 significant figures.

c Copy and complete the reverse function machine to make L the subject of the formula.

… ← […] ← […] ← […] ← T

d Use your answer to part **c** to find the value of L when $T = 1$ second.

Write your answer correct to 3 significant figures.

7 $x = \dfrac{-b \pm \sqrt{b^2 - 4ac}}{2a}$ is a very famous formula which you will learn how to use in your IGCSE.

Show that you can rearrange the formula to give $c = \dfrac{b^2 - (2ax + b)^2}{4a}$

You must show all the steps in your working.

Review

1 $y = \dfrac{3x^2 - 8}{x - 2}$

Find the value of y when:

a $x = 3$ **b** $x = -2$

2 The formula $s = \frac{1}{2}(u + v)t$ is used in physics.

a Find the value of s when $u = 5$, $v = 7$ and $t = 4$

b Find the value of s when $u = -4$, $v = 9$ and $t = 6$

c Find the value of t when $s = 60$, $u = 13$ and $v = -3$

d Find the value of v when $s = 45$, $u = -2$ and $t = 15$

3 Mangos cost $\$x$ each and pineapples cost $\$(x + 1)$ each.

Write a formula for the total cost, T dollars, of 5 pineapples and 10 mangos.

4 Write a formula for:

a the area A of the rectangle

b the perimeter P of the rectangle.

(rectangle with side x and $2x + 1$)

5 Make x the subject of these formulae.

a $y = x - 8$ **b** $y - a = 3x$ **c** $a(2x - 5) = c$ **d** $a - cx = b$

e $h(f - 2x) = 3g^2$ **f** $\dfrac{x}{5p} = q$ **g** $\dfrac{x}{2a + 3b} = 4c$ **h** $\dfrac{x + m}{n} = 5p$

i $\dfrac{10}{x} = y$ **j** $\dfrac{3x}{a} = \dfrac{4b}{c}$ **k** $\dfrac{y}{x - 5} = 2$ **l** $2y = 5 - \dfrac{1}{x}$

m $y = \frac{1}{5}x - 4$ **n** $2 - x^2 = y$ **o** $\dfrac{x^2}{3} - 1 = y$ **p** $\sqrt{x - 4} = 2y$

Examination-style questions

1. $p = 5r$

   Find p when $r = 4.5$ [1]

2. $P = 2L + 2W$

   Find P when $L = 6$ and $W = 9$ [2]

3. $E = \frac{1}{2}mv^2$

   a Find E when $m = 7$ and $v = 12$ [2]

   b Find v when $E = 32$ and $m = 9$ [2]

   c Make m the subject of the formula. [2]

4. $A = 2\pi rh + 2\pi r^2$

   Find A when $r = 3$ and $h = 7$

   Write your answer correct to 3 significant figures. [2]

5. Ice lollies cost $\$x$ each and ice creams cost $\$y$ each.

   Write a formula for the total cost, $\$T$, of 3 ice lollies and 5 ice creams. [2]

6. Make f the subject of the formula $g = 5f - 6$ [2]

7. Make R the subject of $V = IR$ [2]

8. Make x the subject of the formula $2(x - y) = 7y + 8$ [2]

9. Make h the subject of the formula $g = \sqrt{h + 2}$ [2]

10. Make x the subject of the formula $y = \frac{5x}{2} + 3$ [3]

11. Make P the subject of the formula $Q = \frac{2}{P + 1}$ [3]

12. Make r the subject of the formula $A = \frac{3\pi r^2}{2}$ [3]

13 Position and movement

> **Learning outcomes**
>
> - Tessellate triangles and quadrilaterals and relate to angle sums and half-turn rotations; know which regular polygons tessellate, and explain why others will not.
> - Find by reasoning the locus of a point that moves at a given distance from a fixed point, or at a given distance from a fixed straight line.
> - Use the coordinate grid to solve problems involving translations, rotations, reflections and enlargements.
> - Transform 2-D shapes by combinations of rotations, reflections and translations; describe the transformation that maps an object on to its image.
> - Enlarge 2-D shapes, given a centre and positive integer scale factor; identify the scale factor of an enlargement as the ratio of the lengths of any two corresponding line segments.
> - Recognise that translations, rotations and reflections preserve length and angle, and map objects on to congruent images, and that enlargements preserve angle but not length.
> - Know what is needed to give a precise description of a reflection, rotation, translation or enlargement.

Roman mosaics

The original tessellations were inspired by mosaic designs.

The Romans in particular were famous for their mosaic floors and walls.

This picture shows a Roman mosaic in Jordan.

The Latin word tessera, sometimes written as tessella, was the word for one small tile.

The tiling patterns came to be known as tessellations.

13.1 Tessellation

Triangles and quadrilaterals

A pattern of shapes that covers a flat surface with no overlaps or gaps is called a **tessellation**.

Here are some examples of tessellations.

Each pattern uses one shape repeatedly to form the design.

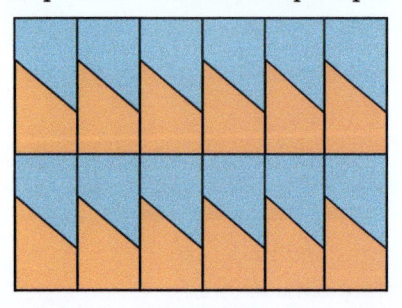

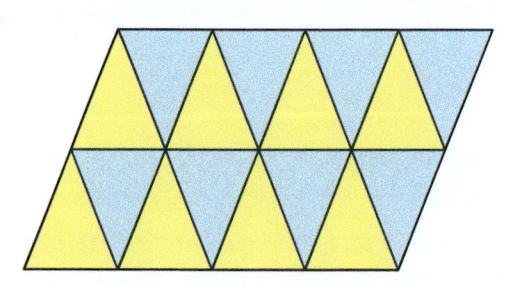

This is how to tessellate any triangle.

Step 1

Draw a triangle – it can be any shape as long as it is a triangle.

Step 2

Draw a 180° rotation of the triangle next to the original.

You now have a parallelogram made up of two triangles.

Step 3

Repeat these two steps to place another parallelogram next to the first one.

You are now building up a row of triangles.

Step 4

Now draw another row of triangles next to the first row.

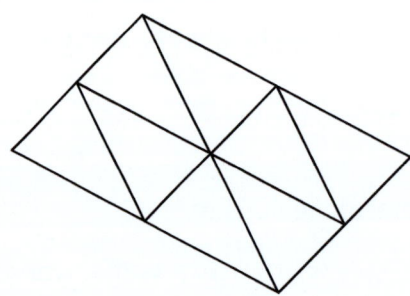

Step 5

You can now see how the tessellation is built up. Colour in the final design.

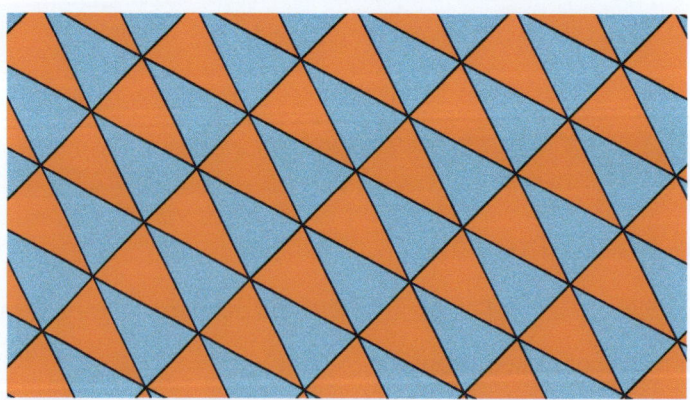

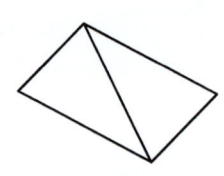

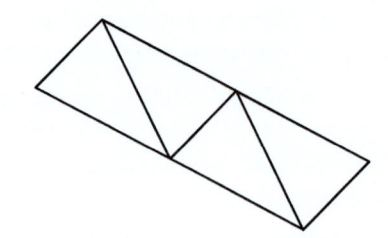

It is possible to see why this works if the angles of the triangle are labelled.

Here is the original triangle with the angles marked a, b and c.

The angles of a triangle add up to 180° so $a + b + c = 180°$

Look now at a corner in the final tessellation.

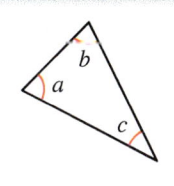

Around the corner circled in this diagram there are two of each angle from the original triangle.

Altogether the angles around that point add up to $2a + 2b + 2c$

$$= 2(a + b + c)$$
$$= 2 \times 180°$$
$$= 360°$$

You know that the angles around a point add up to 360° and this is always the case if a triangle is tessellated like this.

Worked example 1

a Use the same technique to tessellate a quadrilateral.

b Show that the angles at any vertex of the tessellation add up to 360°.

a

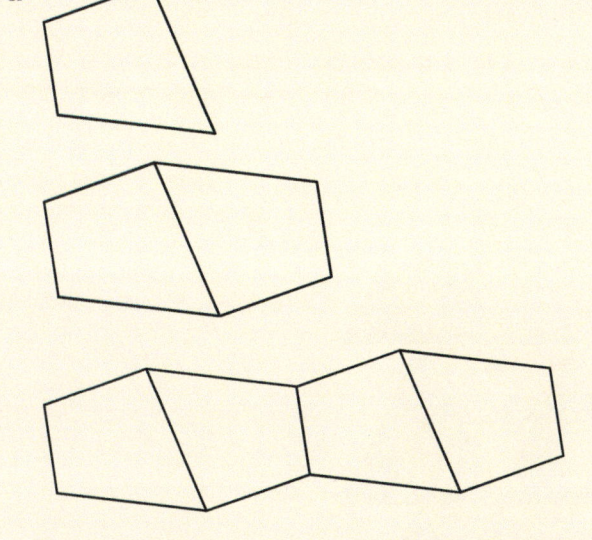

step 1: draw a quadrilateral

step 2: draw a 180° rotation next to the original

step 3: repeat this process to build up a row of quadrilaterals step 4: draw another row next to the first row

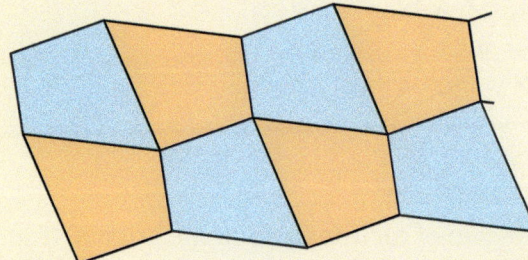

step 5: colour the final tessellation

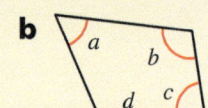

b label the four angles of the original shape a, b, c and d

The angles of a quadrilateral add up to 360° so $a + b + c + d = 360°$

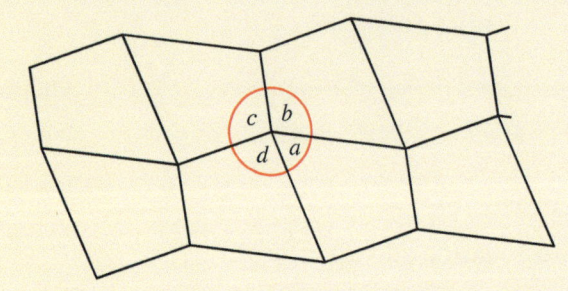

now label the angles at a vertex of the final shape

The four angles at each vertex are a, b, c and d.

But $a + b + c + d = 360°$ so the angles at the vertex add up to 360°.

Regular tessellations

Here is a tessellation of an equilateral triangle.

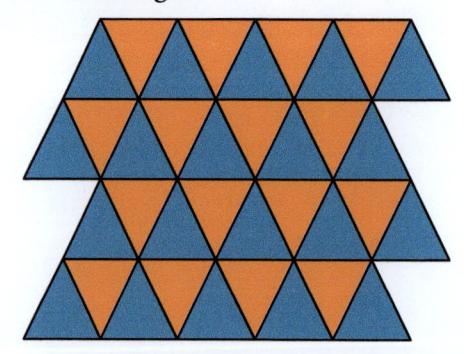

186

An equilateral triangle is a **regular polygon**.

A polygon is regular if all of the sides and angles are equal.

This is called a **regular tessellation**.

Each angle in an equilateral triangle is 60°.

At each vertex of the tessellation there are six angles meeting.

It tessellates because 6 × 60° = 360°, or because the interior angle is a factor of 360°.

Here is a diagram showing some regular octagons.

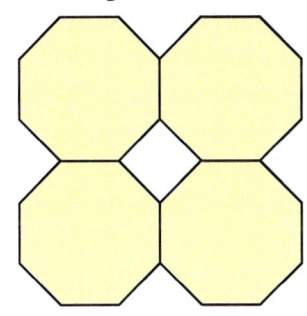

When four octagons are put together there is a square gap.

Regular octagons do not form a regular tessellation.

Check this by looking at the interior angles.

Each interior angle is 135°.

135° × 2 = 270° and 135° × 3 = 405° so 135°, is not a factor of 360°.

It is not possible to arrange 135° angles around a point with no gaps.

Worked example 2

a Draw a diagram showing the tessellation of a regular quadrilateral.

b Write down the interior angle of a regular quadrilateral and justify the tessellation.

a A regular quadrilateral has four equal sides and four equal angles.

This is a square.

Here is a tessellation of squares.

b The interior angle of a square is 90°.

90° is a factor of 360° because 4 × 90° = 360°

There are four 90° corners meeting at each vertex of the tessellation.

Mathematics for Cambridge Secondary 1

Exercise 13.1

1 Copy this triangle on to plain paper.

Leave room around it to draw a tessellation pattern.

Draw a tessellation of the triangle.

2 Copy this quadrilateral on to plain paper.

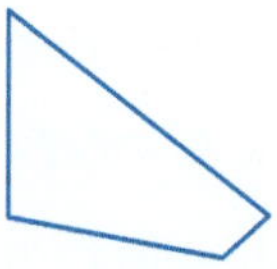

Leave room around it to draw a tessellation pattern.

Draw a tessellation of the quadrilateral.

3 a Draw a tessellation of this quadrilateral.

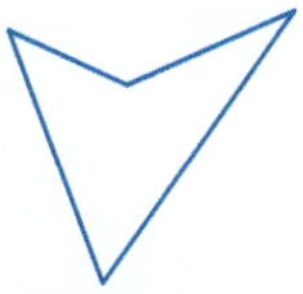

 b By labelling the angles show that the angles at each vertex add up to 360°.

4 a Work out the interior angle of a regular hexagon.

 b Show that 120° is a factor of 360°.

 c Draw a regular tessellation using regular hexagons.

> Look this up in Chapter 3 if you need a reminder.

5 This L shape is made up of three squares.

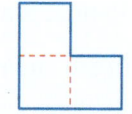

> There are several possible ways of tessellating this shape.

Draw a tessellation of this shape.

You may find it helpful to draw it on squared paper.

6 In Chapter 3 you worked out the interior angle of a series of regular polygons.

Here is the completed table showing the results.

Number of sides	3	4	5	6	7	8	9	10	12	15	20
Interior angle	60°	90°	108°	120°	126.6°	135°	140°	144°	150°	156°	162°

You have already seen that the interior angles of regular polygons with 3, 4 and 6 sides are factors of 360°.

 a Work out if any of the other interior angles in the table are factors of 360°.

 b Confirm that there are only three regular tessellations.

> A factor of 360 must divide exactly into 360.

7 This L shape is made up of four squares.

Draw two different tessellations of this shape.

> It helps if you use squared paper to draw this tessellation.

8 This is a pentagon with interior angles of 90°, 90°, 135°, 90° and 135°.

   a Draw a tessellation of this pentagon.

   b Show that the sum of the angles at each vertex is 360°.

> Draw at least 8 copies of the shape so that you can see that the pattern continues.

9 You have seen how two octagons and a square can make a tiling pattern.

This is called a semi-regular tessellation.

It is made of more than one type of regular polygon.

   a Add together the interior angles of two octagons and one square.

      Confirm that they add up to 360°.

   b Add together the interior angles of two hexagons and two equilateral triangles.

      Confirm that they add up to 360°.

   c Draw a semi-regular tessellation with two hexagons and two triangles at each vertex.

   d See if you can find another set of interior angles that add up to 360°.

> There are 8 altogether but some are harder to draw than others.

      Find another semi-regular tessellation.

13.2 Loci

A **locus** is a set of points that satisfy a particular rule.

A locus can be a line or a region.

The plural of locus is **loci**.

Worked example 1

Find the locus of points that are 3 centimetres from the point X.

X •

By using a ruler it is possible to find some points that are 3 cm from X.

The diagram shows some of these points marked with blue dots.

However, there are many more points that could be drawn.

It should be clear that they all lie on a circle of radius 3 cm.

This circle is the locus of points that are 3 cm from *X*.

Worked example 2

PQ is a straight line 7 cm long.

Draw the locus of all the points that are exactly 2.5 cm from the line *PQ*.

a line can be drawn parallel to *PQ*, 2.5 cm away from it on one side

all of the points on the line are 2.5 cm from *PQ* so this is part of the locus

another line can be drawn on the opposite side of *PQ*

at the left-hand end the locus will be all points that are 2.5 cm from point *P*

that is a semicircle, centre *P*, radius 2.5 cm

similarly, at the other end there is another semicircle

Here is the full locus.

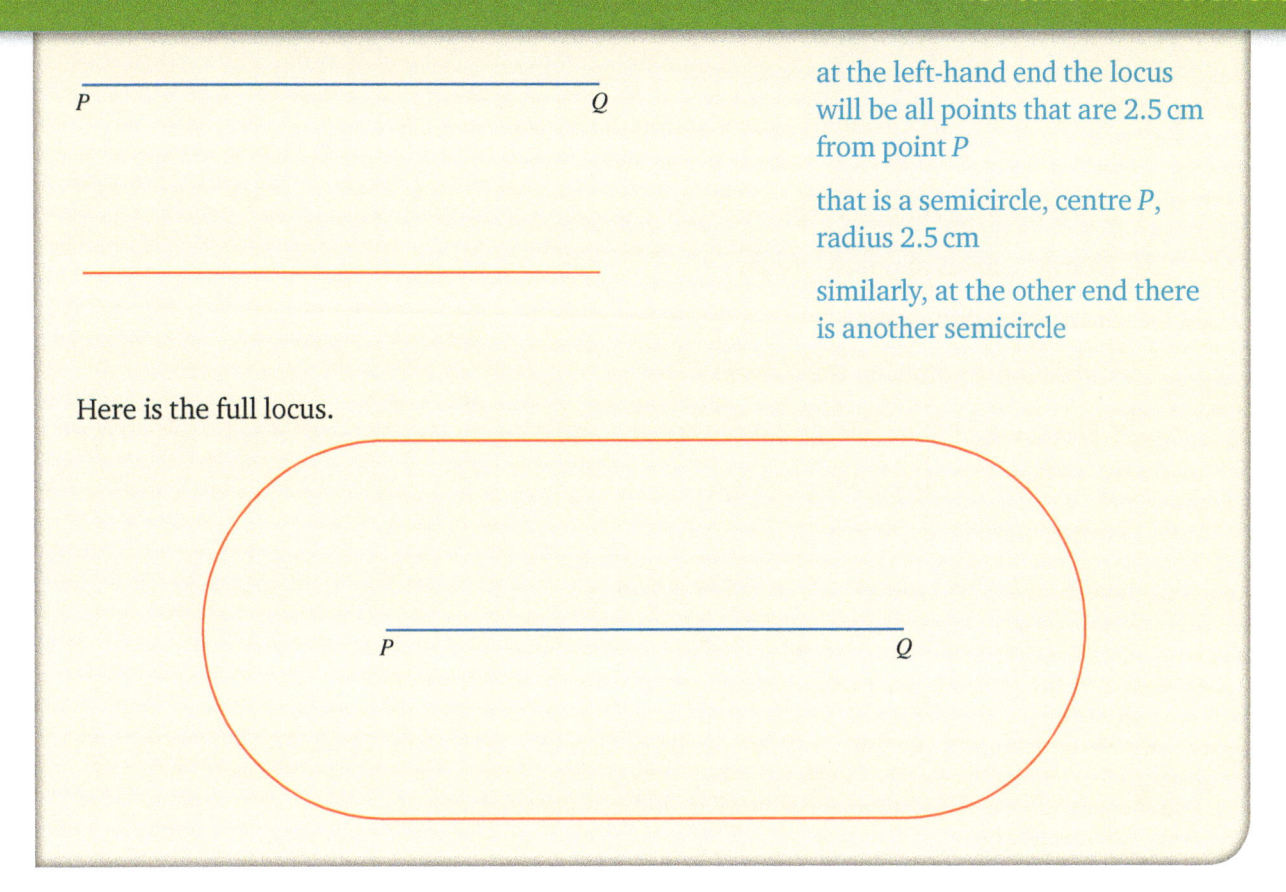

Exercise 13.2

1. A spider is sitting at the end of the hour hand of a clock.

   The hour hand is 4.8 centimetres long.

   Draw the locus of the points that the spider travels through.

2. A painting is hanging on the wall of a gallery.

   This is a plan view of the wall with the painting.

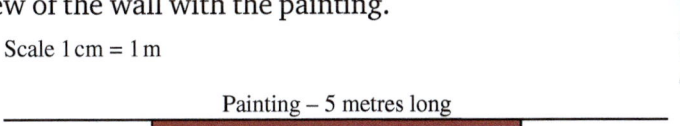

   Scale 1 cm = 1 m

   Painting – 5 metres long

   The public are not allowed within 2 metres of the painting.

   The gallery decide to put a small fence up to keep people away from the painting.

   Draw a diagram to show the locus of all the points 2 metres away from the painting.

3. Mark a point *A* on a piece of plain paper.

   Draw the locus of points that are 3.5 cm from *A*.

4. Draw a line 4.5 centimetres long on a piece of plain paper.

   Draw the locus of points that are exactly 3 cm from the line.

Mathematics for Cambridge Secondary 1

5. Draw two points 6 centimetres apart on plain paper. Label the points *A* and *B*.

   > The locus of the points a fixed distance from a single point is a circle.

   a Draw the locus of points that are 4 cm from *A*.

   b Draw the locus of points that are 4 cm from *B*.

   c How many points are 4 cm from both *A* and *B*?

   Indicate the points on the diagram.

6. The plan shows a rectangular garden labelled *ABCD*.

   The garden is 10 metres by 8 metres.

   A house runs along the side *CD*.

   The owner wants to plant a tree in the garden.

   He wants the tree to be exactly 5 metres from the house.

   It must also be 4 metres from corner *A*.

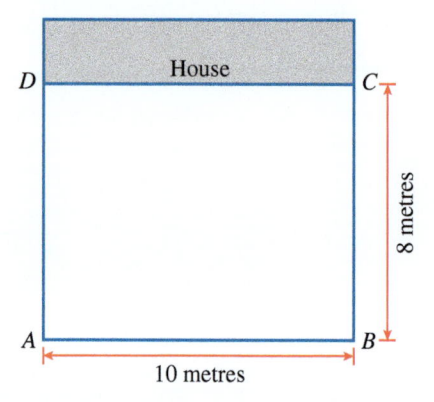

   a Draw a plan of the garden. Use a scale of 1 cm = 1 m

   b Draw the locus of points 5 metres from the house on your diagram.

   c Draw the locus of points 4 metres from *A* on your diagram.

   d Label the position of the tree with the letter *X*.

   > *X* will be the point where the two loci cross.

   e Measure the distance *BX*.

7. Draw a square with sides 5 cm long.

   Draw the locus of points outside the square that are exactly 3 cm from the side of the square.

   > At each corner there will be a set of points that are 3 cm from the vertex. These will make a curve.

13.3 Describing transformations

In your Stages 7 and 8 Student Books you learned about four types of **transformation**.

A transformation changes an **object**, the original shape, into an **image**, the final shape.

The transformations you have met are reflection, rotation, translation and enlargement.

With reflection, rotation or translation the image is congruent to the object.

Congruent means they have the same lengths and the same angles.

If there is a combination of two or more of these transformations the image will still be congruent to the object.

13 Position and movement

Worked example 1

Copy this diagram on to squared paper.

a Reflect triangle A in the y-axis. Label the new triangle B.

b Rotate shape B through 180° about (0, 0). Label the new shape C.

c Describe fully the single transformation that will move shape A to shape C.

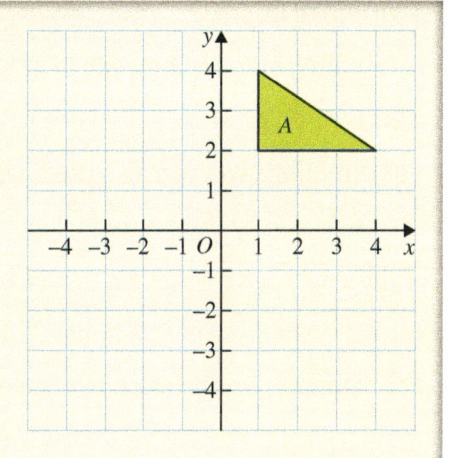

a and b

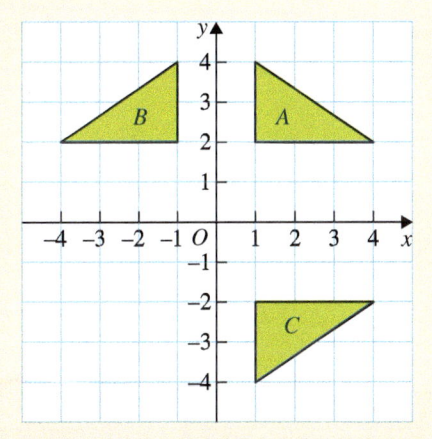

this diagram shows shapes B and C after the reflection and rotation

c The transformation that takes shape A to shape C is a reflection.

   The mirror line is the x-axis.

   The full description is a reflection in the x-axis.

Complete descriptions

In the previous example you were asked for a full description.

Here is what is required to give a full description of each transformation.

Reflection

A description of the mirror line.

This may be given as an equation of a line on a graph.

Rotation

The centre of rotation, often given as coordinates.

The angle of rotation in degrees or fractions of a turn.

The direction (clockwise or anticlockwise). If the rotation is 180° the direction is not required.

193

Translation

A full description of how far to move.

This is often given as a distance to the right (or left) and a distance up (or down).

For directions you can also use a vector like this: $\begin{pmatrix} x \\ y \end{pmatrix}$

A translation by the vector $\begin{pmatrix} x \\ y \end{pmatrix}$ means moving x units right and y units up.

If the x-value is negative it moves to the left.

If the y-value is negative it moves downwards.

For example, a translation by the vector $\begin{pmatrix} -4 \\ -5 \end{pmatrix}$ means moving 4 squares left and 5 squares down.

Worked example 2

The diagram shows shapes P and Q.

Describe fully the transformation that moves shape P to shape Q.

The shape is congruent and it is the same way up.

It has been translated across the page.

From P to Q it has moved 7 squares to the left and 3 squares upwards.

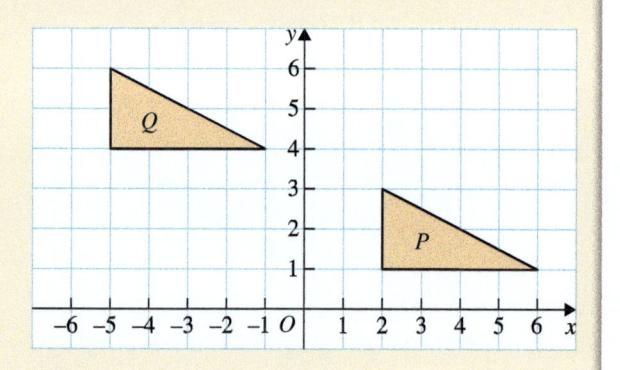

The transformation from P to Q is a translation, 7 squares left and 3 squares up.

Another way to write the answer is a translation by the vector $\begin{pmatrix} -7 \\ 3 \end{pmatrix}$.

Worked example 3

Draw a coordinate grid with x-values from -6 to 6 and y-values from 0 to 6.

Draw a triangle with vertices at $(4, 1)$, $(4, 4)$ and $(6, 1)$. Label the triangle A.

Draw a second triangle with vertices at $(-2, 1)$, $(-2, 4)$ and $(-4, 1)$.

Label the second triangle B.

Describe fully the transformation that moves triangle A to triangle B.

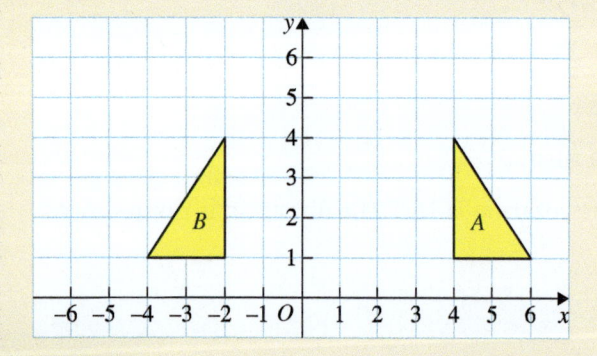

first draw the axes

plot the points to make triangle A

plot the points to make triangle B

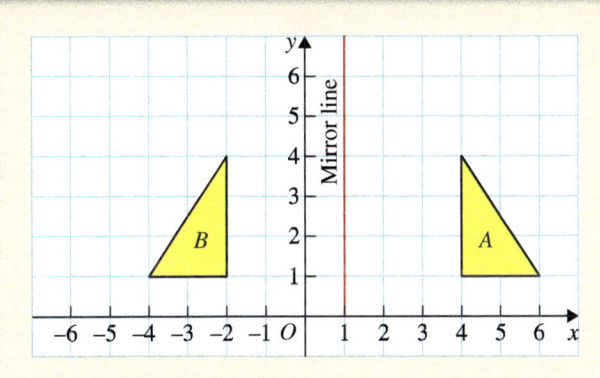

13 Position and movement

the triangles are congruent but one is a reflection of the other

the mirror line is halfway between them

the mirror line is shown here in red

the equation of this straight line is $x = 1$

You may need to revise the work you have done on equations of straight lines.

The transformation from A to B is a reflection in the line $x = 1$.

Exercise 13.3

In questions **1** to **9**, describe fully the transformation that moves shape A to shape B.

1

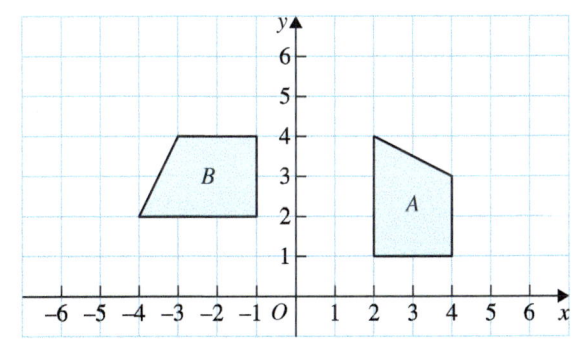

2

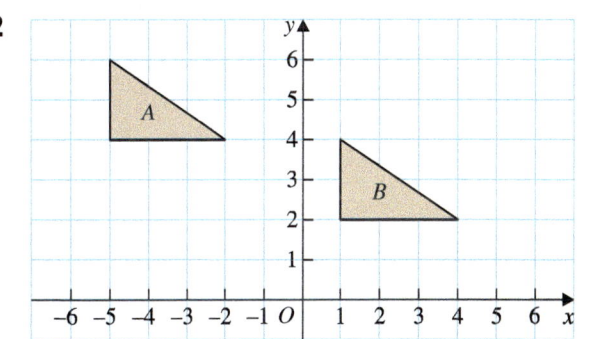

3

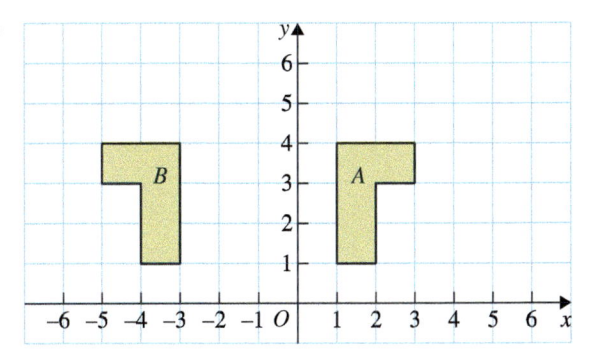

4

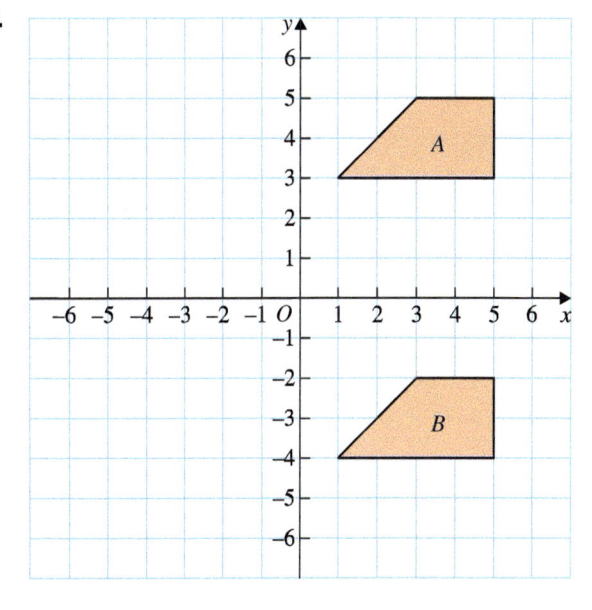

195

Mathematics for Cambridge Secondary 1

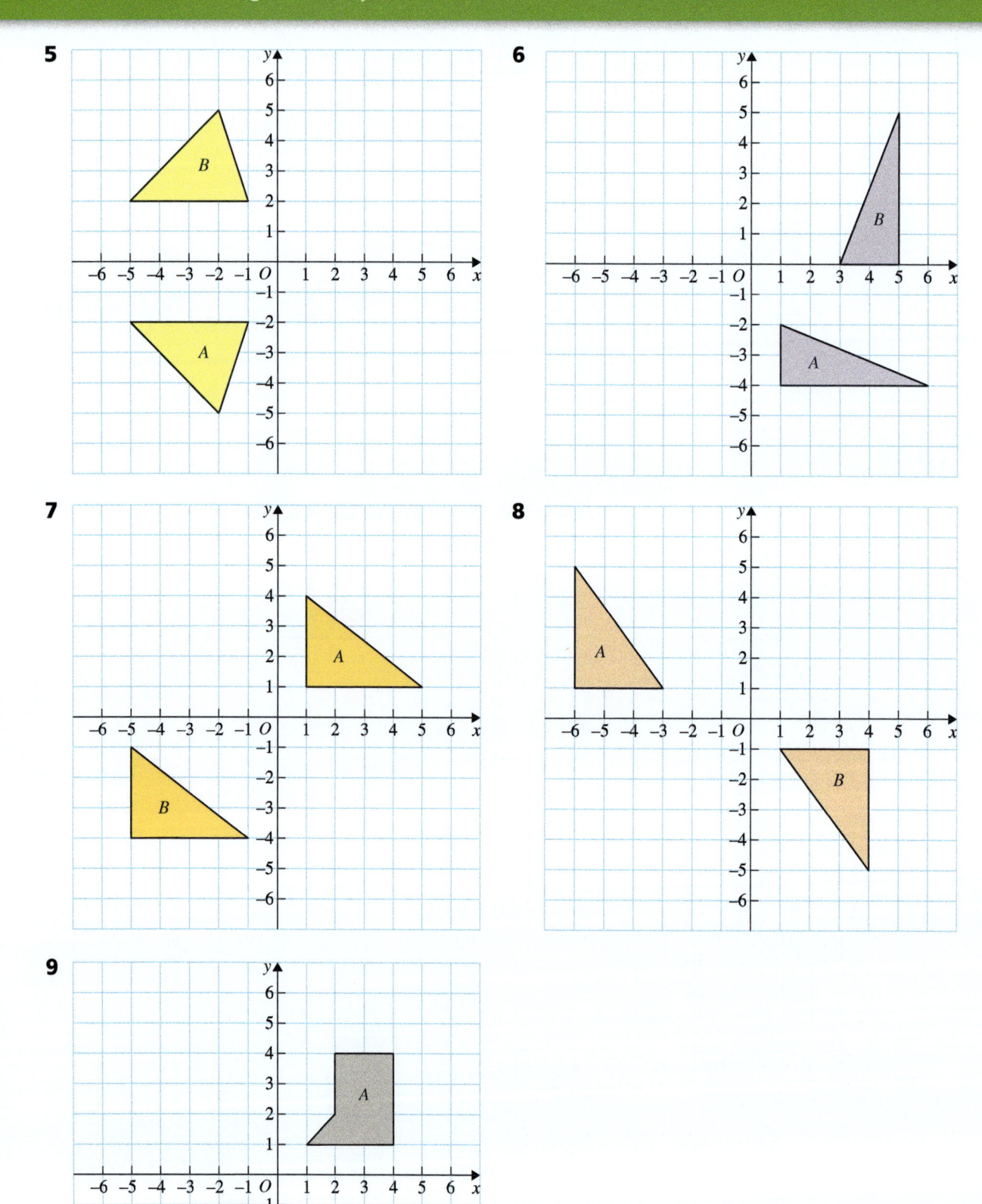

5

6

7

8

9

196

13 Position and movement

10 Copy this diagram on to squared paper.

   a Reflect shape X in the y-axis. Label the new shape Y.

   b Rotate Y 180° about the point (0, 1). Label the shape Z.

   > Use tracing paper to draw the rotation.

   c Describe fully the transformation from X to Z.

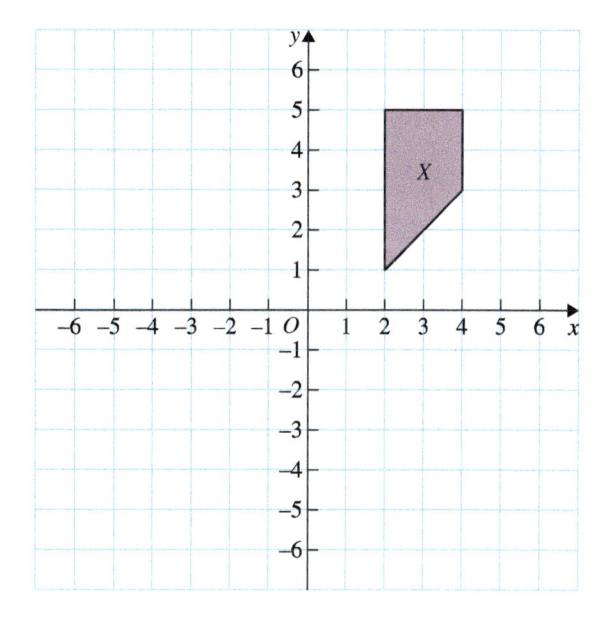

11 a Draw a coordinate grid with axes from −6 to +6.

   b Draw an L shape with vertices at (4, 1), (4, 4), (5, 4), (5, 2), (6, 2) and (6, 1). Label the shape A.

   c Reflect A in the x-axis. Label the new shape B.

   d Reflect B in the line x = 1. Label the new shape C.

   e Translate C by 5 squares upwards. Label the new shape D.

   f Describe fully the single transformation that moves A directly to D.

   > It may help to use tracing paper to find a full description of the transformation.

12 An inverse transformation reverses another transformation.

   For example, the inverse of a translation by 5 squares to the right is a translation by 5 squares to the left.

   Describe fully the inverse of these transformations.

   a A reflection in the x-axis.

   b A translation by the vector $\begin{pmatrix} 3 \\ -4 \end{pmatrix}$.

   > Draw diagrams to find the inverse of these transformations.

   c A rotation of 180° around the point (0, 0).

   d A reflection in the line y = −x

   e A rotation of 90° clockwise around the point (1, 1).

   f A translation by the vector $\begin{pmatrix} 0 \\ 4 \end{pmatrix}$.

197

13.4 Enlargements

Drawing enlargements

Translation, reflection and rotation all keep objects the same size and shape.

Enlargement is different as it changes the size of the object.

After an enlargement the image has the same angles as the object, but the lengths have changed.

In short, the object and the image are the same shape but different sizes.

These two examples will remind you how to draw an enlargement given a centre and a scale factor.

Worked example 1

Copy this diagram on to plain paper.

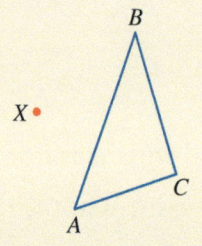

Enlarge triangle ABC by scale factor 2 with point X as the centre of enlargement.

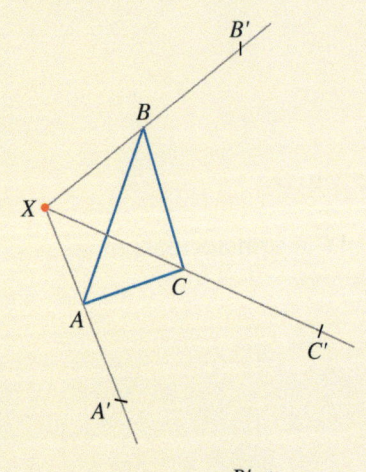

first draw construction lines from X through each vertex of the triangle

mark a point A' so that XA' = 2 × XA

mark B' and C' so that XB' = 2 × XB and XC' = 2 × XC

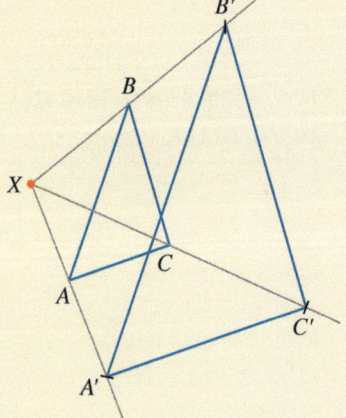

finally, join up the three points A', B' and C' to form the new triangle

13 Position and movement

Worked example 2

Copy this diagram on to squared paper.

a Enlarge PQRS by scale factor 3, centre (0, 0). Label the image P'Q'R'S'.

b Measure the sides and work out:

  i $\dfrac{P'Q'}{PQ}$ **ii** $\dfrac{Q'S'}{QS}$

a

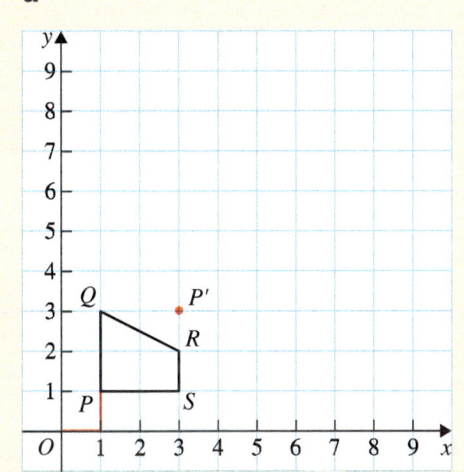

when the object is on a graph or squared paper you can use the squares to help with the enlargement

the vertex P is 1 square across and 1 square up from the centre of enlargement (0,0)

the image of that vertex will be 3 squares across and 3 squares up

you multiply the number of squares by the scale factor

here is the diagram with P' marked

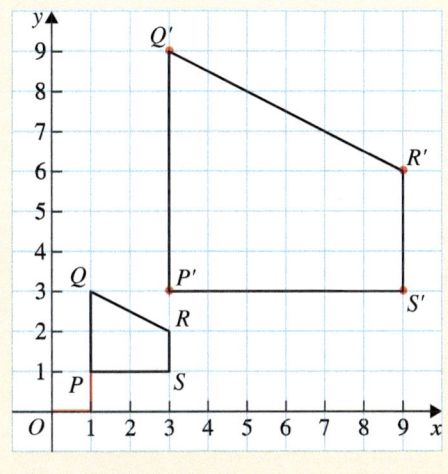

repeat this process for Q, R and S to find the positions of Q', R' and S'

join up the four points to make the enlarged image

b i If the diagram is on centimetre squared paper then P'Q' = 6 cm and PQ = 2 cm

So $\dfrac{P'Q'}{PQ} = \dfrac{6}{2} = 3$

ii Q'S' = 8.49 cm and QS = 2.83 cm   measure the diagonals carefully

$\dfrac{Q'S'}{QS} = \dfrac{8.49}{2.83} = 3$

> The answer may not come out as exactly 3 as it is based on measurement.

199

Describing enlargements

The answers to part **b** of Worked example 2 lead to a general result.

The ratio of corresponding lengths in the image and the object is equal to the scale factor of the enlargement.

You can work out the scale factor by measuring two lengths and dividing one by the other.

To find the centre of enlargement you need to draw faint lines through corresponding vertices.

The centre of enlargement is where the lines cross.

This is what is required to give a full description of an enlargement.

Enlargement

The scale factor. This is a single number, not a ratio.

The centre of enlargement. This may be given as a pair of coordinates.

Worked example 3

Describe fully the transformation that moves triangle *ABC* on to triangle *PQR*.

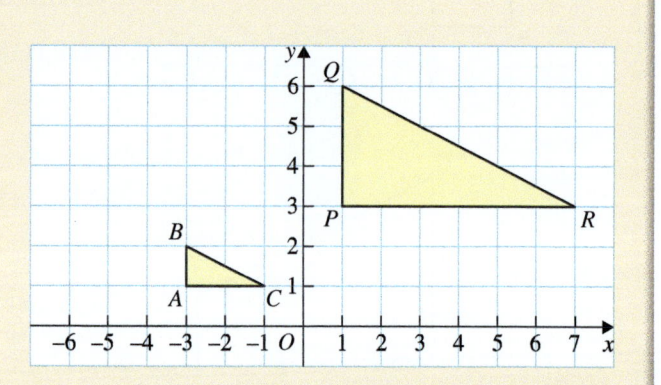

PQ and *AB* are corresponding sides.

$PQ \div AB = 3 \div 1 = 3$

The scale factor is 3.

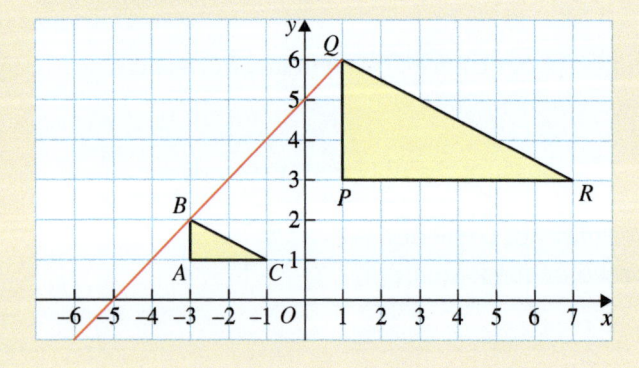

first find the scale factor

now find the centre of enlargement

draw a line through two corresponding vertices, *Q* and *B*

the centre of enlargement is somewhere on this line

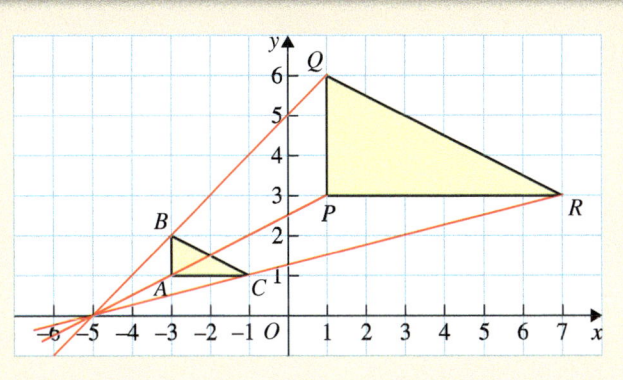

13 Position and movement

repeat through another pair of vertices

the lines all cross at $(-5, 0)$, this is the centre of enlargement

The transformation is an enlargement, centre $(-5, 0)$, scale factor 3.

Exercise 13.4

For questions **1** and **2** you need to copy the diagram on to plain paper.

1 Enlarge *ABCD* by scale factor 2, centre *X*.

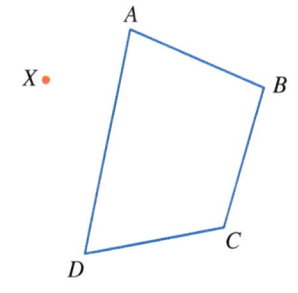

2 Enlarge *ABC* by scale factor 3, centre *O*.

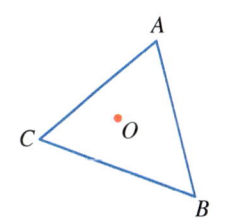

Copy questions **3**, **4** and **5** on to squared paper.

3 Enlarge shape *X* by scale factor 2, centre $(0, 0)$.

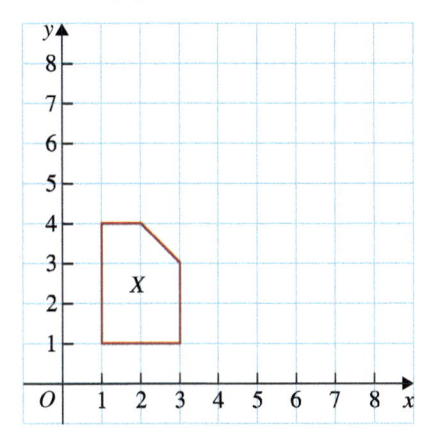

4 Enlarge shape *Y* by scale factor 3, centre $(0, 0)$.

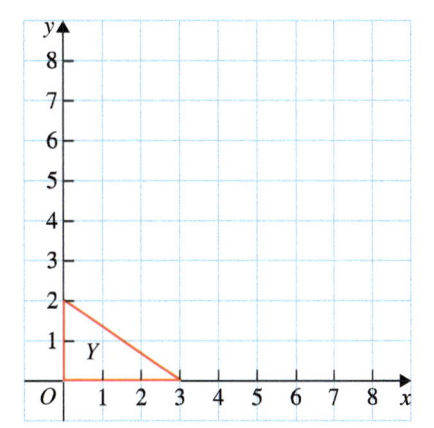

5 Enlarge shape Z by scale factor 3, centre (−5, 4).

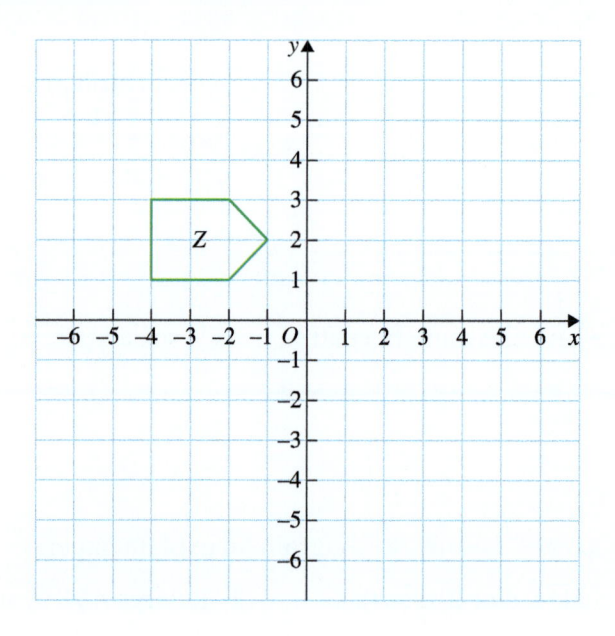

6 Describe fully the transformation that moves shape A on to shape B.

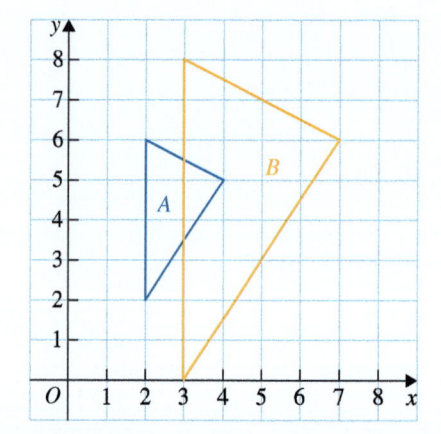

7 Describe fully the transformation that moves shape P on to shape Q.

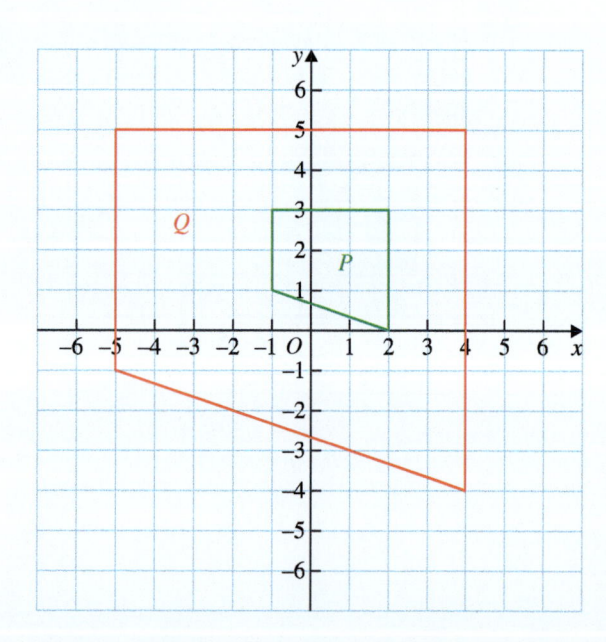

13 Position and movement

8 Describe fully the transformation that moves shape Y on to shape Z.

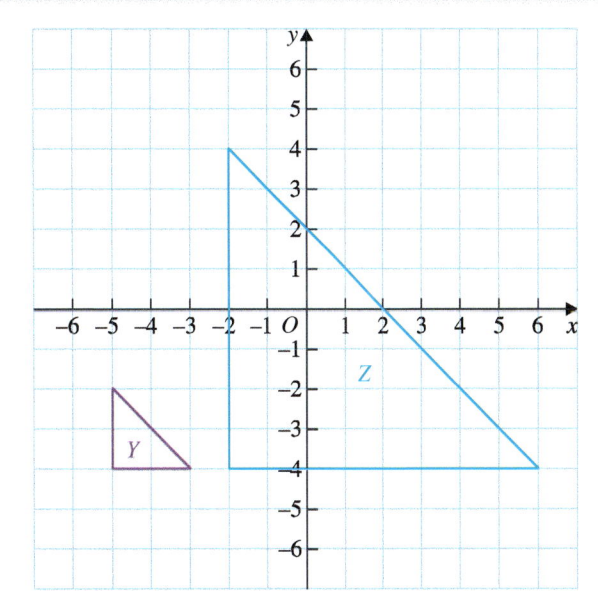

9 a Draw coordinate axes with values from −6 to 6 on both the x-axis and the y-axis.
  b Plot points at (2, 3), (3, 6) and (3, 3). Join them up to form a triangle. Label the triangle T.
  c Enlarge T by scale factor 2, centre of enlargement (5, 6). Label the new triangle U.
  d Enlarge U by scale factor 2, centre of enlargement (3, 6). Label the new triangle V.
  e Describe fully the single transformation that moves triangle T on to triangle V.

> The transformation is an enlargement. Draw lines to find the centre of enlargement.

Review

1 This triangle has sides of length 4.5 cm, 2.5 cm and 4 cm.

Copy the triangle on to squared paper.

Draw a tessellation of this triangle.

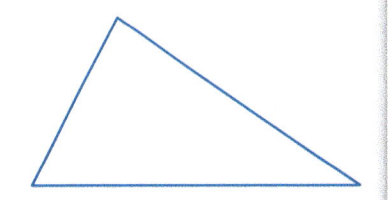

2 Copy this quadrilateral on to squared paper.

Draw a tessellation of the quadrilateral.

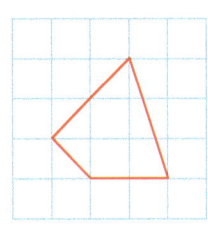

3 Work out the interior angle of a regular 10-sided polygon.

Use your answer to explain why it is not possible to draw a regular tessellation with a decagon.

203

4 The diagram represents a triangular garden *ABC*.

   It is drawn to a scale of 1 cm = 2 m

   a Copy the diagram on to plain paper.

   A tree is to be planted 8 metres from *A* and 6 metres from *BC*.

   b Draw the locus of points 8 metres from *A*.

   c Draw the locus of points 6 metres from the wall *BC*.

   d Mark the position of the tree with the letter *T*.

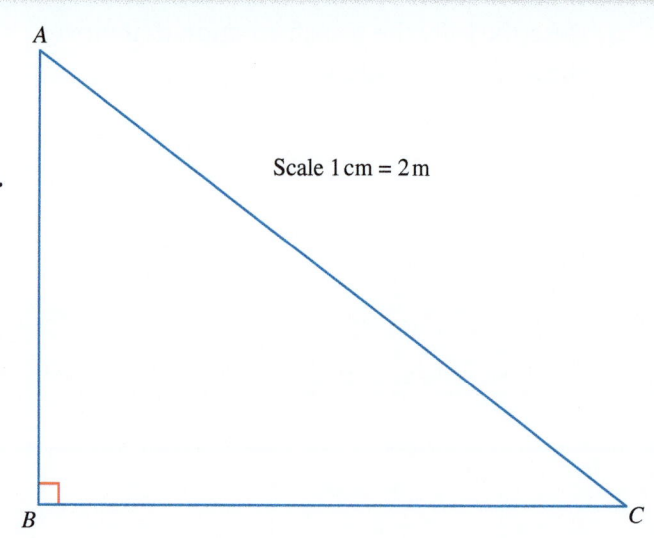

   Scale 1 cm = 2 m

In questions **5** to **12** describe fully the single transformation that moves shape *A* on to shape *B*.

5

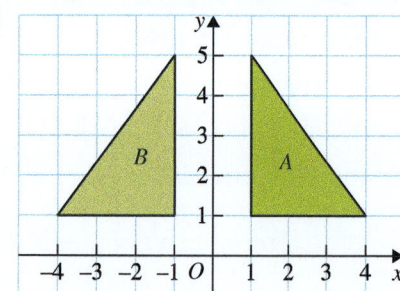

6

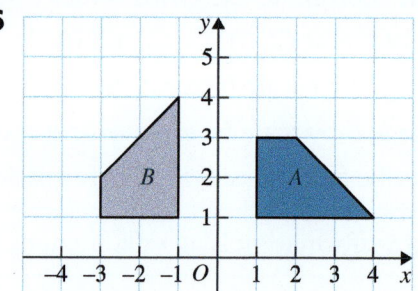

7

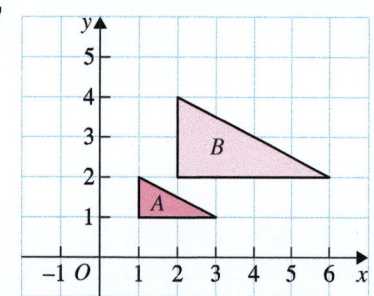

8

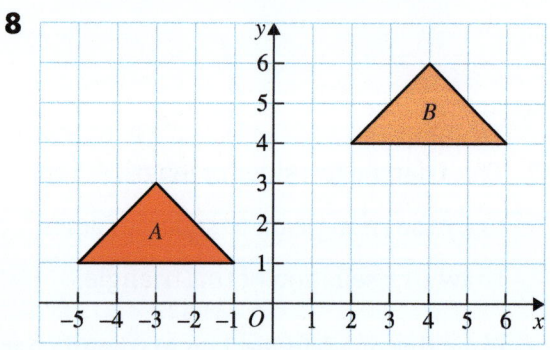

9

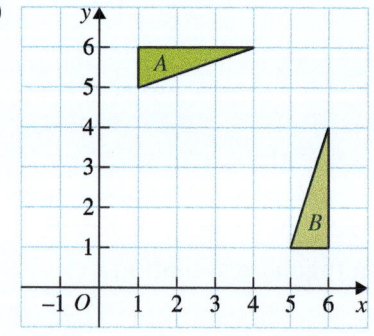

10

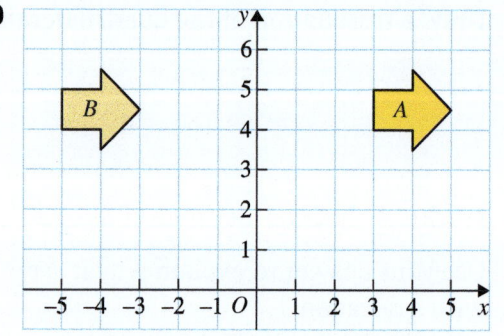

13 Position and movement

11

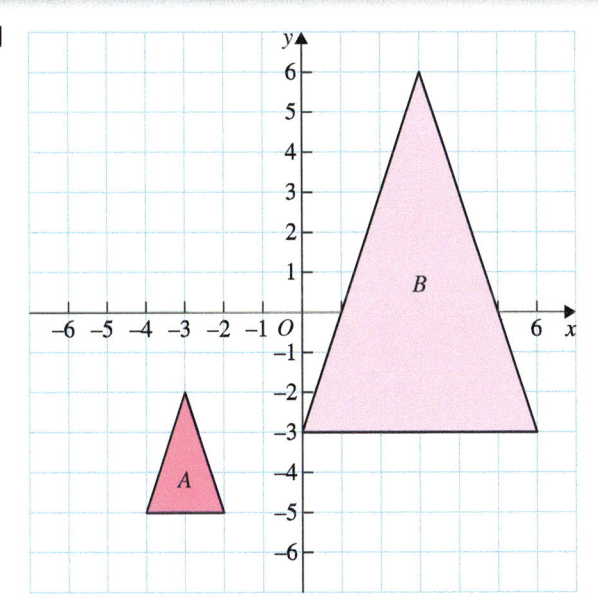

12

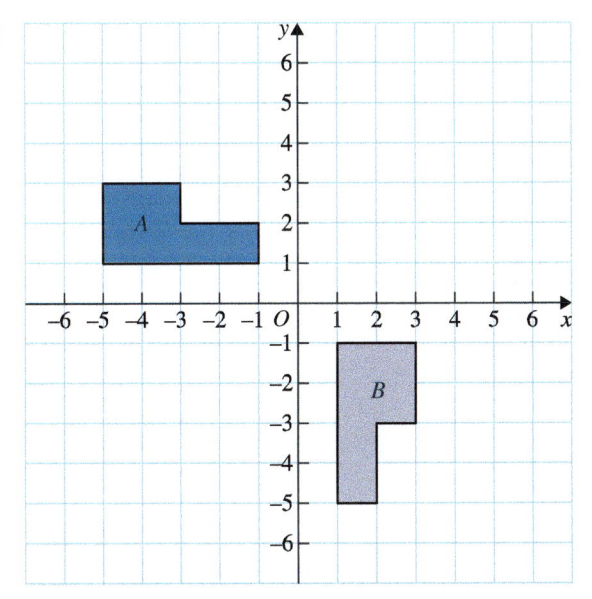

13 Copy this diagram on to squared paper.

   a Rotate shape A 180° about the point (1, 0).

   Label the new shape B.

   b Reflect shape B in the y-axis.

   Label the new shape C.

   c Translate shape C by the vector $\begin{pmatrix} 2 \\ -2 \end{pmatrix}$.

   Label the new shape D.

   d Describe fully the single transformation that moves shape D on to shape A.

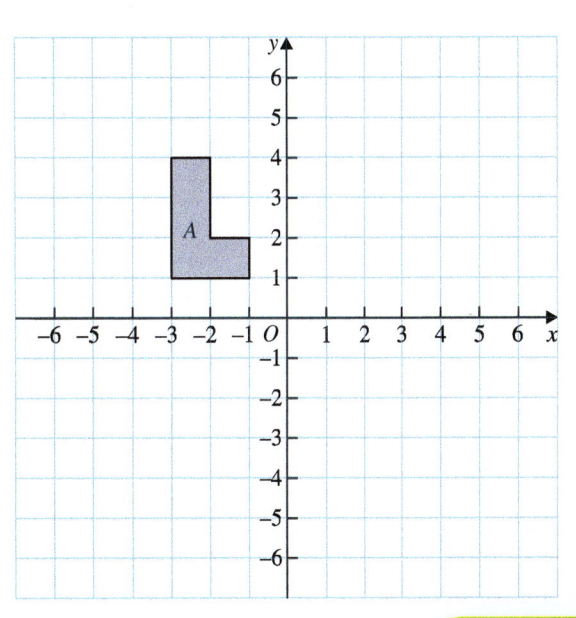

14 Copy this diagram on to squared paper.

  a Enlarge triangle A by scale factor 2, centre (−4, 1).

  Label the new triangle B.

  b Translate triangle B by the vector $\begin{pmatrix} 2 \\ 1 \end{pmatrix}$.

  Label the new triangle C.

  c Describe fully the single transformation from A to C.

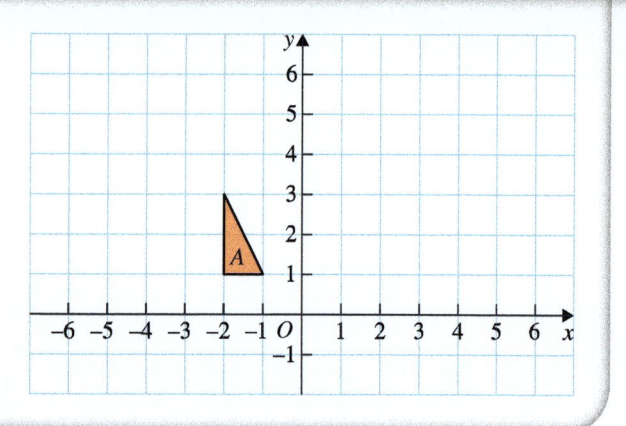

Examination-style questions

Copy the coordinate grid shown and use it to answer questions **1** and **2**.

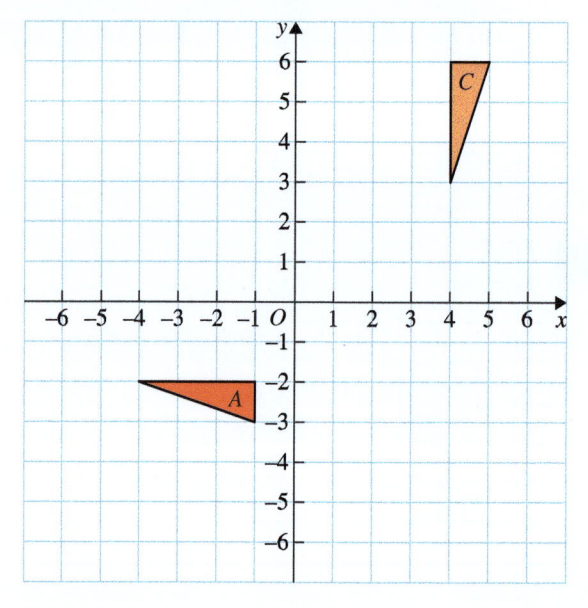

1 **a** Plot the point (4, 2). Label it P. [1]

  b Plot the point (−4, 4). Label it Q. [1]

  c Write down the coordinates of the midpoint of PQ. [1]

2 **a** Rotate triangle A through 90° anticlockwise about (0, 0).
    Label the new triangle B. [2]

  b Describe fully the single transformation that moves triangle B on to triangle C. [2]

3 Copy this quadrilateral on to squared paper.
  Draw a tessellation of the quadrilateral.
  You must show at least five more shapes. [2]

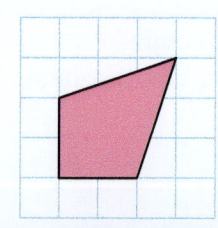

13 Position and movement

4   Copy this diagram on to squared paper.

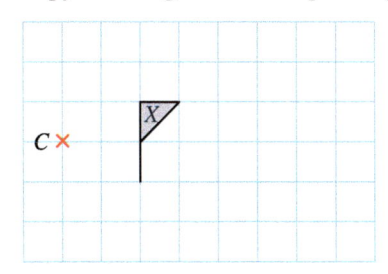

   Enlarge shape *X*, scale factor 3, centre of enlargement *C*. [2]

5   *PQR* is an isosceles triangle with *QR* = 13 cm and *PQ* = *PR* = 9 cm

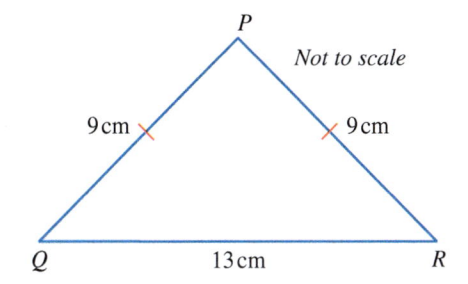

   a   Construct an accurate drawing of triangle *PQR* [2]
   b   Draw the locus of points that are 3.5 cm from side *PQ*. [1]
   c   Draw the locus of points that are 5 cm from the point *P*. [1]
   d   Mark the point where the loci intersect as *X*.
       Measure distance *XR*. [1]

6   The diagram shows three triangles *A*, *B* and *C*.

   Describe fully the transformation from:

   a   triangle *A* to triangle *B* [2]
   b   triangle *B* to triangle *C*. [2]

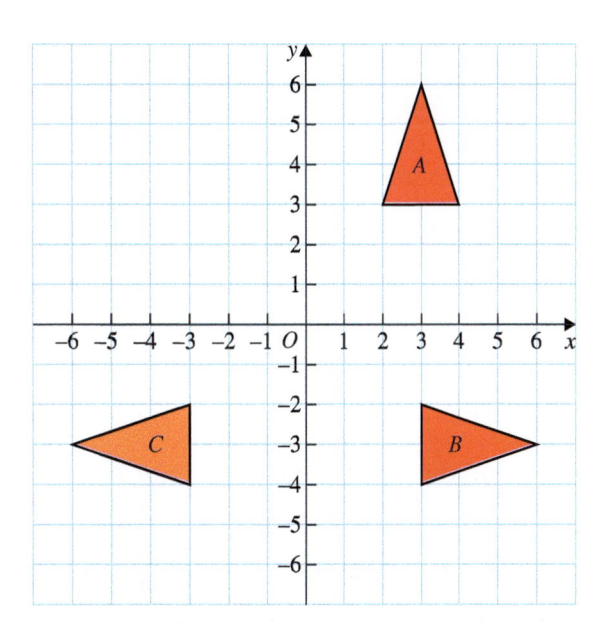

207

14 Sequences

> **Learning outcomes**
> - Generate terms of a sequence using term-to-term rule.
> - Generate terms of a sequence using position-to-term rule.
> - Derive an expression for the nth term of an arithmetic sequence.

Natural beauty

Snowflakes are amazing creations of nature.

If you look very carefully at a snowflake you will see some very intricate patterns. These patterns can be modelled using fractals.

The fractal that has been used to make this picture is called the Koch snowflake.

In this chapter you will learn how the Koch snowflake fractal is formed using a sequence of diagrams.

You may wish to use the internet to find out more about other famous fractals and sequences.

14.1 Using the term-to-term rule

A **sequence** is a list of numbers or diagrams that are connected by a rule.

8, 2, −4, −10, … is a linear sequence.

It is called a **linear sequence** (or arithmetic sequence) because the differences between the terms are all the same.

$$8 \xrightarrow{-6} 2 \xrightarrow{-6} -4 \xrightarrow{-6} -10$$

The rule to find the next term (the **term-to-term rule**) is '−6' or 'subtract 6'.

The sequence 3, 6, 11, 18, 27, … is called a **non-linear sequence** because the differences between the terms are not the same.

$$3 \xrightarrow{+3} 6 \xrightarrow{+5} 11 \xrightarrow{+7} 18 \xrightarrow{+9} 27$$

14 Sequences

The differences form their own number sequence.

You can use the differences sequence to find the next terms in the sequence.

3, 6, 11, 18, 27, (38), (51)
+3, +5, +7, +9, +11, +13

Worked example 1

15, 18, 21, 24, …

Write down: **a** the term-to-term rule **b** the next two terms of the sequence.

a 15, 18, 21, 24
+3, +3, +3

The term-to-term rule is 'add 3'.

b 15, 18, 21, 24, (27), (30)
+3, +3, +3, +3, +3

The next two terms are 27 and 30.

Worked example 2

12, 17, 25, 36, 50, …

Write down the next two terms of the sequence.

You must first look at the number pattern for the differences and then use the pattern to write down the next two terms.

12, 17, 25, 36, 50, (67), (87)
+5, +8, +11, +14, +17, +20

The next two terms in the sequence are 67 and 87.

Exercise 14.1

1 Write down for each sequence: **i** the term-to-term rule **ii** the next three terms.

    a 3, 5, 7, 9, …
    b 7, 11, 15, 19, …
    c 1, 6, 11, 16, …

    d 7, 8.5, 10, 11.5, …
    e 16, 13, 10, 7, …
    f 5, −4, −13, −22, …

    g 1, 3, 9, 27, …
    h 1, 10, 100, 1000, …
    i 32, 16, 8, 4, …

2 Look at the differences between terms in each of these sequences and write down the next three terms.

    a 1, 2, 4, 7, 11, …
    b 5, 6, 9, 14, 21, …
    c 2, 5, 11, 20, 32, …

    d 4, 9, 19, 34, 54, …
    e 20, 19, 17, 14, 10, …
    f −4, −2, 2, 8, 16, …

3 The diagram below shows how the first four triangle numbers are formed.

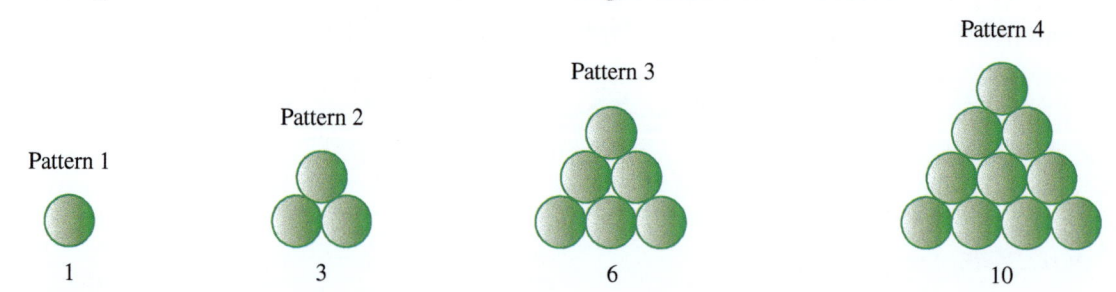

Pattern 1 — 1
Pattern 2 — 3
Pattern 3 — 6
Pattern 4 — 10

a Write down the next three triangle numbers.

b Copy and complete the table.

Pattern	1	2	3	4	5	6	7
Number of circles	$1 = \frac{1 \times 2}{2}$	$3 = \frac{2 \times 3}{2}$	$6 = \frac{3 \times 4}{2}$	$10 = \frac{4 \times 5}{2}$			

c Use the patterns in your table for part **b** to write down an expression for finding the nth triangle number.

d Use your answer to part **c** to write down the 100th triangle number.

4 The sequence 1, 1, 2, 3, 5, 8, 13, 21, ... is called the Fibonacci sequence.

Each term in a Fibonacci sequence is the sum of the two previous terms in the sequence.

Write down the next three terms in each of these Fibonacci sequences:

a 5, 7, 12, 19, 31, ...   **b** 2, 2, 4, 6, 10, 16, ...   **c** 3, −4, −1, −5, −6, ...

5 The diagram shows the growth of a plant over three years.

Each year a flower is replaced by three stems and three flowers.

Year 1   Year 2   Year 3

Copy and complete the table.

Year	1	2	3	4	5
Number of flowers	1	3	9		
Number of stems	1	4			

6

Diagram 1   Diagram 2   Diagram 3   Diagram 4

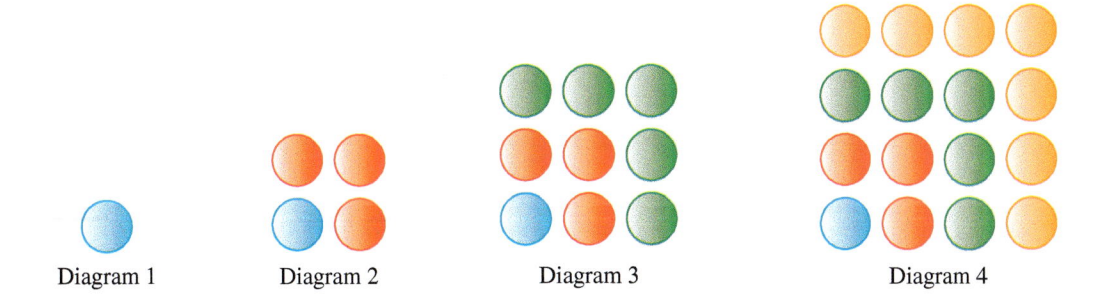

The diagrams show how the following number sequence is formed:

Diagram 1:   1   = 1   = 1^2

Diagram 2:   1 + 3   = 4   = 2^2

Diagram 3:   1 + 3 + 5   = 9   = 3^2

Diagram 4:   1 + 3 + 5 + 7 = 16   = 4^2

a Write down the next two rows in this number sequence.

b Use the number patterns to help you write down the sum of the first 20 odd numbers.

7 The diagram below shows how a pyramid of oranges is formed.

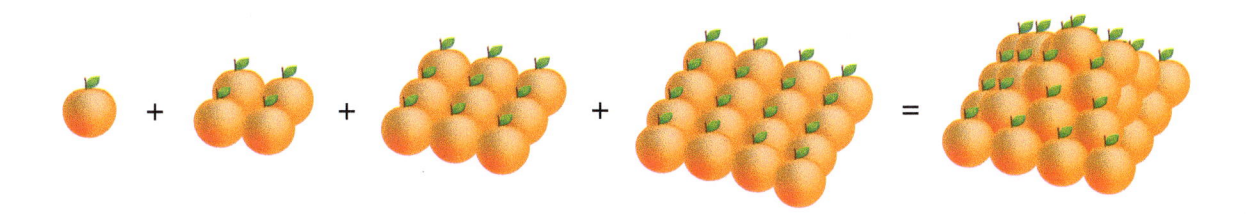

This pyramid has four layers. The number of oranges in the pyramid = 1 + 4 + 9 + 16 = 30

a How many oranges are in a pyramid with three layers?

b How many oranges are in a pyramid with five layers?

c How many oranges are in a pyramid with six layers?

8 83, 39, 42, 23, …

The rule to find the next term in the sequence is 'multiply the digits of the last number and then add 15'.

a Write down the first 10 terms in the sequence.

b What is the 50th term in the sequence?

9 Here is a number pattern:

Row 1: 1^3 $= 1$ $= 1$ $= \left(\dfrac{1 \times 2^2}{2}\right)$

Row 2: $1^3 + 2^3$ $= 1 + 8$ $= 9$ $= \left(\dfrac{2 \times 3^2}{2}\right)$

Row 3: $1^3 + 2^3 + 3^3$ $= 1 + 8 + 27$ $= 36$ $= \left(\dfrac{3 \times 4^2}{2}\right)$

Row 4: $1^3 + 2^3 + 3^3 + 4^3 = 1 + 8 + 27 + 64 = 100 = \left(\dfrac{4 \times 5}{2}\right)^2$

a Write down the next two rows in this number pattern.

b Write down an expression for the sum of the first n cube numbers.

10 William is investigating the Fibonacci sequence 1, 1, 2, 3, 5, 8, 13, 21, …

He investigates the ratio of adjacent numbers.

He writes each ratio in the form $1 : n$.

> Ratio of 1st number to 2nd number = 1 : 1
>
> Ratio of 2nd number to 3rd number = 1 : 2
>
> Ratio of 3rd number to 4th number = 2 : 3 = 1 : 1.5
>
> Ratio of 4th number to 5th number = 3 : 5 = 1 : 1.666…

a Write down the next six ratios and describe what happens to the ratio.

b Use the internet to find out more about this 'golden ratio'.

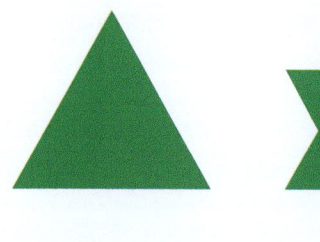

11

Diagram 1   Diagram 2   Diagram 3   Diagram 4

The diagrams show the first four steps in drawing a Koch snowflake.

The Koch snowflake is a fractal pattern.

a Use the internet to find out about the Sierpinski triangle.

Draw the first four diagrams for this fractal.

b Use the internet to investigate other fractal patterns.

14 Sequences

14.2 Using the position-to-term rule

You can also use a **position-to-term rule** to find the numbers in a sequence.

If the position-to-term rule for a sequence is: term = 3 × position number then add 5, then the first four terms are:

1st term = 3 × 1 + 5 = 3 + 5 = 8
2nd term = 3 × 2 + 5 = 6 + 5 = 11
3rd term = 3 × 3 + 5 = 9 + 5 = 14
4th term = 3 × 4 + 5 = 12 + 5 = 17

The position-to-term rule can be written algebraically as: nth term = $3n + 5$

Worked example 1

The rule for finding the nth term of a sequence is nth term = $7n - 2$

a Write down the first four terms in the sequence.
b Which term in the sequence has a value of 327?

a 1st term = 7 × 1 − 2 = 7 − 2 = 5
2nd term = 7 × 2 − 2 = 14 − 2 = 12
3rd term = 7 × 3 − 2 = 21 − 2 = 19
4th term = 7 × 4 − 2 = 28 − 2 = 26
The first four terms are: 5, 12, 19, 26

b Solve the equation $7n - 2 = 327$
$7n = 329$ add 2 to both sides
$n = 47$ divide both sides by 7

The 47th term has a value of 327.

Worked example 2

The rule for finding the nth term of a sequence is nth term = $2n^2 + 5$
Write down the first four terms in the sequence.

1st term = 2 × 1^2 + 5 = 2 × 1 + 5 = 7
2nd term = 2 × 2^2 + 5 = 2 × 4 + 5 = 13
3rd term = 2 × 3^2 + 5 = 2 × 9 + 5 = 23
4th term = 2 × 4^2 + 5 = 2 × 16 + 5 = 37
The first four terms are: 7, 13, 23, 37

Exercise 14.2

1 Write down the first four terms of the linear sequence whose nth term is:

- **a** $n + 2$
- **b** $3n - 1$
- **c** $5n + 3$
- **d** $2n + 5$
- **e** $4n - 3$
- **f** $40 - n$
- **g** $80 - 5n$
- **h** $5 - 3n$
- **i** $0.5n + 3$
- **j** $12 - 0.5n$
- **k** $\dfrac{2n - 7}{2}$
- **l** $\dfrac{6 - 5n}{4}$

2 Write down the first five terms of the non-linear sequence whose nth term is:

- **a** n^2
- **b** $n^2 + 2$
- **c** $n^2 - 1$
- **d** $3n^2$
- **e** $n(n + 1)$
- **f** $n^2 - 3n$
- **g** n^3
- **h** $n^3 + 1$
- **i** $2n^3$
- **j** $n(n + 1)(n + 2)$
- **k** $\dfrac{n}{n + 1}$
- **l** $\dfrac{n + 5}{2n}$

3 nth term $= 6n + 3$

Which term in the sequence has a value of 405?

4 nth term $= 2n - 7$

Which term in the sequence has a value of 301?

5 nth term $= 10 - 4n$

Which term in the sequence has a value of -258?

6 nth term $= \dfrac{5n + 3}{2}$

Which term in the sequence has a value of 374?

7 nth term $= n^2 + 4$

Which term in the sequence has a value of 488?

8 nth term $= 2^n - 1$

Which term in the sequence has a value of 511?

9

1   4   9   16

This sequence of diagrams gives the first four terms in the sequence of square numbers.

The rule for finding the nth term in the sequence of square numbers is: nth term $= n^2$

The following sequences are all connected with the sequence of square numbers.

Find the connection and then write down the rule for the nth term of each sequence.

The first sequence has been done for you.

a 4, 7, 12, 19, …

Square numbers: 1　　4　　9　　16
　　　　　　　　↓　　↓　　↓　　↓　　The rule is add 3.
Sequence:　　　 4　　7　　12　 19

So the rule is: nth term = $n^2 + 3$

b 2, 5, 10, 17, …　　**c** 0, 3, 8, 15, …　　**d** 7, 10, 15, 22, …

e 3, 6, 11, 18, …　　**f** 2, 8, 18, 32, …　　**g** 1, 7, 17, 31, …

10

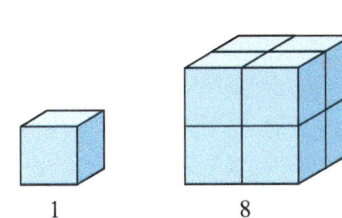

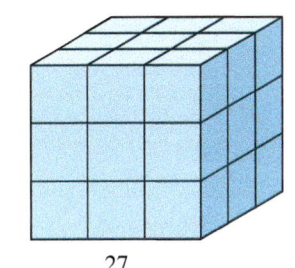

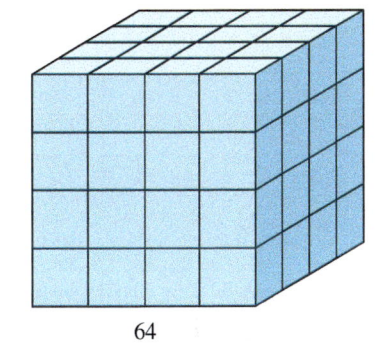

　　1　　　8　　　27　　　64

This sequence of diagrams gives the first four terms in the sequence of cube numbers.

The rule for finding the nth term in the sequence of cube numbers is: nth term = n^3

The following sequences are all connected with the sequence of cube numbers.

Find the connection and then write down the rule for the nth term of each sequence.

a 2, 9, 28, 65, …　　**b** 0, 7, 26, 63, …　　**c** 6, 13, 32, 69, …

d 4, 11, 30, 67, …　　**e** 2, 16, 54, 128, …　　**f** 3, 17, 55, 129, …

11 The sequence 2, 4, 8, 16, … is formed from:

$2^1 = 2$　　$2^2 = 2 \times 2 = 4$　　$2^3 = 2 \times 2 \times 2 = 8$　　$2^4 = 2 \times 2 \times 2 \times 2 = 16$

So the rule for finding the nth term in the sequence is: nth term = 2^n

The following sequences are all connected with the sequence 2, 4, 8, 16, …

Find the connection and then write down the rule for the nth term of each sequence.

a 3, 5, 9, 17, …　　**b** 0, 2, 6, 14, …　　**c** 4, 6, 10, 18, …

d 1, 3, 7, 15, …　　**e** 6, 12, 24, 48, …　　**f** 7, 13, 25, 49, …

14.3 The nth term of an arithmetic sequence

When working with number sequences you are often asked to find the value of one of the terms in the sequence.

For example, you might be asked to find the 40th term in the sequence: 3, 10, 17, 24, 31, …

Writing down a long list of terms until you reach the term that you need is not an efficient method.

The most efficient method is to use the position-to-term rule.

The position-to-term rule is often written using algebra. This is known as finding the **nth term**.

To find the position-to-term rule, first look at the differences between the terms of the sequence.

Position, n	1	2	3	4	5
Term	3	10	17	24	31

+7   +7   +7   +7

In this case the terms increase by 7.

The numbers in the 7 times table also increase by 7.

To find the rest of the rule you must compare the terms with the 7 times table.

Position, n	1	2	3	4	5
7 times table	7	14	21	28	35
Term	3	10	17	24	31

×7
−4

The position-to-term rule is:

   term = 7 × position number then subtract 4

This can be written using algebra as:

   nth term = $7n - 4$

So the 40th term in this sequence is found using $n = 40$ in the formula.

   40th term = $7 \times 40 - 4$

   $= 280 - 4$

   $= 276$

Worked example

Find the nth term and 50th term of the linear sequence 37, 33, 29, 25, 21, ...

37   33   29   25   21
  −4   −4   −4   −4

The terms decrease by 4. So compare the terms with the −4 times table.

Position, n	1	2	3	4	5
−4 times table	−4	−8	−12	−16	−20
Term	37	33	29	25	21

×−4
+41

The position-to-term rule is:

   term = −4 × position number then add 41

So nth term = $-4n + 41 = 41 - 4n$

   50th term = $41 - 4 \times 50$

   $= 41 - 200$

   $= -159$

216

14 Sequences

Exercise 14.3

1 Find: **i** the nth term **ii** the 50th term for each of these increasing arithmetic sequences.

Show your working in a table. (Part **a** has been done for you.)

a 3, 7, 11, 15, …

Position, n	1	2	3	4	50	n
Term	3	7	11	15	199	$4n - 1$

$\times 4$ and subtract 1

+4 +4 +4

b 2, 5, 8, 11, …   **c** 2, 9, 16, 23, …   **d** 9, 14, 19, 24, …

e 1, 3, 5, 7, …   **f** 5, 16, 27, 38, …   **g** 10, 18, 26, 34, …

h −2, 4, 10, 16, …   **i** 3, 7, 11, 15, …   **j** −12, −5, 2, 9, …

2 Find: **i** the nth term **ii** the 30th term for each of these decreasing arithmetic sequences.

Show your working in a table. (Part **a** has been done for you.)

a 5, 2, −1, −4, …

Position, n	1	2	3	4	30	n
Term	5	2	−1	−4	−82	$8 - 3n$

$\times -3$ and add 8

−3 −3 −3

b 30, 27, 24, 21, …   **c** 56, 51, 46, 41, …   **d** 88, 77, 66, 55, …

e 40, 34, 28, 22, …   **f** 13, 9, 5, 1, …   **g** 15, 7, −1, −9, …

h 7, 5, 3, 1, …   **i** −2, −7, −12, −17, …   **j** 3, −7, −17, −27, …

3 Fatimath is asked to find the nth term of the sequence −11, −4, 3, 10, …

Fatimath writes:

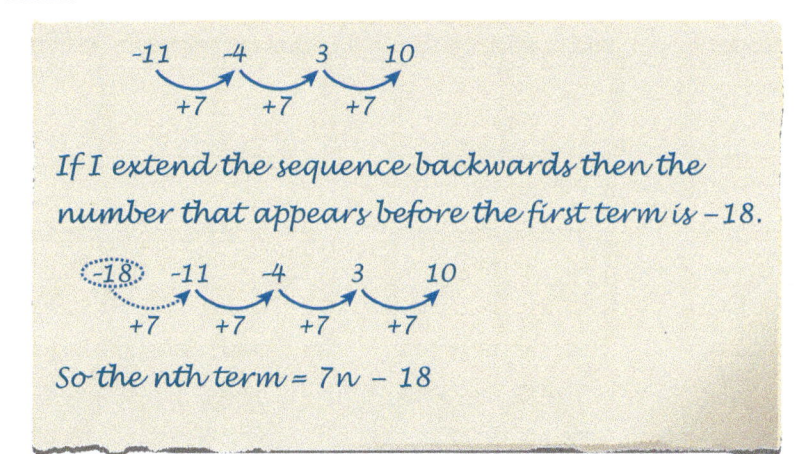

> −11   −4   3   10
> +7   +7   +7
>
> *If I extend the sequence backwards then the number that appears before the first term is −18.*
>
> (−18)   −11   −4   3   10
> +7   +7   +7   +7
>
> *So the nth term = $7n - 18$*

a Copy and complete these checks to show that the nth term of the sequence −11, −4, 3, 10, … is $7n - 18$.

1st term = $7 \times 1 - 18 = -11$ 2nd term = $7 \times 2 - 18 = \ldots$

3rd term = $7 \times 3 - 18 = \ldots$ 4th term = $7 \times 4 - 18 = \ldots$

217

b Use Fatimath's method to find the nth term of these linear sequences.

　　i 6, 13, 20, 27, …　　　**ii** 9, 20, 31, 42, …　　　**iii** 40, 42, 44, 46, …

　　iv 33, 37, 41, 45, …　　**v** 8, 3, −2, −5, …　　　**vi** 83, 76, 69, 62, …

4 Below is a sequence of diagrams made from crosses.

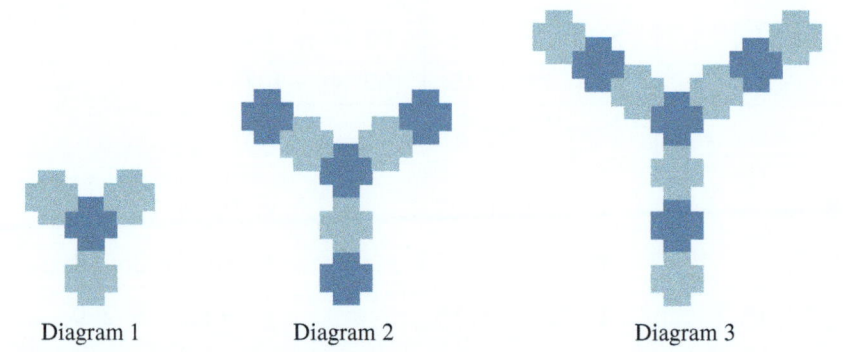

　　　　Diagram 1　　　　Diagram 2　　　　Diagram 3

　a Find the number of crosses needed to make:

　　i diagram 4　　　**ii** diagram 5　　　**iii** diagram 6.

　b Copy and complete the table for the pattern.

Diagram	1	2	3	4	5	6
Number of crosses	4	7				

　c Write down the rule for finding the number of crosses in diagram n.

5 The diagrams below show four different patterns made with sticks.

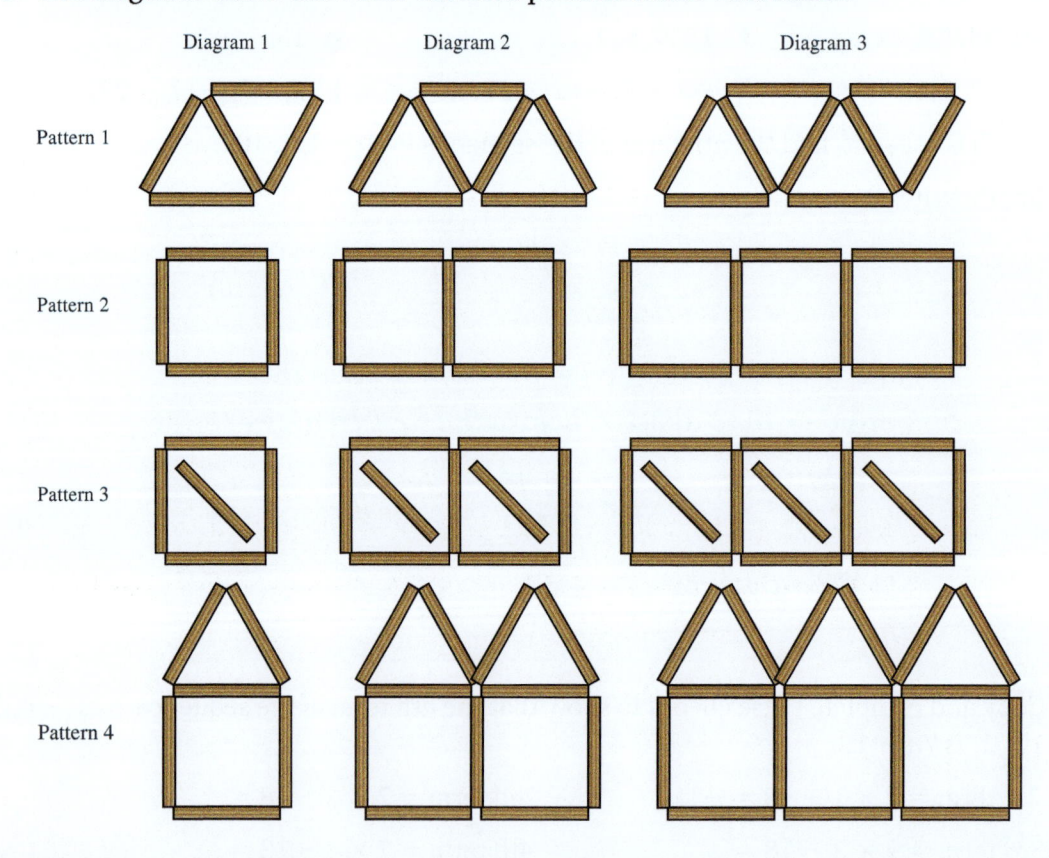

　　　　Diagram 1　　　　Diagram 2　　　　Diagram 3

Pattern 1

Pattern 2

Pattern 3

Pattern 4

14 Sequences

  a Draw diagram 4 for each pattern.

  b Copy and complete the table for each pattern.

Diagram	1	2	3	4
Number of sticks				

  c Write down the rule for finding the number of sticks in diagram n for each pattern.

Review

1 Write down the next three terms for each of these sequences.

  a 13, 19, 25, 31, …
  b −14, −11, −8, −5, …
  c 7, 14, 28, 56, …
  d 7, 5, 3, 1, …
  e 400, 200, 100, 50, …
  f 3, 6, 11, 18, 27, …
  g −7, −5, −1, 5, 13, …
  h 0, 1, 1, 2, 3, 5, 8, …
  i 16, 8, 4, 2, 1, …
  j $\frac{1}{3}, \frac{2}{5}, \frac{3}{7}, \frac{4}{9}, …$
  k $\frac{3}{4}, \frac{4}{9}, \frac{5}{16}, \frac{6}{25}, …$
  l $5a^5, 8a^7, 11a^9, 14a^{11}, …$

2 Write down the first four terms of the sequence whose nth term is:

  a $n + 3$
  b $2n − 1$
  c $4n + 3$
  d $4n − 5$
  e $23 − 2n$
  f $15 − 4n$
  g $n^2 + 1$
  h $n^2 − 2$
  i $2n^2$
  j $n(n + 1)$
  k $n^2 + 5n$
  l $n^3 + 2$

3 Here is a number pattern:

  Row 1: 2   = 2 = $1^2 + 1$
  Row 2: 2 + 4   = 6 = $2^2 + 2$
  Row 3: 2 + 4 + 6   = 12 = $3^2 + 3$
  Row 4: 2 + 4 + 6 + 8   = 20 = $4^2 + 4$

  a Write down the next two rows in this number pattern.

  b Write down an expression for the sum of the numbers in row n.

4 The sequence 3, 9, 27, 81, … is formed from:

  $3^1 = 3$ $3^2 = 3 \times 3 = 9$ $3^3 = 3 \times 3 \times 3 = 27$ $3^4 = 3 \times 3 \times 3 \times 3 = 81$

  So the rule for finding the nth term in the sequence is: nth term = 3^n

  The following sequences are all connected with the sequence 3, 9, 27, 81, …

  Find the connection and then write down the rule for the nth term of each sequence.

  a 4, 10, 28, 82, …
  b 7, 13, 31, 85, …
  c 1, 7, 25, 79, …
  d 2, 8, 26, 80
  e 6, 18, 54, 162, …
  f 7, 20, 57, 166, …

5

Diagram 1   Diagram 2   Diagram 3   Diagram 4

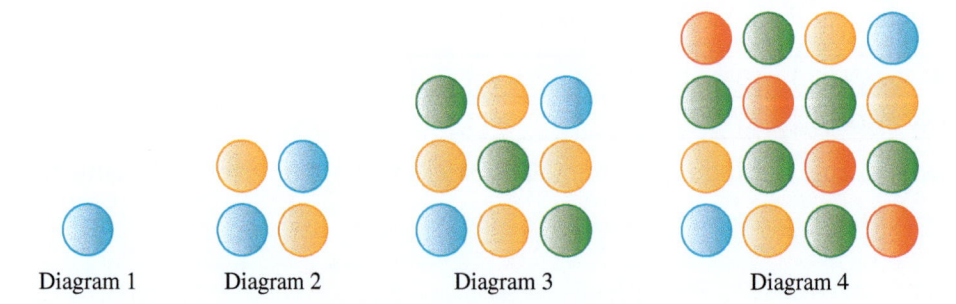

The diagrams show how the following number sequence is formed:

Diagram 1: $\quad\quad\quad\quad\quad\quad 1 \quad\quad\quad\quad\quad\quad\quad = \quad 1 = 1^2$

Diagram 2: $\quad\quad\quad\quad 1 + 2 + 1 \quad\quad\quad\quad\quad = \quad 4 = 2^2$

Diagram 3: $\quad\quad 1 + 2 + 3 + 2 + 1 \quad\quad\quad = \quad 9 = 3^2$

Diagram 4: $\; 1 + 2 + 3 + 4 + 3 + 2 + 1 = 16 = 4^2$

  a Write down the next two rows in this number sequence.

  b Use the number patterns to help you find the total number of dots in:

    i diagram 20   **ii** diagram 50.

6 Find: **i** the nth term   **ii** the 40th term for each of these arithmetic sequences.

  a 12, 14, 16, 18, …   **b** 5, 11, 17, 23, …   **c** −9, −5, −1, 3, …

  d 41, 45, 49, 53, …   **e** 3, 14, 25, 36, …   **f** 2, 11, 20, 29, …

  g 17, 14, 11, 8, …   **h** 38, 32, 26, 20, …   **i** −11, −19, −27, −35, …

7 Chairs are arranged around tables as shown.

1 table   2 tables   3 tables

  a Copy and complete the table.

Number of tables	1	2	3	4
Number of chairs	6			

  b Find the number of people that can be seated at 10 tables arranged in this way.

  c Find the number of people that can be seated at n tables arranged in this way.

  d Find the number of tables needed to seat 50 people in this way.

14 Sequences

Examination-style questions

1   **a**   Write down the next term in the sequence 4, 7, 10, 13, … [1]

    b   Write down the next term in the sequence 25, 18, 11, 4, … [1]

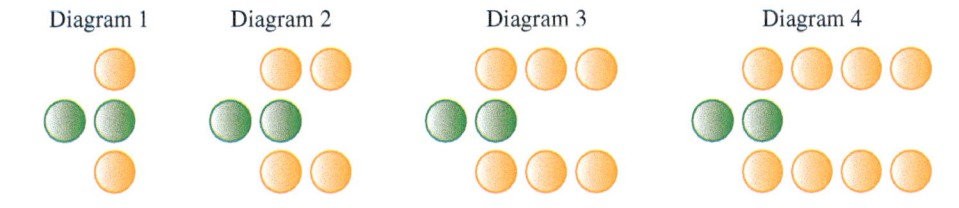

2   Diagram 1   Diagram 2   Diagram 3   Diagram 4

Draw the next diagram in the sequence. [1]

3   Diagram 1   Diagram 2   Diagram 3   Diagram 4

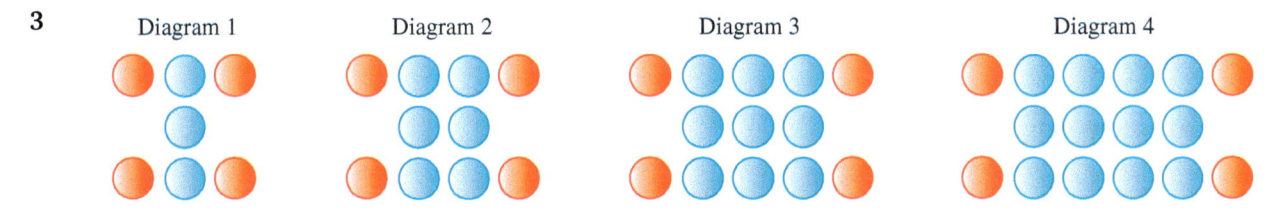

a Copy and complete the table for the sequence of diagrams.

Pattern, n	1	2	3	4	5	6
Number of dots	7	10				

[2]

b   Write down the term-to-term rule for the number of dots. [1]

c   Write down an expression for the number of dots in the nth diagram. [2]

d   Find the number of dots in diagram 40. [1]

e   A diagram in the sequence has 232 dots.

    Find the value of n for this diagram. [2]

4   5, 10, 20, 40, …

    a   Write down in words the term-to-term rule for continuing the sequence. [1]

    b   Write down the next term in the sequence. [1]

5   Write down the next two terms in the sequence 29, 26, 21, 14, 5, … [2]

6   Write down the first three terms in the sequence whose nth term is $n^2 + 5$ [2]

7 Write down the *n*th term for each of these sequences.

  a 3, 6, 9, 12, … [1]

  b 17, 14, 11, 8, … [2]

8 Write down the *n*th term for each of these sequences.

  a 1, 4, 9, 16, … [1]

  b 1, 8, 27, 64, … [1]

9 Diagram 1   Diagram 2   Diagram 3   Diagram 4

  a Draw diagram 5. [1]

  b Copy and complete the table for the sequence of diagrams.

Diagram, *n*	1	2	3	4	5
Number of dots	2 = 1 × 2	6 = 2 × 3	12 = 3 × 4		

  [2]

  c Write down the rule for finding the number of dots in diagram *n*. [2]

15 Probability

> **Learning outcomes**
> - Know that the sum of all mutually exclusive outcomes is 1.
> - Find and record all outcomes for two successive events in a sample space diagram.
> - Understand relative frequency as an estimate of probability.
> - Use relative frequency to compare outcomes of experiments.

Buffon's needle

This idea was in a paper written in 1777 by Georges Louis Leclerc, Comte de Buffon.

A 'needle' is needed. This could be a straw or a matchstick.

Some parallel lines are drawn, all the same distance apart. This distance should be twice the length of the 'needle'.

The 'needle' is dropped on to the lines. It either touches a line or it does not. This is recorded.

This experiment is then repeated.

The relative frequency of 'touching a line' is then found.

If the experiment is repeated enough times the relative frequency can be used to estimate π.

If you try this experiment yourself, you will be doing one of the earliest geometric probability problems to be solved.

15.1 Mutually exclusive outcomes

Mutually exclusive outcomes

When a coin is thrown it can land either 'heads' or 'tails'. Both outcomes cannot occur at the same time. Outcomes are called **mutually exclusive** outcomes if only one of them can happen.

> **Worked example 1**
>
> A dice is rolled.
>
> State whether the outcomes A and B are mutually exclusive or not.
>
> Give a reason for each answer.
>
> **a** A A 6 is rolled.   B A 3 is rolled.
>
> **b** A A 2 is rolled.   B An even number is rolled.
>
> **a** These are mutually exclusive outcomes.
>
>   A 6 and a 3 cannot be rolled at the same time.
>
> **b** These are not mutually exclusive outcomes.
>
>   The even numbers are 2, 4 and 6. If a 2 is rolled then it is also an even number.

The sum of the probabilities of mutually exclusive outcomes

The sum of the probabilities of all mutually exclusive outcomes for any event is 1, as one of the outcomes is certain to happen.

The spinner has six equally sized sectors.

The three colours are not equally likely.

The probabilities are:

$P(red) = \frac{2}{6}$ $P(white) = \frac{1}{6}$ $P(blue) = \frac{3}{6}$

> P(red) means the probability of the spinner landing on red.

There are no other colours on the spinner so one of these colours is certain to happen when the spinner is spun.

So these probabilities add up to 1.

If you know the probability of two of these outcomes you can find the third.

15 Probability

Worked example 2

Adrienne has a spinner with four colours.

The table shows the probability of spinning three of the colours.

Colour	Red	White	Yellow	Green
Probability	0.1	0.25	0.4	

What is the probability of spinning green?

The outcomes are mutually exclusive.

There are no other colours possible.

So the probabilities add up to 1.

P(red, white or yellow) = 0.1 + 0.25 + 0.4 = 0.75

P(green) = 1 − 0.75

= 0.25

Exercise 15.1

1 A dice is rolled.

State whether the outcomes A and B are mutually exclusive or not.

Give a reason for each answer.

a A A 4 is rolled.　　B A number less than 5 is rolled.

b A An even number is rolled.　　B An odd number is rolled.

c A An even number is rolled.　　B A number greater than 4 is rolled.

2 Brad has a bag of coloured marbles.

The bag only contains red, yellow, blue and green marbles.

The table shows the probability of taking a red, yellow or blue marble.

Colour	Red	Yellow	Blue	Green
Probability	0.1	0.4	0.2	

What is the probability of taking a green marble?

3 A box contains some discs.

Each disc is numbered 1, 2, 3, 4 or 5.

A disc is taken from the box at random.

> 'At random' means that each disc is equally likely to be chosen.

a Copy and complete the table to show the probability of each number being chosen.

Number	1	2	3	4	5
Probability	0.15	0.25	0.1		0.35

b Is the disc more likely to have an odd number or an even number?

225

4 A bag contains only red, yellow and orange sweets.

A sweet is picked at random from the bag.

The table shows the probability of picking each colour.

Sweet	Red	Yellow	Orange
Probability	0.2	0.35	

What is the probability of picking an orange sweet?

5 When a coin is tossed, it can land either 'heads' (H) or 'tails' (T).

Two coins are tossed at the same time.

There are four possible outcomes: HH, HT, TH, TT.

   a How many outcomes are there when three coins are tossed?

   b What is the probability that all three coins show the same?

6 Two bags contain red and black counters.

Both bags contain the same number of counters.

The probability of taking a red counter from bag A is 0.5

The probability of taking a red counter from bag B is 0.2

Both bags are emptied and all the counters are placed in bag C.

What is the probability of taking a red counter from bag C?

> Choose a value for the number of counters in each bag. Make sure that you get a whole number answer when you multiply it by 0.5 and 0.2

7 Every day Charlie's mother gives him a fruit from a fruit bowl.

The table shows the probability of each type of fruit being in the fruit bowl.

Fruit	Apple	Pear	Banana	Orange
Probability	$\frac{1}{4}$	$\frac{1}{3}$	$\frac{3}{8}$	

What is the smallest number of oranges there could be in the bowl?

Show how you decide.

> First find the probability of getting an orange. Then find the lowest common multiple of the denominators.

8 A pile of picture cards is used in a children's game.

The table shows the probability of each picture being chosen.

Picture	Bus	Car	Bike	Scooter
Probability	$\frac{1}{16}$	$\frac{1}{4}$		$\frac{1}{2}$

   a What is the probability of getting a bike?

   b Which outcome is most likely?

   c What is the probability of not getting a bus?

   d After a number of games, the scooter had appeared 15 times.

   How many games do you think had been played?

15 Probability

15.2 Sample space diagrams

When you list all the possible outcomes for two successive events it can be helpful to use a table.

This table is called a **sample space diagram**.

Worked example 1

A red dice and a white dice are rolled.

Show all the possible outcomes on a sample space diagram.

		\multicolumn{6}{c	}{Red dice}
		1	2	3	4	5	6
White dice	1	1, 1	2, 1	3, 1	4, 1	5, 1	6, 1
	2	1, 2	2, 2	3, 2	4, 2	5, 2	6, 2
	3	1, 3	2, 3	3, 3	4, 3	5, 3	6, 3
	4	1, 4	2, 4	3, 4	4, 4	5, 4	6, 4
	5	1, 5	2, 5	3, 5	4, 5	5, 5	6, 5
	6	1, 6	2, 6	3, 6	4, 6	5, 6	6, 6

Worked example 2

A fair red dice and a fair blue dice are rolled.

The numbers showing are added.

> 'Fair' means the number on each face is equally likely.

a Show all the possible outcomes in a sample space diagram.

b Use your table to find the probability of getting:

  i a total of 3

  ii a total of 12

  iii a total greater than 6.

c Which total score is most likely?

a

		\multicolumn{6}{c	}{Blue dice}
		1	2	3	4	5	6
Red dice	1	2	③	4	5	6	7
	2	③	4	5	6	7	8
	3	4	5	6	7	8	9
	4	5	6	7	8	9	10
	5	6	7	8	9	10	11
	6	7	8	9	10	11	12

227

b **i** P(a total of 3) = $\frac{2}{36}$ = $\frac{1}{18}$ count how many times 3 occurs in the table (shown circled)
   ii P(a total of 12) = $\frac{1}{36}$ divide by the total number of possible outcomes
   iii P(a total greater than 6) = $\frac{21}{36}$ = $\frac{7}{12}$ simplify the fraction where possible
c A total score of 7 is most likely. There are six ways in which a total of 7 may be scored.

Exercise 15.2

1 A coin is thrown and a dice is rolled.

Copy and complete the sample space diagram to show all the possible outcomes.

		Dice					
		1	2	3	4	5	6
Coin	H	H, 1					
	T						

2 Two fair dice are rolled.

The numbers showing are multiplied together to make a final score.

  a Show the scores in a sample space diagram.
  b What is the probability of scoring 10 or less?
  c What is the probability of scoring more than 20?
  d What is the probability of getting an even score?

3 Two fair dice are rolled in a game.

The score is the difference between the numbers shown.

The sample space diagram shows some of the scores.

		Dice 1					
		1	2	3	4	5	6
Dice 2	1	0	1	2			
	2	1	0				
	3	2					
	4						
	5						
	6						

  a How can a score of 0 be obtained?
  b Copy and complete the table.
  c What is the probability of getting a score of 1?
  d What is the probability of getting an even number?
  e What is the probability of getting a score less than 4?
  f Which score has a probability of $\frac{1}{18}$?

4 A fair three-sided spinner has the numbers 1, 3, 5.

The spinner is spun once and a fair dice is thrown.

The score is the sum of the two numbers.

The sample space diagram shows some of the scores.

		Dice					
		1	2	3	4	5	6
Spinner	1	2	3				
	3	4					
	5						11

  a Copy and complete the sample space diagram.

  b What is the probability of scoring 9?

  c What is the probability of scoring less than 6?

5 A bag contains five discs.

The numbers on the discs are 2, 4, 6, 8 or 10.

One disc is taken from the bag without looking.

It is then replaced.

A second disc is taken from the bag without looking.

The score is the sum of the numbers on the discs.

The sample space diagram shows some of the scores.

		Disc 1				
		2	4	6	8	10
Disc 2	2	4	6	8		
	4	6	8			
	6	8				
	8					18
	10				18	20

  a Copy and complete the sample space diagram.

  b What is the probability that the score is 16?

  c Which is greater – the probability that the score is a square number or the probability that the score is a cube number?

  Show how you decide.

> 4 is a square number because 4 × 4 = 16.
> 27 is a cube number because 3 × 3 × 3 = 27.

6 Two fair spinners each show the numbers 0, 1 and 2.

Each spinner is spun in turn.

The score is obtained by multiplying the two numbers together.

The sample space diagram shows some of the possible scores.

		Spinner 1		
		0	1	2
Spinner 2	0	0		
	1			2
	2		2	4

a Copy and complete the sample space diagram.

b Maria says, 'There are only four scores, one of which is 0. So the probability of scoring 0 is $\frac{1}{4}$.'

Explain why Maria is wrong.

> Think about the number of ways of getting each score.

15.3 Experimental probability

Relative frequency

When a drawing pin is dropped there are two possible outcomes.

It may land point up or point down.

These two outcomes may be equally likely, or they may not be.

You could do an **experiment** to estimate the probability.

You could drop a drawing pin lots of times and record the results.

Each time the drawing pin is dropped is called a **trial**.

You could then use the results to find relative frequencies using this formula.

Relative frequency $= \dfrac{\text{number of successful outcomes}}{\text{total number of trials}}$

Relative frequency is sometimes called **experimental probability**.

It is used as an estimate of the probability of the outcome of the event.

The larger the number of trials used to estimate the probability, the better the estimate.

The number of trials that you need to find the experimental probability depends on the number of possible outcomes.

It should be about 50 times the number of outcomes.

For example, for a biased coin with two possible outcomes about 100 throws should be enough.

> When a coin is fair, a 'head' and a 'tail' are equally likely. When a coin is biased, a 'head' and a 'tail' are not equally likely.

When an experiment is repeated you would not expect to get the same results.

Using probability

When an ordinary dice is thrown a large number of times, you would expect each score to come up the same number of times.

In practice, this does not happen.

Each score comes up roughly the same number of times.

For example, if you throw a dice 300 times you would expect each score to come up:

$\frac{1}{6} \times 300 = 50$ times

In practice, you would be likely to get frequencies close to 50.

If the frequencies are very different, you might think the dice is biased.

Worked example 1

Julia spins a spinner 150 times. It shows red 40 times.

a What is the experimental probability that the spinner shows red?

b What is the experimental probability that the spinner does not show red?

c The spinner is spun 200 times.

Estimate the number of times the spinner will show red.

a $\frac{40}{150}$ $\dfrac{\text{number of times red is shown}}{\text{total number of trials}}$

b $1 - \frac{40}{150}$ P(outcome not occurring) = 1 − P(outcome)

$= \frac{110}{150}$

c $\frac{40}{150} \times 200$ probability × number of spins

$= 53.\dot{3}$

A reasonable estimate is about 53 times.

> The number you would expect should be given as a whole number.

Worked example 2

Tomas plants a bag of mixed tulip bulbs in his garden.

All the bulbs grow and produce one flower each.

The table shows the colours produced by the bulbs when they flower.

Colour	Red	Yellow	Orange	Purple
Frequency	8	32	6	4

a How many bulbs does Tomas plant?

b Work out the relative frequency for each colour.

c Use your answer to part b to estimate the number of purple flowers that will be produced by a bag containing 240 mixed bulbs.

231

a 8 + 32 + 6 + 4 = 50 bulbs

b

Colour	Red	Yellow	Orange	Purple
Frequency	8	32	6	4
Relative frequency	$\frac{8}{50}$	$\frac{32}{50}$	$\frac{6}{50}$	$\frac{4}{50}$

c $\frac{4}{50} \times 240 = 19.2$

An estimate of the number of purple flowers is 19.

Worked example 3

A dice is thrown 240 times.

The table shows the frequencies of each score.

Score	1	2	3	4	5	6
Frequency	10	20	50	60	40	60

Is the dice likely to be biased?

Give a reason for your answer.

$\frac{1}{6} \times 240 = 40$ so you would expect about 40 of each score.

10 is quite different from 40, so:

yes, the dice is likely to be biased.

More throws would give more evidence.

Exercise 15.3

1 A garage sells cars.

The table shows the number of cars of each colour.

Colour	White	Blue	Yellow	Black	Green	Grey
Frequency	11	4	2	12	5	6

a Find the relative frequency for each colour of car.

b The manager selects a car at random. Estimate the probability that the car chosen is yellow.

2 Deema records the number of times her bus is late. Out of 45 occasions, the bus is late 6 times.

a Estimate the probability that the bus is late tomorrow.

b She catches the bus 10 times each week. How many times can she expect the bus to be late next week?

3 A dice is rolled 20 times.

The frequency table shows the number of times each number is rolled.

Number	1	2	3	4	5	6
Frequency	2	6	4	7	0	1

15 Probability

  a What is the relative frequency of a 4 for this dice?

  b If this dice is biased, use the results in the table to estimate the probability of getting a 4 on the next roll.

  c If this dice is a fair dice, write down the probability of getting a 4 on the next roll.

4 The probability a plane will land late is 0.3.

Out of 300 flights, how many times would you expect the plane to land late?

5 A company makes parts for computers.

It tests 300 to see how long they last.

25 last less than 500 hours.

220 last between 500 hours and 2000 hours.

The rest last more than 2000 hours.

  a Estimate the probability that a component lasts less than 500 hours.

  b Estimate the probability that a component lasts over 2000 hours.

  c A computer maker orders 4000 components.

  Estimate how many of them will last longer than 500 hours.

6 Sara has a dice that does not roll as many 6s as she would like.

She does an experiment to see if the dice is biased against 6s.

She rolls the dice 100 times and records whether the dice shows a 6 or not.

Here are her results in groups of 10.

'6' means a 6 is showing, 'N' means a 6 is not showing.

N	N	N	N	6	N	N	N	6	6
N	N	6	N	N	N	N	N	6	6
6	N	N	N	N	N	N	6	N	N
N	N	N	N	N	6	N	N	N	N
N	6	N	N	N	N	N	N	N	6
N	N	N	N	N	N	N	N	6	N
N	N	N	N	N	N	N	N	N	N
N	6	N	N	N	N	N	N	N	N
N	N	N	6	N	N	N	N	N	N
N	N	N	N	N	N	6	N	N	6

  a Calculate the relative frequency of a 6 after 10 rolls.

  b Calculate the relative frequency of a 6 after 20 rolls.

  c Continue for 30, 40, 50, 60, 70 80, 90, 100 rolls.

  Copy and complete the table to show your results.

  Where necessary, give the relative frequency to 2 decimal places.

Number of rolls	10	20	30	40	50	60	70	80	90	100
Frequency of 6	3									
Relative frequency of 6	$\frac{3}{10}$ = 0.3									

233

d Draw axes on graph paper, using the scales shown here.

Make the horizontal axis go up to 100.

Plot the points from your table.

e What does your graph show about the relative frequency as the number of rolls increases?

f Do you think the dice is biased against 6s?

Give a reason for your answer.

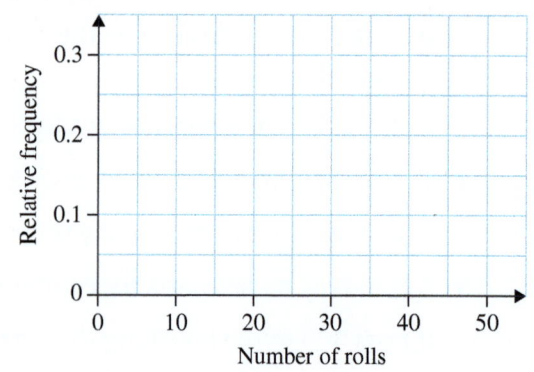

7 This spinner has five equal sections.

Aiden and Claude want to find out whether the spinner is fair.

They are estimating the probability of getting red.

a Aiden spins the spinner 10 times.

It lands on red three times.

What is the relative frequency of red after these spins?

b Claude spins the spinner 100 times.

It lands on red eight times.

Explain why Claude's results should give a better estimate of the probability.

c Combine Aiden and Claude's results.

Use the combined results to estimate the probability of getting red with this spinner.

d Do you think the spinner is fair?

Use your answer to part **c** to help you explain.

8 A four-sided spinner is biased.

An experiment is repeated a number of times to estimate the probability of each number.

The table shows some of the results.

Number	1	2	3	4
Relative frequency	0.1			0.3

The spinner lands on 2 twice as often as it lands on 3.

The spinner is spun 50 times.

How many times would you expect it to land on 2? Show how you decide.

> Remember that the four probabilities must add up to 1.

9 Two spinners are each numbered from 1 to 4

The tables show the results after spinning each spinner 120 times.

Spinner A	
Score	Frequency
1	28
2	32
3	29
4	31

Spinner B	
Score	Frequency
1	24
2	23
3	25
4	48

Only one of the spinners is fair.

Which spinner do you think is fair? Give a reason for your answer.

Review

1 A car is chosen at random from a car park.

State whether the outcomes A and B are mutually exclusive or not.

Give a reason for each answer.

　a A　The car is left-hand drive.　　B　The car is right-hand drive.

　b A　The car is red.　　B　The car is blue.

　c A　The car has seats in the front.　B　The car has seats in the back.

2 A dice is biased.

The table shows the probability of getting a 1, 2, 3, 4, 5 with the dice.

Score	1	2	3	4	5	6
Probability	0.15	0.1	0.05	0.1	0.2	

Find the probability of getting a 6.

3 Yusuf is making sandwiches to sell in his shop.

He has a table showing the probability of selling each type of sandwich he makes.

Sandwich	Haloumi	Tuna	Salad	Egg	Salmon
Probability	0.3	0.41	0.1	0.25	0.2

　a How can you tell there is a mistake in the table?

　b The probability for tuna is not correct.

　　What should this probability be?

　c Yusuf makes 200 sandwiches to sell in the shop.

　　How many of these would you expect to be haloumi?

4 A spinner has three equal sections.

Each section has one of the numbers 3, 4, 5.

The spinner is spun twice.

The score is found by adding the two numbers.

235

a Copy and complete the sample space diagram to show the possible totals.

	2nd spin		
	3	4	5
1st spin 3			
4			
5			

b What is the probability of getting a score of 8?

c What is the probability of getting a score that is an odd number?

5 Kerry drops a drawing pin 120 times.

It lands point up 35 times.

a What is the relative frequency of the pin landing point up?

b Kerry is planning on dropping the pin 500 times.

How many times would you expect it to land point up?

6 The students in a class make their own dice.

They roll the dice to test the fairness, and record the results.

The table shows the results for three of the students in the class.

	1	2	3	4	5	6
Anna	20	20	20	20	20	20
Bella	9	11	12	8	9	11
Cara	14	16	12	11	12	25

Answer each of the following questions. Give a reason for each answer.

a Which student produced a dice that appears to be fair?

b Which student produced a dice that appears to be biased?

c Which student appears to have cheated and not rolled the dice at all?

Examination-style questions

1 A bag contains only red, green, yellow and blue beads.

The probability of picking each colour bead at random is shown in the table.

Colour	Red	Green	Yellow	Blue
Probability	0.2	0.1	0.4	

What is the probability of picking a blue bead? [1]

15 Probability

2 The face of a fair spinner is a regular hexagon.

The probability of getting an even number is twice that of getting an odd number.

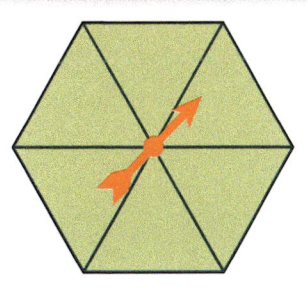

Copy the diagram and write a whole number in each section to make this correct. [1]

3 The diagrams show two fair spinners.

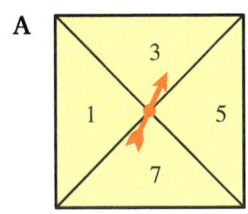

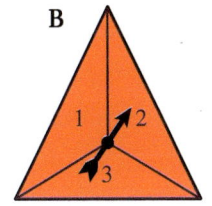

a Complete the sample space diagram to show all the possible outcomes of spinning the triangular and the square spinner. [1]

		Spinner B		
		1	2	3
Spinner A	1	1, 1	1, 2	1, 3
	3	3, 1		
	5		5, 2	
	7		7, 2	

b What is the probability of scoring an **odd total** when the numbers on the two spinners are added? [1]

4 Jamal uses two fair five-sided spinners in a game.

His score is the total of the two numbers shown on the spinners.

a Complete the table to show all of his possible scores. [1]

		Spinner 1				
		1	2	3	4	5
Spinner 2	1	2	3	4	5	6
	2		4	5	6	7
	3			6	7	8
	4				8	9
	5					10

b Find the probability that Jamal gets:

  i a score of 2   **ii** a score of 11. [2]

c Find the probability that Jamal gets a score more than 6.

Give your answer as a fraction in its lowest terms. [1]

237

16 Functions and graphs

Learning outcomes

- Construct tables of values and plot the graphs of linear functions.
- Find the gradient of a straight-line graph and know the significance of m in $y = mx + c$
- Find the approximate solutions of a simple pair of simultaneous linear equations by finding the point of intersection of their graphs.
- Find the inverse of a linear function.
- Construct functions arising from real-life problems; draw and interpret their graphs.
- Use algebraic methods to solve problems involving direct proportion, relating solutions to graphs of the equations.

Modelling change

Scientists use graphs to see trends and tendencies between two or more parameters. If the scientists can then find a function to fit the trend then it is possible for them to predict what the future may be like.

One example of this kind of work is recognising the effect that climate change has on rising sea levels.

16.1 Gradients

$$\text{Gradient} = \frac{\text{vertical distance}}{\text{horizontal distance}}$$

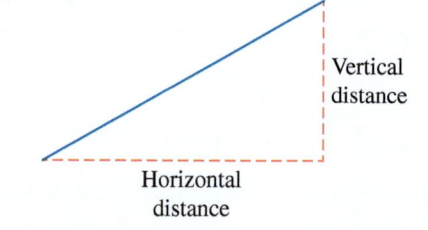

The **gradient** of a straight line is a measure of how steep the line is.

A line can have positive, negative or zero gradient.

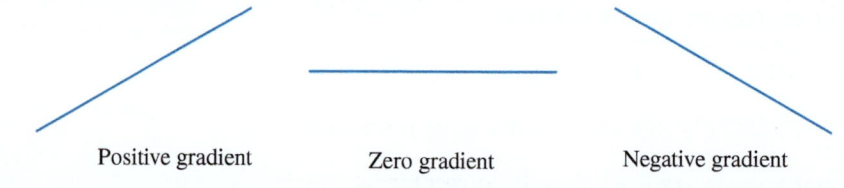

Positive gradient   Zero gradient   Negative gradient

238

16 Functions and graphs

Worked example

Find the gradient of the line joining the points (1, 3) and (6, 1).

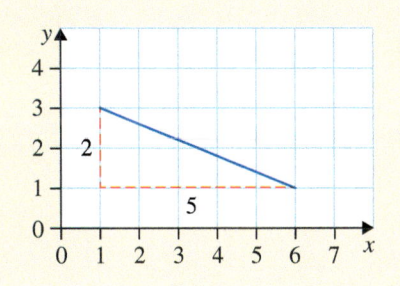

draw the line joining the two points

$$\text{Gradient} = \frac{\text{vertical distance}}{\text{horizontal distance}}$$

$$\text{gradient} = -\frac{2}{5}$$

Note that the gradient is negative.

Exercise 16.1

1 Find the gradient of each of these lines.

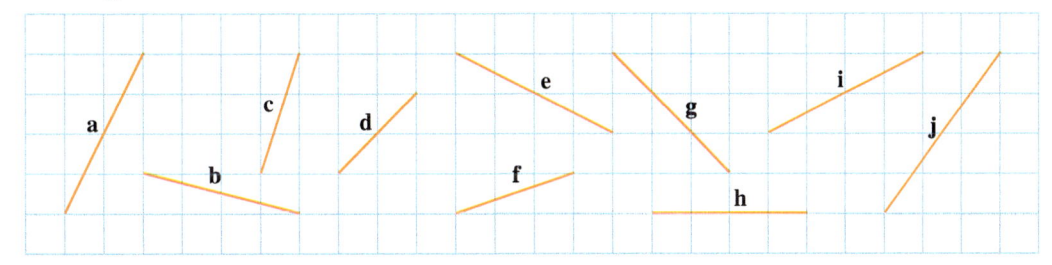

2 Draw the straight lines joining these pairs of points.

Find the gradient of each line.

   a (1, 2) and (3, 6)   **b** (6, 3) and (3, 4)

   c (2, 1) and (4, 2)   **d** (4, −2) and (6, 1)

   e (1, −3) and (6, −4)   **f** (−2, −6) and (2, −4)

   g (−3, 2) and (−2, −4)   **h** (−6, −6) and (−3, −3)

3 The line joining the points (1, 3) and (6, a) has gradient 2.

Find the value of a.

4 The line joining the points (2, 0) and (6, −6) has the same gradient as the line joining (−1, 2) and (7, b).

Find the value of b.

5 The gradient of a line AB is $\frac{4}{5}$.

Find the gradient of a line that is perpendicular to AB.

239

16.2 Straight-line graphs of the form $ax + by = c$

If you are asked to draw the graph of $3x + 4y = 12$, it is usually easiest to find points where the line crosses the axes.

Putting $x = 0$ into the equation gives you the y-axis crossing point.

If $x = 0$, then $4y = 12$, $y = 3$

So the graph crosses the y-axis at 3.

Putting $y = 0$ into the equation gives you the x-axis crossing point.

If $y = 0$, then $3x = 12$, $x = 4$

So the graph crosses the x-axis at 4.

The table of values is:

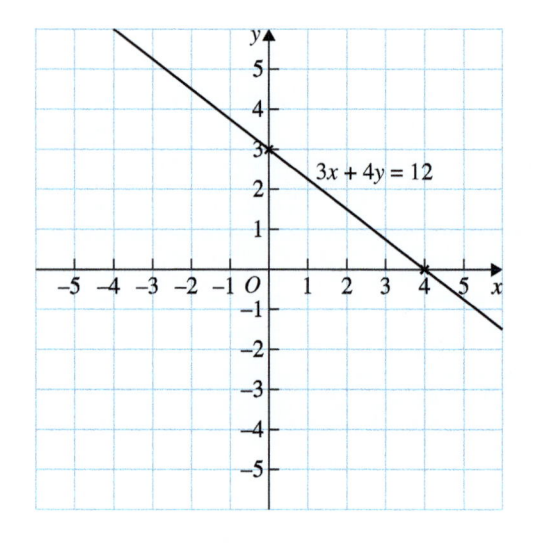

x	0	4
y	3	0

Worked example

Draw the graph of $3x - 7y = 21$

If $x = 0$, then $-7y = 21$, $y = -3$ find the y-axis crossing point

So the graph crosses the y-axis at -3.

If $y = 0$, then $3x = 21$, $x = 7$ find the x-axis crossing point

So the graph crosses the x-axis at 7.

The table of values is:

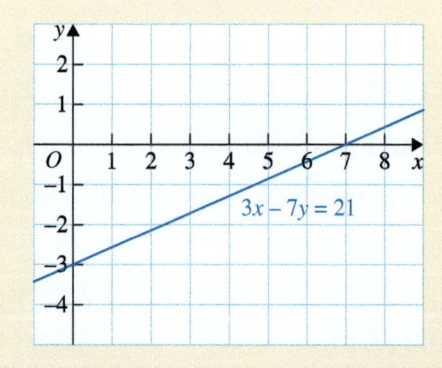

x	0	7
y	-3	0

Exercise 16.2

1. Find where each graph crosses the axes and then draw the graph.

   a $x + y = 5$ **b** $x + y = 2$ **c** $x + y = -3$ **d** $2x + y = 8$

   e $x + 3y = 9$ **f** $4x + 3y = 12$ **g** $2x + 3y = 12$ **h** $3x + 8y = 24$

   i $2x + 4y = -8$ **j** $5x + 4y = -20$ **k** $6x + 7y = 21$ **l** $2x + 7y = -21$

2 Find where each graph crosses the axes and then draw the graph.

   a $x - y = 4$ **b** $x - y = 2$ **c** $2x - y = 4$ **d** $x - 3y = 9$

   e $3x - 4y = 24$ **f** $5x - 2y = 10$ **g** $2x - 5y = -10$ **h** $7x - 2y = 21$

16.3 Using graphs to solve simultaneous equations

The equations $x + y = 5$ and $x + 2y = 8$ can be represented by straight lines on a grid as shown.

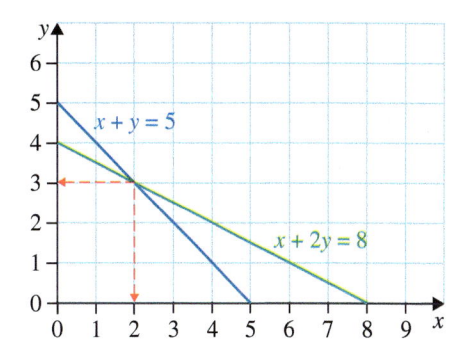

The lines intersect at the point (2, 3).

The solution to the **simultaneous equations**

$$x + y = 5$$
$$x + 2y = 8$$

is $x = 2, y = 3$

Check: substitute the values for x and y into the original equations.

$2 + 3 = 5$ ✓   and $2 + (2 \times 3) = 8$ ✓

Exercise 16.3

1 Use the graph to solve these simultaneous equations.

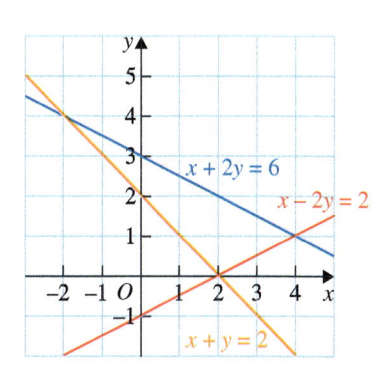

   a $x + 2y = 6$
       $x - 2y = 2$

   b $x + y = 2$
       $x - 2y = 2$

   c $x + y = 2$
       $x + 2y = 6$

2 Solve these simultaneous equations by drawing a graph.

   $5x + 2y = 25$
   $4x - 2y = 2$

3 Use the graph to solve these simultaneous equations.

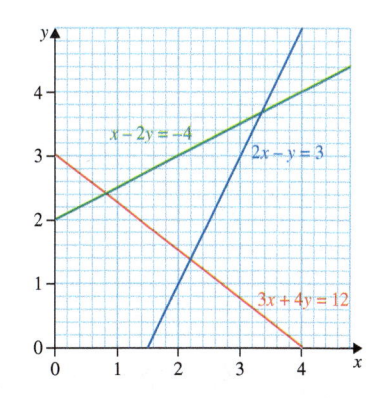

   a $2x - y = 3$
       $x - 2y = -4$

   b $x - 2y = -4$
       $3x + 4y = 12$

   c $3x + 4y = 12$
       $2x - y = 3$

Give your answers correct to one decimal place.

4 Solve these simultaneous equations by drawing a graph.

$3x + 7y = 21$

$2x - 3y = -3$

Give your answers correct to one decimal place.

16.4 The general equation of a straight line

In your Stages 7 and 8 Student Books you learnt how to draw graphs of the form $y = mx + c$

For example, to draw the graph of $y = 2x + 1$ you make a table of values, plot the points on a grid and then join them with a straight line.

x	−3	0	2
y	−5	1	5

The gradient of the line $y = 2x + 1$ can be found by using any two points on the line.

Using the points (0, 1) and (2, 5):

$$\text{gradient} = \frac{\text{vertical distance}}{\text{horizontal distance}} = \frac{4}{2} = 2$$

So the gradient of the line $y = 2x + 1$ is 2.

Similarly, it can be shown that the gradient of the line $y = 3x - 5$ is 3.

This leads to the general rule that:

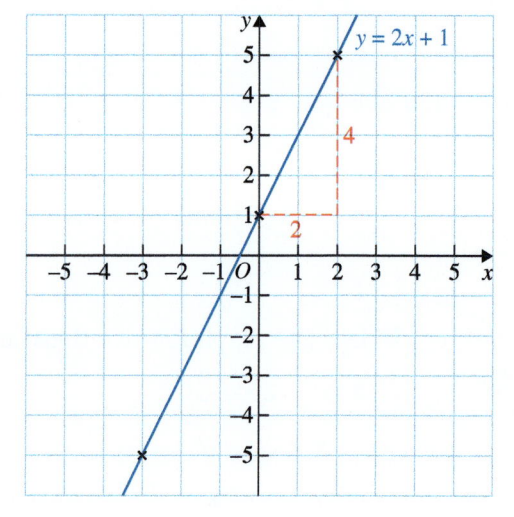

> The graph of $y = mx + c$ is a straight line, where m = the gradient of the line.

Worked example 1

Write down the gradient of each of these lines.

a $y = 6x - 1$ **b** $y = 5 - 3x$ **c** $x + 2y - 6 = 0$

a $y = 6x - 1$ gradient = 6

b $y = 5 - 3x$ rearrange the equation to the form $y = mx + c$

   $y = -3x + 5$ gradient = −3

c $x + 2y - 6 = 0$ rearrange the equation to the form $y = mx + c$

   $2y = -x + 6$

   $y = -\frac{1}{2}x + 3$ gradient = $-\frac{1}{2}$

Worked example 2

Find the equation of the line with gradient 3 that passes through the point (2, 5).

The gradient is 3, so substitute $m = 3$ into the equation $y = mx + c$

$y = 3x + c$

The point (2, 5) lies on the line, so substitute $x = 2$ and $y = 5$ into the equation.

$5 = (3 \times 2) + c$

$c = -1$

The equation of the line is $y = 3x - 1$

Exercise 16.4

1 Draw each of these lines and calculate their gradients.
   a $y = x + 4$
   b $y = 3x - 1$
   c $y = \frac{1}{2}x + 3$
   d $y = 5 - 2x$

2 Write down the gradient of each of these lines.
   a $y = 2x + 5$
   b $y = 3x + 7$
   c $y = x + 4$
   d $y = \frac{1}{2}x + 7$
   e $y = -2x + 3$
   f $y = -\frac{1}{2}x + 1$
   g $y = 2x$
   h $y = 6$

3 Jacques says that the gradient of the line $y = 3 - 2x$ is 3.

   Explain why he is wrong and write down the correct answer.

4 Write down the gradient of each of these lines.
   a $y = 1 - 3x$
   b $y = 5 - x$
   c $y = 4 + 2x$
   d $y = 5 - \frac{1}{2}x$

5 Write each of these equations in the form $y = mx + c$ and then write down the gradient of each line.
   a $y - 3x = 2$
   b $y - 2x + 1 = 0$
   c $4x + y = 3$
   d $5x + 2y = 10$
   e $3y - x = 9$
   f $2y = 4x + 8$
   g $2y = 6x - 2$
   h $3y - 5x = -15$

6 Which of these lines are parallel?

   A $y = -2x + 2$ B $4x + 2y = 6$ C $2y - 2x = 7$ D $x = -\frac{1}{2}y + 5$

   > Parallel lines have the same gradient.

7 Find the equation of the line with gradient -2 that passes through the point (1, 5).

8 Find the equation of the line that passes through the points (5, 4) and (9, −1).

243

16.5 Finding the inverse of a function

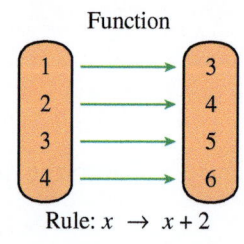

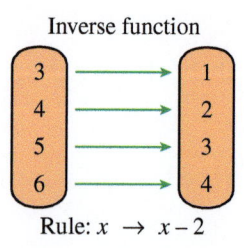

Function: 1→3, 2→4, 3→5, 4→6 Rule: $x \to x+2$

Inverse function: 3→1, 4→2, 5→3, 6→4 Rule: $x \to x-2$

The diagram shows a function 'add 2' and its **inverse function** 'subtract 2'.

The inverse of a function is the function that 'undoes' what the original function has done.

Similarly, the inverse of the function $x \to 3x$ is $x \to \frac{x}{3}$

This can also be written as:

the inverse of the function $y = 3x$ is $y = \frac{x}{3}$

You can find the inverse of any linear function using a reverse function machine.

The function machine for the function $x \to 2x + 3$ is:

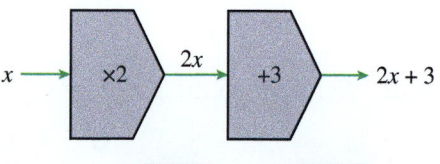

$x \to \times 2 \to 2x \to +3 \to 2x+3$

The reverse function machine is:

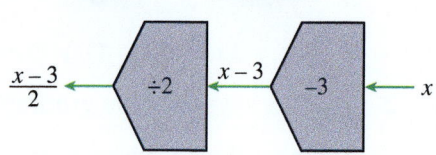

$\frac{x-3}{2} \leftarrow \div 2 \leftarrow x-3 \leftarrow -3 \leftarrow x$

So the inverse of the function $x \to 2x + 3$ is $x \to \frac{x-3}{2}$

If you draw the function $y = 2x + 3$ and its inverse function $y = \frac{x-3}{2}$ on a graph you will see a special relationship.

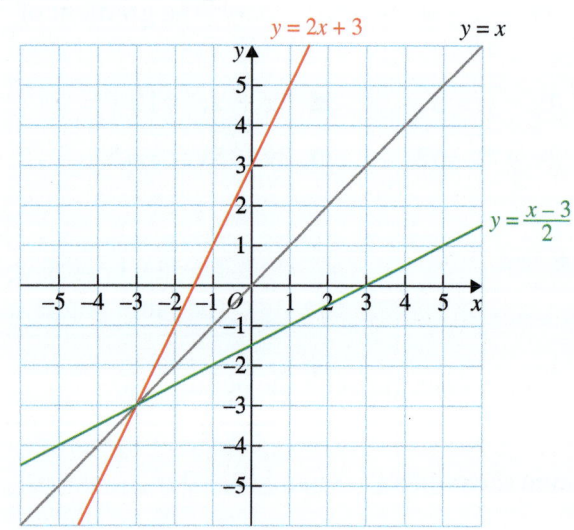

A function and its inverse function are reflections of each other in the line $y = x$

16 Functions and graphs

Worked example 1

Find the inverse of the function $x \to \dfrac{3x-5}{2}$

The function machine for the function is: $x \to [\times 3] \to 3x \to [-5] \to 3x-5 \to [\div 2] \to \dfrac{3x-5}{2}$

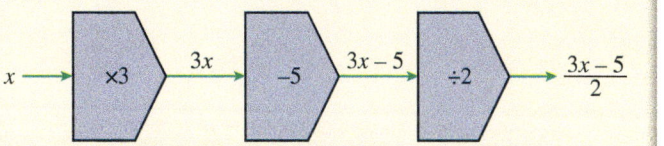

The reverse function machine is: $\dfrac{2x+5}{3} \leftarrow [\div 3] \leftarrow 2x+5 \leftarrow [+5] \leftarrow 2x \leftarrow [\times 2] \leftarrow x$

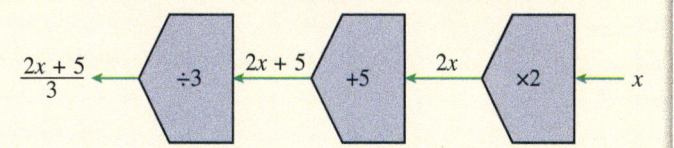

So the inverse function is $x \to \dfrac{2x+5}{3}$

Worked example 2

Find the inverse of the function $x \to 5 - 2x$

First, rewrite the function as $x \to -2x + 5$

So the rule for the function is 'multiply by -2' and then 'add 5'.

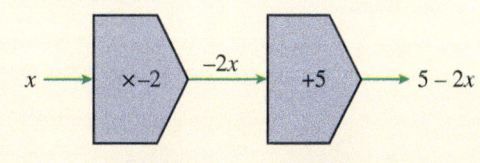

$x \to [\times -2] \to -2x \to [+5] \to 5-2x$

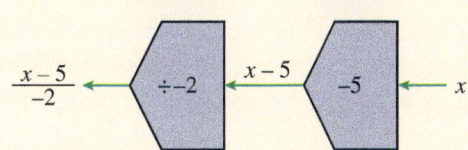

$\dfrac{x-5}{-2} \leftarrow [\div -2] \leftarrow x-5 \leftarrow [-5] \leftarrow x$

So the inverse function is $x \to \dfrac{x-5}{-2}$, which simplifies to $x \to \dfrac{5-x}{2}$

Exercise 16.5

1 Write down the inverse of each of these functions.

  a $x \to 2x$ **b** $x \to x+3$ **c** $x \to x-6$ **d** $x \to \dfrac{x}{5}$

2 Use function machines to find the inverse of each of these functions.

  a $x \to 3x+1$ **b** $x \to 5x-3$ **c** $x \to 4x+3$ **d** $x \to 7x-1$
  e $x \to 2(x+3)$ **f** $x \to 4(x-7)$ **g** $x \to \dfrac{x+2}{4}$ **h** $x \to \dfrac{x-8}{3}$
  i $x \to \dfrac{x}{4}-2$ **j** $x \to \dfrac{x}{6}+5$ **k** $x \to \dfrac{3x+1}{2}$ **l** $x \to \dfrac{2x-3}{5}$
  m $x \to 5(2x-1)$ **n** $x \to \dfrac{2x}{3}+1$ **o** $x \to \dfrac{4x+7}{2}$ **p** $x \to \dfrac{3x-4}{7}$

3 Find the inverse of each of these functions.

  a $x \to 7-x$ **b** $x \to 8-2x$ **c** $x \to 15-3x$ **d** $x \to \dfrac{3-2x}{4}$

245

Mathematics for Cambridge Secondary 1

4 Roberto is asked to find the inverse of the function $x \to 2x - 5$

He writes:

If I follow these steps I can find the inverse function:

STEP 1 Write the function as $\quad y = 2x - 5$

STEP 2 Interchange the x's and y's $\quad x = 2y - 5$

STEP 3 Rearrange to get 'y =' $\quad y = \dfrac{x + 5}{2}$

STEP 4 Write out the inverse function $\quad x \to \dfrac{x + 5}{2}$

Use Roberto's method to find the inverse of each of these functions.

a $x \to 2x - 9$ b $x \to 4x + 2$ c $x \to \dfrac{2x - 1}{4}$ d $x \to \dfrac{3 - 2x}{5}$

5 Copy the graph and draw the inverse function.

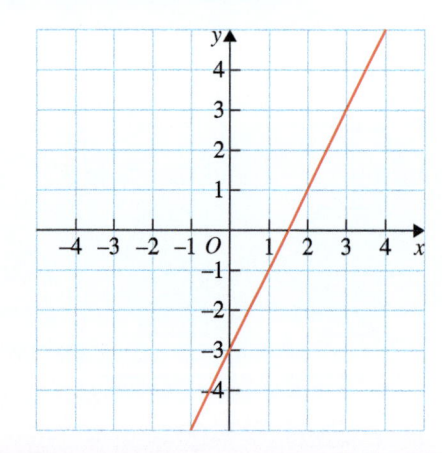

16.6 Functions arising from real-life problems

The rule for calculating the cooking time, t minutes, for a chicken of mass m kilograms is:

'allow 45 minutes per kilogram and then an extra 20 minutes'

This rule can be written algebraically as:

$t = 45m + 20$

A table of values can be made and a graph drawn to show the relationship between m and t.

m	1	2	3	4
t	65	110	155	200

The graph can then be used to read off information.

For example, the mass of a chicken that takes 125 minutes cooking time is 2.3 kg (shown by the red line on the graph).

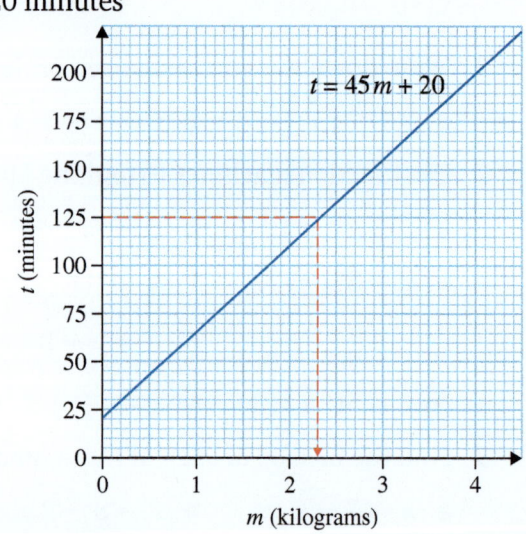

Worked example

Zainab can hire a boat from Company A or Company B.

The cost C, in $, for hiring a boat for n hours from each company is shown in the table.

Company A	C = 5 + 3n
Company B	C = 10 + 2n

Show the costs for each company on a graph and find the values of n for which Company A is more expensive than Company B.

Company A

n ($)	1	2	3	4
C (hours)	8	11	14	17

Company B

n ($)	1	2	3	4
C (hours)	12	14	16	18

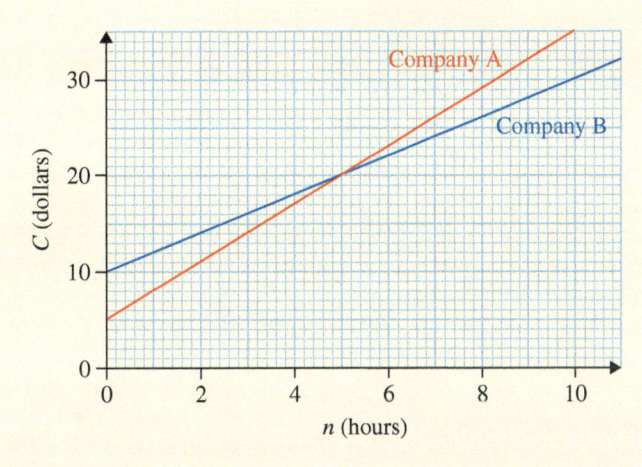

From the graph you can see that Company A is more expensive if $n > 5$

Exercise 16.6

1. Ali is heating a tank of water.

   He records the temperature every 5 minutes.

   The table shows his results.

Time, m (minutes)	5	10	15	20
Temperature, t (°C)	10	13	16	19

   a Draw a straight-line graph to show this information.

   b What was the temperature of the water when he started to heat the water?

   c Find the gradient of the graph.

   What information does the gradient represent?

   d Use your graph to find the value of t when the temperature was 17°C.

247

2 The rule for changing a temperature, C, in degrees Celsius (°C) to a temperature, F, in degrees Fahrenheit (°F) is:

'multiply by 1.8 and then add 32'

a Write this rule in the form '$F = \ldots\ldots\ldots$'

b Represent this rule on a graph using $-50 \leq C \leq 50$

c Use your graph to find the temperature:

  i in °F when it is 18 °C   **ii** in °C when it is -32 °F.

d Use your graph to find the temperature which has the same value in both degrees Celsius and degrees Fahrenheit.

16.7 Direct proportion

If fabric is sold for $6 per metre, then the cost, C, for n metres is given by the formula $C = 6n$

The table shows some possible values of n and C.

n (m)	1	2	3	4	5	6
C ($)	6	12	18	24	30	36

This means that:

if n doubles, then C will also double

if n trebles, then C will also treble.

The quantities n and C are said to be in **direct proportion** to each other and the graph of n against C will be a straight line through the origin.

> If y is directly proportional to x then $y = kx$, where k is called the constant of proportionality

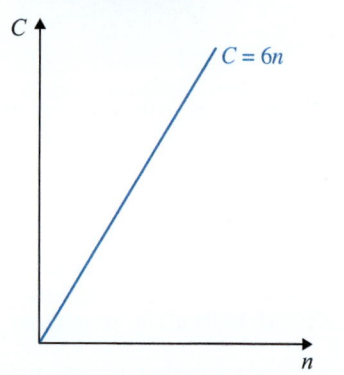

Worked example

y is directly proportional to x.

When $y = 4, x = 5$.

a Find a formula connecting y and x.

b Find y when $x = 3$

a y is directly proportional to x so

$y = kx$

$4 = 5k$ substitute $y = 4$ and $x = 5$ into the equation

$k = \frac{4}{5}$ divide both sides by 5

The formula is $y = \frac{4}{5}x$

b When $x = 3$, $y = \frac{4}{5} \times 3 = 2.4$ substitute $x = 3$ into the formula from part **a**

The last example can also be illustrated on a graph.

Since y is directly proportional to x, the graph will be a straight line passing through the origin and the point (5, 4).

The line has gradient $\frac{4}{5}$,

and the equation of the line is $y = \frac{4}{5}x$

The dashed red lines show that when $x = 3$, $y = 2.4$

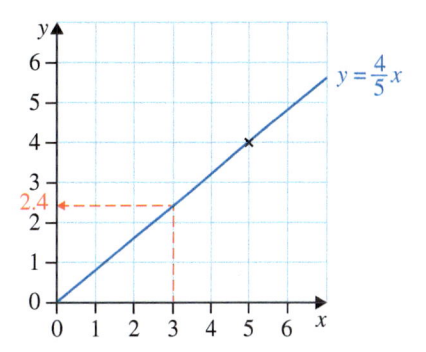

Exercise 16.7

1 y is directly proportional to x.

$y = 18$ when $x = 6$

  a Find a formula connecting y and x.

  b Find the value of y when $x = 12$

2 y is directly proportional to x.

$y = 15$ when $x = 20$

  a Find a formula connecting y and x.

  b Find the value of y when $x = 34$

3 p is directly proportional to q.

$p = 0.5$ when $q = 8$

  a Find a formula connecting p and q.

  b Find the value of p when $q = 20$

  c Find the value of q when $p = 0.2$

4 The distance travelled, d m, by a cyclist is directly proportional to the number of turns, n, made by his bicycle wheel.

Jonny travels 21 m for 10 turns of his bicycle wheel.

   a Find a formula connecting d and n.
   b Find the distance travelled when $n = 27$

5 The extension of a spring, e cm, is directly proportional to the mass, m kg, which hangs on the end of the spring.

When $e = 3$, $m = 5$

   a Find a formula connecting e and m.
   b Find e when $m = 6$
   c Find m when $e = 2.6$

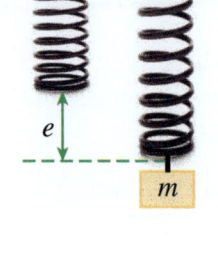

6 The time for one complete swing of a pendulum, t seconds, is directly proportional to the square root of the length, l metres, of the pendulum.

$t = 1.6$ when $l = 0.64$

   a Find a formula connecting t and l.
   b Find t when $l = 0.81$

Review

1 Find the gradients of the lines joining:

   a (2, 5) and (6, 7)   b (2, −3) and (−2, 6).

2 Draw the line $x - 3y = 6$

3 Write down the gradient of each of these lines.

   a $y = 2x + 9$ b $y = 7 - 3x$ c $3x + 6y = 12$ d $2y - 5x = 15$

4 $x + y = 5$

   $x - 7y = -7$

   Use a graph to solve the simultaneous equations.

5 Find the inverse of each of these functions.

   a $x \to 5x$ b $x \to x - 8$ c $x \to 6x - 4$ d $x \to \dfrac{3x - 5}{2}$

6 w is directly proportional to h.

   $w = 8$ when $h = 12$

   Find w when $h = 15$

Examination-style questions

1. Find the gradient of the line joining the points (5, −2) and (3, 8). [2]

2. Complete the table of values for $y = 2x + 3$ and draw the graph.

x	−2	0	1
y			

   [3]

3. On a grid draw the lines:

   a $y = 3$ [1]

   b $y = 2x$ [1]

   c $5x + 2y = 10$ [2]

4. Find the coordinates of the point where the line $y = 3x + 4$ crosses the y-axis. [1]

5. Write down the gradient of the line $y = 4x + 2$ [1]

6. The line $y = 2x + 5$ intersects the line $y = 2$ at the point P.

   Find the coordinates of P. [2]

7. Find the inverse of the function $x \to 2x + 7$ [2]

8. $2x + 5y = 15$

   $2x − 3y = −6$

   Use a graph to solve the simultaneous equations, giving your answers correct to one decimal place. [3]

9. y is directly proportional to x.

   $y = 5$ when $x = 25$

   Find the value of y when $x = 9$ [3]

17 Fractions, decimals and percentages

> **Learning outcomes**
> - Solve problems involving percentage changes, choosing the correct number to take as 100%.
> - Solve problems involving personal and household finance.
> - Recognise when fractions or percentages are needed to compare different quantities.

Tax and other uses of percentages

The word tax comes from the Latin word 'taxo' meaning 'rate'.

Taxes are paid by individual people and also by companies.

People pay tax on the money they earn and the money they spend. Tax is expressed as a percentage, so that the same proportion is paid on different amounts.

17.1 Percentage change

Percentage increase and decrease

To increase by a percentage, you find the percentage and then add it on.

To decrease by a percentage, you find the percentage and then subtract it.

Sometimes decreasing by a percentage is called **percentage reduction**.

> **Worked example 1**
>
> The population of a village is 260.
>
> It is expected to increase by 5% in the next year.
>
> Work out the expected population of the village.
>
> $\frac{5}{100} \times 260 = 13$ this is the increase
>
> $260 + 13 = 273$ add it on
>
> The expected population is 273.

Worked example 2

A car is priced at $12 500.

It is reduced by 15% in a sale.

What is the price of the car in the sale?

$\frac{15}{100} \times \$12\,500 = \$1875$ this is the reduction

$\$12\,500 - \$1875 = \$10\,625$ subtract it

The car costs $10 625 in the sale.

Uses of percentages

When a shop buys a product and sells it for more than they paid, the difference is called a **profit**. When they sell it for less than they paid, the difference is called a **loss**.

When you have money in the bank you may be paid interest. This is often paid at the end of the year. An interest rate of 5% each year is often described as 5% **per annum**. If the interest is not added to the account, it is called **simple interest**.

Sometimes prices are shown including tax. Sometimes prices are shown not including tax. Rates of tax may be different, depending on what is being sold and which country it is being sold in.

Worked example 3

Nikki sells some cakes at the school fête.

The cakes cost $5.80 to make.

She wants to make a profit of 80%.

How much must she sell the cakes for?

$\frac{80}{100} \times \$5.80 = \$4.64$ this is the profit she needs to make 80% of the cost price

$\$5.80 + \$4.64 = \$10.44$

She must sell the cakes for $10.44.

Worked example 4

Tiago has $500 in the bank.

He is paid 6% per annum simple interest on his savings.

How much interest does he receive if he leaves his money in the bank for two years?

$\frac{6}{100} \times \$500 = \$30$ this is the interest he receives each year

$\$30 \times 2 = \$60$

The total interest is $60.

Finding a percentage

Sometimes you need to write one number as a fraction or percentage of another.

To do this you write the numbers as a fraction, then convert this fraction to a percentage.

When you find a percentage increase or decrease you compare the change with the *original* value.

> ### Worked example 5
>
> Tania buys a bike for $750.
>
> She sells it for $650.
>
> Calculate her percentage loss.
>
> Give your answer correct to 1 decimal place.
>
> $750 - $650 = $100   this is the loss she makes on selling the bike
>
> $\frac{100}{750} \times 100 = 13.\dot{3}\% = 13.3\%$ (to 1 d.p.)   this is the loss as a percentage of the original cost
>
> Tania's loss is 13.3% (to 1 d.p.).
>
> ### Worked example 6
>
> Roger takes some tests in maths.
>
> He scores 25 out of 30 in his mental arithmetic test.
>
> He scores 34 out of 40 in his calculator test.
>
> He scores 48 out of 60 in his non-calculator test.
>
> Which test did Roger do best in?
>
> In which test did he do least well?
>
> The tests are out of different amounts, so you need to find percentages to compare them.
>
> Mental arithmetic: $\frac{25}{30} \times 100 = 83.\dot{3}\%$ (to 1 d.p.)
>
> Calculator test: $\frac{34}{40} \times 100 = 85\%$
>
> Non-calculator test: $\frac{48}{60} \times 100 = 80\%$
>
> Roger did best in the calculator test. He did least well in the non-calculator test.

Exercise 17.1

1 A travel company increases all its prices by 15%.

  Calculate the new price for tickets that previously cost:

  a $12   **b** $24   **c** $36

2 A shop advertises 25% off all marked prices in a sale.

Find the sale price of:

   a a fridge with a marked price of $280

   b a kettle with a marked price of $14

   c a toaster with a marked price of $60.

3 A bank pays 3% per annum simple interest on its accounts.

Calculate the interest paid at the end of the year to:

   a Miriam, who has a balance of $250

   b Lina, who has a balance of $320

   c Klaus, who has a balance of $28.

4 Bricks are advertised as $386 plus tax.

Tax is 20%.

How much do the bricks cost to buy?

5 This is the menu at a restaurant.

The restaurant adds a 10% service charge to everything.

Work out the cost of each type of pizza including service.

Mushroom pizza $18.00
Tomato pizza $17.00
Tuna pizza $24.00
Meat feast $28.00

6 Imran has $360.

He spends $30 on shoes and $24 on a shirt.

What percentage of his money has he spent?

7 Leroy sells lemonade to raise money.

His costs in making lemonade are $1.20 per litre.

He sells the lemonade for $1.50 per litre.

   a What is his percentage profit?

   b He sells 20 litres of lemonade. How much profit does he make?

8 The table shows the total area and the area that is water for five countries.

Country	Total area (square km)	Area water (square km)
Australia	7 741 220	58 920
Canada	9 984 670	891 163
Belgium	30 528	250
Uruguay	176 215	1 200
Ghana	238 533	11 000

Source: World Factbook https://www.cia.gov/library/publications/the-world-factbook/

Find the percentage of each country that is water.

Give your answers to 2 decimal places.

9   Moko buys a car costing $24 000.

    She pays a deposit of 35% and then makes 24 equal monthly payments.

    How much are her monthly payments?

    > Take care to subtract the deposit first.

10 The table shows information about the number of men and women students in full-time further education in the UK between 1970 and 2008.

| Men (thousands) |||| Women (thousands) ||||
1970/71	1980/81	1990/91	2007/08	1970/71	1980/81	1990/91	2007/08
116	154	219	520	95	196	261	534

Source: Office for National Statistics

   a Calculate the percentage changes from 1970/71 to 1980/81 for:

      i men   ii women.

   b Calculate the percentage changes from 1990/91 to 2007/08 for:

      i men   ii women.

   c Calculate the percentage of the students that are men for each of the four years shown in the table.

   d Comment on any changes in the percentages in part **c**.

   > Compare the change with the original value.

17.2 More calculating with percentages

Sometimes you need to find amounts before a percentage change was made.

You need read the question carefully to decide whether the amount in the question is before or after the percentage change.

To calculate the amount before the change:

- work out the total percentage
- divide to find 1%
- then multiply by 100 to find the amount before the percentage change.

Worked example 1

Chang receives a bill in a restaurant for $52.80.

This includes a 10% service charge.

What is the bill before the service charge is added?

The original bill is 100%.

10% + 100% = 110%   this is the original plus the service charge

$52.80 is 110%.

$1\% = \frac{52.80}{110} = \$0.48$ divide by 110 to find the value of 1%

$100\% = 100 \times \$0.48 = \$48$ multiply by 100 to find the value of 100%

The bill is $48.

Worked example 2

A bike is reduced in a sale by 15%.

It is now priced at $595.

What was the original price?

The original price is 100%.

$100\% - 15\% = 85\%$ this is the original less the reduction

$595 is 85%.

$1\% = \frac{595}{85} = \$7$ divide by 85 to find the value of 1%

$100\% = 100 \times \$7 = \$700$ multiply by 100 to find the value of 100%

The original price was $700.

Exercise 17.2

1. A cooker is reduced in a sale by 25%.

   The sale price is $315.

   What was the original price?

2. A freezer is bought in a sale.

   The label says '20% reduction'.

   The saving is $49.

   What was the original price?

3. 14 400 people visited a museum in 2011.

   This was an increase of 20% compared with 2010.

   How many people visited the museum in 2010?

4. A jacket is advertised in a sale for $96.

   It is reduced by 40% from its original price.

   Calculate its original price.

5. Bob receives $30 interest on his savings at the end of the year.

   The interest rate advertised is 6%.

   Tax at 20% is taken off before he receives his payment.

   Show that his savings were $625.

   > Start by finding the amount of interest before tax is taken off.

6 A set of chairs is reduced in a sale by 30%.

The reduction is $45.

What is the sale price?

Review

1 Elizabeth pays a fuel bill of $320 plus tax at 6%.

How much does she pay altogether?

2 A table is reduced by 12.5% in a sale.

Its original price is $360.

What is its sale price?

3 Dimitri receives $30 interest on his savings.

He has $480 saved.

What is the percentage interest rate?

4 The number of people going to a cinema was 25 000 in 2012.

In 2013, this increased to 27 000.

What percentage increase was this?

5 The table shows Fay's scores in some tests.

Subject	Mark	Total for test
Maths	25	40
History	21	30
Geography	48	60
Music	52	80

In which subject did Fay do best? Show how you decided.

6 A desk is reduced in a sale by 40%.

The sale price is $570.

What was the price before the sale?

7 Train tickets are increased by 25%.

A train ticket now costs $36.

What was the cost of the train ticket before the increase?

Examination-style questions

1. Kumar measures the height of a plant.

   At the start of the first week it is 36 mm.

   A week later it is 63 mm.

   Work out the percentage increase in the height of the plant. [2]

2. Lola puts $150 into a bank account.

   The account pays 4% per annum simple interest.

   Work out the total amount of money in her account at the end of the year. [2]

3. A 15% service charge is added to a restaurant bill.

   What is the service charge on a bill of $45? [2]

4. Samira buys a car.

   The car loses 23% of its value in the first year.

   At the end of the first year her car is worth $6575.80

   How much did Samira pay for her car?

   Show your working. [3]

18 Planning and collecting data

Chapter 6 covers processing data
Chapter 10 covers presenting data

Learning outcomes

- Suggest a question to explore.
- Identify the sets of data needed and how to collect them.
- Identify sample sizes and degree of accuracy.
- Identify primary and secondary sources of suitable data.
- Design, trial and refine data collection sheets.
- Collect and tabulate discrete and continuous data, choosing suitable equal-width classes.

The first questionnaire

Sir Francis Galton was born in Birmingham in 1822.

He introduced the use of surveys and questionnaires for collecting information on human communities.

Questionnaires and surveys are now carried out widely.

They gain a lot of useful information that helps with planning and forming policies.

18.1 Planning

Suggest a question to explore

The question you ask before you actually start a statistical investigation will help you decide what data to collect and how to collect it.

You must make sure that your question is clear.

It is much better to ask two simple questions than to combine them into one big question.

Worked example 1

Debbie wants to do a statistics investigation on football.

Suggest:

    a two questions she could ask

    b the data needed for each one.

18 Planning and collecting data

Question 1
a Debbie could ask 'Do football teams score more goals when there are more people watching?'
b She would need attendance figures and the final score for a number of football matches.

Question 2
a Debbie could ask 'Are goalkeepers taller than other players?'
b She would need information about heights of footballers.

Worked example 2

Simon wants to do a statistical investigation on holidays.

Suggest:

a two questions he could ask
b the data needed for each one.

Question 1
a Simon could ask 'Are holidays in hotter countries more expensive than holidays in colder countries?'
b He would need to find average temperatures for a number of countries, and the cost of holidays there.

Question 2
a Simon could ask 'Do people prefer holidays in hot countries?'
b He would need to ask a number of people where they prefer to go and ask if there is a reason.

Data sources

You can use **primary data**. This is data that you collect yourself.

You can use **secondary data**. This is data that someone else has already collected.

There are lots of sources of secondary data. Many governments publish national statistics. The internet has a lot of data available. Almanacs containing data are also available (an almanac is a reference book that is normally published annually).

Sampling

The whole group you are asking about is called the **population**.

It may not be possible to ask everyone in the population.

When you ask part of the population, the part you ask is called a **sample**.

About 10% of the population will often give a sensible **sample** size. Usually, a minimum of about 30 is enough.

Methods of data collection

When you need to collect data yourself you can carry out a **survey**.

You can ask people questions by using a **questionnaire**, or by using an **interview**.

A questionnaire has printed questions and people complete the answers themselves.

In an interview, you ask them the questions yourself.

When you need to record the results of something happening, you can carry out an **experiment** or record **observations**.

When you record measurements you need to make sure you use a suitable **degree of accuracy**. For example, if you are measuring the time it takes to run in a 100-metre race and you record the time to the nearest second, you may find that a lot of the times are the same. Measuring to a tenth of a second would be a better degree of accuracy.

Worked example 3

How would you collect data to answer these questions?

   a What is the most popular type of music for the students in your class?

   b What is the most popular colour of car?

   c Is a slice of toast more likely to land butter side down when dropped?

   a Carry out a survey.

   You could ask the students in your class. Either interview each person and ask them or write a questionnaire to give them.

   b Record observations.

   You could record the colour of cars in the school car park, or as they drive past the school gates.

   c Carry out an experiment.

   You could drop a piece of buttered toast and record whether it lands butter side down or not.

Worked example 4

Juana wants to find out the height of the 20 members of her netball team.

   a Should she ask everyone or a sample?

   b What degree of accuracy should she use?

   a She should ask everyone as there are only 20.

   b Measurements to the nearest centimetre will be sufficient.

Exercise 18.1

1 Suggest one question that could be asked and the data that would be needed for:
  a an investigation into cricket
  b an investigation into buying music
  c an investigation into weather
  d an 18-year-old buying a car.

2 Would you use primary data or secondary data to find out about:
  a heights of students in your class
  b recycling habits in your village
  c recycling habits in your country
  d area of countries around the world
  e weather patterns?

3 Choose from Survey, Experiment or Observation to decide which method would be best for collecting data in each of the following:
  a finding out which type of film is most popular with students in your class
  b investigating the number of people in cars
  c investigating the number of pets people have in your village
  d finding out whether teenagers have quicker reactions than adults
  e finding out whether one fertiliser works better than another
  f finding out whether the new swimming pool is better than the old one
  g finding out how many people use the new swimming pool each day.

4 The manager of a football club wants to find out whether supporters like the catering at matches.
  a Give a reason why the manager should ask a sample of supporters.
  b How many supporters should the manager ask?

5 A company makes 1000 light bulbs every day.

They want to test them to find out how long they last.
  a Give two reasons why the company should use a sample.
  b How many should be in the sample?
  c Suggest a degree of accuracy they could use in measuring the life of the bulbs.

6 What degree of accuracy would be suitable for investigating:

  a the length of time a group of people spend exercising

  b the length of time an Olympic athlete takes to run 100 metres

  c the distance between villages

  d the amount of water used each day by families?

7 Michelle owns a beauty salon in a large town.

She is thinking about opening a second one in the same town.

In groups of two or three, discuss and then write down answers to the following:

  a Suggest a question that she should ask.

  b What data does she need to help her make the right decision?

  c Is this data primary or secondary?

  d What method of collection should she use to get the data?

  e What other factors should she consider?

> Remember the town is large, so there will be a lot of people who live or work there.

8 Kurt wants to find out whether his favourite basketball team is improving.

In groups of two or three, discuss and then write down answers to the following:

  a Suggest a question that he could ask.

  b What data does he need?

  c Is this data primary or secondary?

  d What other factors should he consider?

> There are two groups of people that may be involved – spectators and players. Both may be considered.

18.2 Data collection sheets

Observation sheets

Data collection sheets must be easy to use and allow the required data to be collected.

When you make your own data collection sheet, you may want to test it before you use it to collect data. To do this you will carry out a trial to make sure it:

- is easy to use
- actually collects the data you want.

A tally chart is most often used to collect numerical data.

- If there is a lot of different values then it is sensible to use groups.
- Groups should be equal in width.

18 Planning and collecting data

- Generally five to eight groups should be enough. More than that will give too much detail. Fewer than five will not give enough detail.
- The groups must allow the largest and smallest values to be recorded.

When you have decided what the groups should be, you need to make sure that:

- there are no gaps
- there are no overlaps.

Worked example 1

Farid is collecting information about how much pocket money students in his class receive.

He expects the smallest amount to be $3 and the largest amount to be about $21.

Make a data collection sheet that he could use.

A simple tally table will be suitable.

Pocket money, $p	Tally	Frequency
$0 < p \leq 5$		
$5 < p \leq 10$		
$10 < p \leq 15$		
$15 < p \leq 20$		
$20 < p \leq 25$		

There are lots of other groups that could be used.

For example, $2 < p \leq 7$, $7 < p \leq 12$, but it is much easier to work with the numbers in the groups given in the table.

Worked example 2

Nicola is collecting information about the height of students in her class.

She decides to use the following table.

Height, cm	Tally	Frequency
140–150		
150–160		
165–180		

Make four criticisms of the table.

There is a gap, as there is no place to record a height of 162 cm, for example.

There is an overlap, as a height of 150 cm could be placed in two groups.

The classes (groups) are not all the same width.

Three groups will not give enough detail.

Questionnaires

Questions must be easy to understand, otherwise answers will not give the information needed.

For example, for the question 'How much sport do you play?' some people may reply 'a lot' and others may say 'a few hours a week'.

Questions must not give offence.

For example, many people do not like to give their exact age. An alternative could be to offer age ranges, such as 20 to 29, 30 to 39, and so on.

Questions must not be leading.

For example, 'What do you think of the superb new swimming pool in …?' will lead most people to agree. The point of a questionnaire is to find out people's opinions, not to lead them into agreement.

Questions must be quick to answer.

People do not like to spend a lot of time answering questionnaires. Yes/no questions and questions with tick boxes as response sections are quick to answer.

Tick boxes should:

- have no gaps
- have no overlaps
- include the full range of answers
- be 'balanced' so that there are as many positive answers as negative ones for opinions.

Questionnaires are usually trialled before use.

This tests the questions to make sure they:

- are understood
- obtain the information needed.

Worked example 3

Morris wants to find out how much time students in his class spend watching television.

He considers two possible questions:

  i 'How much do you watch television?'
  ii 'How many hours do you spend watching television?'
  a Identify one problem with each of these questions.
  b Write a better question that he could use.

- **a** **i** It is not clear what sort of answer is required. Some students may reply 'lots' or 'not much'.

  ii There is no time period in the question. Some may answer for one evening, or one week, or a different period of time. Some may not remember exactly.

- **b** How many hours did you watch television last night?
  - ❏ Not at all.
  - ❏ Up to 1 hour.
  - ❏ 1 hour up to 2 hours.
  - ❏ 2 hours or more.

Exercise 18.2

1. Lucas wants to know how many people there are in cars that drive past his school.
   - **a** Design a data collection sheet he could use to collect this information.
   - **b** If you can do so safely supervised by your teacher, fill in your data sheet by counting the number of people in cars passing your school in a set amount of time.

2. Nissa wants to know whether her favourite television programme has more adverts than other programmes.
   - **a** Design a data collection sheet she could use to collect this information.
   - **b** Choose some other programmes and fill in your data sheet comparing the number of adverts in them to your favourite programme.

3. Manza grows two different types of bean plant in his garden.

   He wants to find out which type has most beans per plant.

   He knows that no plant had fewer than 9 beans and no plant had more than 51.

   > Remember to use suitable groups and that there are two types of bean plant.

   Design a data collection sheet that he could use.

4. Karen designs this data collection sheet to help her investigate how often people use the local bus service.

   How often do you use the local bus?
   - ❏ Never
   - ❏ Sometimes
   - ❏ Often
   - ❏ Always

   - **a** Make two criticisms of this data collection sheet.
   - **b** Design a better data collection sheet that Karen could use.

5 The manager at a park wants to know what visitors think of the facilities.

He asks this question in his survey.

What do you think of the facilities?

❑ Excellent

❑ Very good

❑ Good

❑ Satisfactory

❑ Poor

  a Make one criticism of the response section.

  b Write an improved question that could be used.

6 Joy wants to know how often teenagers in her village use the local swimming pool.

Write a few suitable questions that could be used as part of a questionnaire.

7 Sukina is investigating the mass of tomatoes from her greenhouse.

She uses this table to record her results.

Mass, m (g)	Tally	Frequency
5–10		
10–20		
25–30		
30–50		

Make two criticisms of the table.

8 Oliver measures the heights of some of the plants in his garden.

The heights, in centimetres, are:

23   45   43   31   50   49   12   32   59   18

16   32   37   38   61   45   37   14   26   53

Record these heights in a grouped frequency table using suitable groups.

> You can use a tally column to help.

9 A new manager starts work at a restaurant. She wants to know how often people living locally go out to eat and what type of food they like to eat.

In groups of two or three, discuss questions that she could ask.

Write down some questions that she could use in a questionnaire.

> Try not to ask too many questions, as long questionnaires are often not completed.

10 A new sports centre is to be opened in a town.

The manager wants to provide facilities that the local people will use. He uses a survey to find out what they want.

In groups of two or three, discuss what information the manager needs to know.

Write down some questions that he could use in a questionnaire.

> It will be helpful to find out whether people will take part in or watch sports, as well as which sports are required.

11 Adam is investigating how often people go to the cinema.

One of his questions is 'How often do you go to the cinema?'

   a Make two criticisms of this question.

   b Write an improved version of the question that could be used.

12 Bernice is investigating the length of tracks on her CD collection.

   a Make a data collection sheet that she could use.   *Use suitable classes.*

   b Fill in the data collection sheet for some of your CDs.

Review

1 Anton is investigating buying a bike.

   a Suggest a question that he could ask.

   b What data would be needed?

   c Would this data be primary or secondary?

2 What is the difference between primary and secondary data?

3 A company employs 451 staff.

   The manager wants find out what people think of the canteen.

   a Give one reason why she should use a sample.

   b How many people should be in her sample?

   c What data collection method should she use?

   d Would she be collecting primary or secondary data?

   e She uses a questionnaire to collect her data.

      Write two questions that she could use.

4 Hannah is investigating how much time students in her class spend reading books.

   One of the questions she asks is 'How long do you spend reading?'

   a Make two criticisms of this question.

   b Write an improved version of the question that could be used.

5 A farmer records the number of eggs laid by her hens over a 28-day period:

25	31	42	12	21	31	45	31	20	17
6	15	27	38	41	52	19	26	24	18
15	16	18	27	34	36	41	50		

Record this information in a grouped frequency table using suitable classes.

Examination-style questions

1 30 students take a maths test.

The marks are out of 50.

Lanika makes up this table to show the marks.

Marks	Frequency
0–20	3
20–25	12
25–40	14
40–50	1
Total	30

Make two criticisms of the table. [2]

2 Wesley is carrying out a survey to find out how much time students spend listening to the radio.

He designs a questionnaire.

This is one of the questions:

How many hours do you spend listening to the radio per week?

Please tick

❑ 1 to 5

❑ 5 to 10

❑ 10 to 20

❑ More than 20

Make two criticisms of Wesley's question. [2]

19 Ratio and proportion

> **Learning outcomes**
> - Interpret and use ratio in a range of contexts; compare two ratios.
> - Recognise when two quantities are directly proportional; solve problems involving proportionality, e.g. converting between different currencies.

All change

Historically, countries have developed their own currencies, or money systems.

Some, like the US dollar ($) or the British pound (£), are well known and have widely recognised symbols.

Most countries have banks or offices where money can be changed from one currency to another. This exchange is based on the currencies being directly proportional to each other.

These banks or offices are often found at ports, airports and main railway stations where people are likely to arrive in a country.

19.1 Ratio review

Ratio is used to compare one quantity with another.

You have already learned how to calculate with ratios by using shares or portions.

Red beads and white beads mixed in the ratio 3 : 7 means 3 portions of red beads and 7 portions of white beads.

There are 10 portions altogether.

$\frac{3}{10}$ of the beads are red and $\frac{7}{10}$ of the beads are white.

Mathematics for Cambridge Secondary **1**

These three examples will remind you of the methods you have learned.

Worked example 1

Divide $2352 in the ratio $3:4:1$

$3 + 4 + 1 = 8$	add the ratios to find the total number of portions
$2352 \div 8 = 294$	divide to find the size of each portion
$294 \times 3 = 882$	multiply to find the size of each share
$294 \times 4 = 1176$	
$294 \times 1 = 294$	

The three amounts are $882, $1176 and $294.

> Check that the answer is correct by adding the three amounts.
> $882 + 1176 + 294 = 2352$ ✓

Worked example 2

The ratio of boys to girls on a school trip is $3:2$

There are 46 girls on the school trip.

How many boys are there?

The ratios add up to 5 so there are 5 portions altogether.

The girls represent 2 of these portions.

2 portions = 46	start with what you know
1 portion = $46 \div 2 = 23$	divide to find one portion
3 portions = $23 \times 3 = 69$	multiply to find the required total

There are 69 boys on the trip.

Worked example 3

A bag contains red beads and blue beads in the ratio $4:5$

What fraction of the beads is red?

$4 + 5 = 9$	add the ratios to find the total number of portions

4 out of the 9 portions are red.

$\frac{4}{9}$ of the beads are red.

Exercise 19.1

1. Cancel these ratios down to their simplest form.

   a 15 : 25   b 36 : 54   c 81 : 45   d 56 : 48   e 80 : 32

2. Simplify these ratios. Give the answers in their simplest form.

   a 400 g : 1.2 kg
   b 1.5 m : 40 cm
   c 2 days : 8 hours
   d 600 ml : 2.4 l
   e 25 mm : 4 cm
   f 450 m : 1 km

   Remember to convert both sides of the ratio to the same units.

3. Write these ratios in their simplest form.

   a 36 : 45 : 27   b 12 : 24 : 18   c 0.5 : 1.5 : 2   d 150 : 240 : 570

4. A cook has three bags of flour.

   Bag A has a mass of 500 g, bag B has a mass of 750 g and bag C has a mass of 1.5 kg.

   Work out the ratio of the mass of bag A to bag B to bag C.

   Simplify your answer.

5. Divide these quantities in the given ratio.

   a $400 in the ratio 4 : 1
   b 120 cm in the ratio 3 : 2
   c $38.70 in the ratio 5 : 4
   d $350 in the ratio 3 : 2
   e 576 km in the ratio 3 : 5
   f 660 m in the ratio 10 : 1

6. Chris, Sue and Liz share $1050 in the ratio 3 : 2 : 1

   Work out how much each receives.

7. The angles of a triangle are in the ratio 5 : 4 : 3

   Work out the size of the smallest angle in degrees.

8. The angles of a quadrilateral are in the ratio 2 : 3 : 3 : 7

   Work out the size of the largest angle in degrees.

9. The ratio of adults to children on a bus is 1 : 9

   What fraction of the people on the bus are children?

10. $\frac{5}{8}$ of the children at a school are girls.

    Work out the ratio of girls to boys.

11. A piece of ribbon is cut in the ratio 5 : 3

    The shorter piece is 45 cm long.

    How long is the longer piece?

12 A bag contains black beads and white beads in the ratio 7 : 4

There are 42 black beads in the bag.

Work out how many beads there are altogether in the bag.

13 Axel and Mimi's ages are in the ratio 3 : 5

Mimi is 6 years older than Axel.

   a Work out how old they each are.

   b What will be the ratio of their ages in one year's time? Simplify your answer.

   > They will both be one year older than the answer to part **a**.

14 There is a mixture of red, green and blue balls in a bag.

There is the same number of green and blue balls.

The probability of picking a red ball from the bag is $\frac{3}{7}$

Work out the ratio of red balls : green balls : blue balls.

   > Work out the probability of picking blue and green balls. These probabilities must be equal.

15 The scale of a map is given as 1 : 25 000

   a A road is 6.4 cm long on the map.

   Work out the actual length of the road in kilometres.

   > Work out the distance in centimetres first.

   b An airport runway is 2.5 km long.

   How long is it on the map?

   > First change the kilometres into centimetres.

   c A rectangular plantation is 3 cm by 1.6 cm on the map.

   Work out the area of the plantation in hectares.

   > 1 hectare (ha) = 10 000 m^2

19.2 Comparing ratios

Equivalent ratios

It is often useful to be able to compare ratios.

To show that ratios are equivalent, you can cancel them down to their simplest form.

If they have the same simplest form they are **equivalent ratios**.

> **Worked example 1**
>
> Compare the ratios 35 : 84 and 60 : 144 to see if they are equivalent.
>
> Simplifying each ratio in turn gives:
>
>   35 : 84 = **5 : 12**   divide both numbers by 7
>
> 60 : 144 = 30 : 72   divide both numbers by 2
>
>   = **5 : 12**   divide both numbers by 6
>
> > This simplification could be done in one step by dividing by 12.
>
> The two ratios are equivalent as they are both equal to 5 : 12

19 Ratio and proportion

Unitary ratios

If the ratios are different, the simplest forms are not always easy to compare.

In these cases there are two methods to use.

The first of these is by converting to a **unitary ratio**.

A unitary ratio is a ratio in the form $1:n$ or $n:1$

So far you have only simplified ratios using whole numbers.

For unitary ratios, the value of n is often not a whole number.

Worked example 2

Bidtown School has 565 students and 27 teachers.

Clayford School has 840 students and 37 teachers.

Which school has the lower student-to-teacher ratio?

The student-to-teacher ratio at Bidtown School is $565:27$

This cannot be simplified as 565 and 27 have no common factors.

> The prime factors of the two numbers are $565 = 113 \times 5$ and $27 = 3 \times 3 \times 3$. They have no factors in common.

The student-to-teacher ratio at Clayford School is $840:37$

This cannot be simplified as 840 and 37 have no common factors.

> The prime factors of 840 are $2 \times 2 \times 2 \times 3 \times 5 \times 7$ and 37 is a prime number. They have no factors in common.

The ratios are not equivalent as they have different simplest forms.

Convert both ratios to unitary ratios.

Bidtown School: $565:27 = \frac{565}{27}:1$ divide both numbers by 27

$\qquad\qquad\qquad = \mathbf{20.9:1}$ to 1 decimal place

Clayford School: $840:37 = \frac{840}{37}:1$ divide both numbers by 37

$\qquad\qquad\qquad = \mathbf{22.7:1}$ to 1 decimal place

Bidtown School has 20.9 students per teacher and Clayford School has 22.7 students per teacher.

Bidtown School has the lower student-to-teacher ratio.

Using percentages

Another method is to convert the ratios to percentages.

This can be helpful when you need to know one quantity as a percentage of another.

To do this you write them as fractions and convert the fractions to percentages.

Worked example 3

Raymond mixes 240 ml red paint and 560 ml of white paint.

Georgia mixes 0.75 l red paint and 1.25 l white paint.

Who has the higher proportion of red paint?

Raymond's mixture has 240 ml of red paint out of a total of 800 ml.

This is $\frac{240}{800}$ as a fraction.

$\frac{240}{800} = 240 \div 800 = 0.3$

$0.3 = 0.3 \times 100 = \mathbf{30\%}$

> Remember – convert the fraction to a decimal, then change that to a percentage.

Georgia's mixture has 0.75 l of red paint out of a total of 2 l.

This is $\frac{0.75}{2}$ as a fraction.

$\frac{0.75}{2} = 0.75 \div 2 = 0.375$

$0.375 = 0.375 \times 100 = \mathbf{37.5\%}$

Georgia's mixture has the higher proportion of red paint.

Exercise 19.2

1. Simplify each of these pairs of ratios and state whether they are equivalent.

   a 15 : 25 and 12 : 18   b 36 : 16 and 45 : 20   c 25 : 40 and 35 : 55

   d 21 : 56 and 18 : 48   e 20 : 30 and 32 : 48   f 48 : 60 and 28 : 35

2. Write each of these ratios in the form 1 : n

   a 4 : 7   b 5 : 8   c 6 : 3

   d 8 : 10   e 10 : 8

3. On school trips the student-to-teacher ratio must be less than 9 : 1

   A day trip has 128 students and 15 teachers.

   a Work out the student-to-teacher ratio for this trip in the form n : 1

   b Are there enough teachers on the trip?

4. Deniz mixes 28 kg soil with 7 kg fertiliser.

   Hamza mixes 63 kg soil with 17 kg fertiliser.

   a Work out the percentage of fertiliser in each mixture.

   b Who has the higher proportion of fertiliser?

5. Two gardeners mix seeds for their lawns.

   Danielle mixes 5 kg grass with 2 kg ryegrass.

   Henry mixes 3.6 kg grass with 1.4 kg ryegrass.

a Work out the percentage of ryegrass in each mixture.

b Whose mixture has the higher proportion of ryegrass?

6 Alex and Milo are mixing fruit drinks for a party.

Alex mixes 1.8 litres of fruit juice with 12 litres of water.

Milo mixes 2.5 litres of fruit juice with 17.5 litres of water.

Work out the proportion of juice in each of the drinks.

7 Jennifer is 10 years old and Andrew is 7 years old.

Their grandmother has some pocket money for them to share.

She suggests two ways they can share the money.

One way to share it is in the ratio of their ages.

The other way to share it is in the ratio of the number of letters in their name.

Which method do you think Jennifer would prefer? Explain why.

> First work out what fraction she would receive using each method. Then use decimals or percentages to compare them.

8 Here are four ratios:

21 : 35 : 14   36 : 60 : 24   27 : 45 : 15   45 : 75 : 30

Three of these ratios are equivalent and one is different.

> Simplify each ratio.

Work out which ones are equivalent and which is the odd one out.

You must show all your working.

19.3 Proportionality

Direct proportion

If two variables are always in the same ratio, they are said to be in **direct proportion**.

For example, if a car uses fuel at a constant rate, the distance travelled is directly proportional to the amount of fuel used. When you drive four times as far, you use four times as much fuel.

You can use multiplication or division to solve most problems involving direct proportion.

In Chapter 16, you used graphs to solve these types of problems.

Sometimes it is more convenient to use the ratio of the two variables.

Worked example 1

Alphonso drives 144 km and uses 9 litres of fuel.

Work out how much fuel he would need when driving 400 km.

144 km uses 9 litres of fuel.

1 km uses $\frac{9}{144}$ litres of fuel.

> The unitary method: divide to find the amount used for 1 km then multiply to find the answer.

277

400 km uses $\frac{9}{144} \times 400$

$$= \frac{9 \times 400}{144}$$
$$= \frac{3600}{144}$$
$$= 25$$

He would need 25 litres of fuel.

Worked example 2

Here are two boxes of breakfast cereal.

Crunchios Cereal
Standard pack
300 grams
Contains
45 g of nuts

Crunchios Cereal
Value pack
440 grams
Contains
66 g nuts

Is the mass of the nuts in each pack directly proportional to the mass of the pack?

The ratio of nuts to pack size for each is:

| Standard pack | 45 : 300 = 3 : 20 | divide by 15 to simplify the ratio |
| Value pack | 66 : 440 = 3 : 20 | divide by 22 to simplify the ratio |

The ratio is the same for each pack.

The mass of nuts is directly proportional to the mass of the pack.

Worked example 3

Anderson is paid $62.80 for 8 hours' work.

Savannah is paid $39.25 for 5 hours' work.

Work out whether they are on the same rate of pay.

For each of them the pay is directly proportional to the hours worked.

The pay per hour for each person is:

| Anderson | 62.80 ÷ 8 = 7.85 | Anderson is paid $7.85 per hour. |
| Savannah | 39.25 ÷ 5 = 7.85 | Savannah is paid $7.85 per hour. |

They are both on the same rate of pay.

19 Ratio and proportion

Exchange rates

In the previous example, the word 'rate' was used to describe the proportionality between two different units.

A common type of rate is an exchange rate between two different currencies.

The amounts of money in two different currencies are in direct proportion.

Worked example 4

Alison changes $120 into euros and receives €91.20.

a Work out the exchange rate from dollars to euros.

At the same time, Kevin changes $250 into euros.

b Work out how much he receives.

a The exchange rate is the amount received for 1 dollar.

$120 = €91.20

$1 = €91.20 ÷ 120

  = €0.76

The exchange rate is $1 = €0.76

b Using the exchange rate from part **a**:

250 × 0.76 = 190

Kevin receives €190.

Exercise 19.3

1 7 metres of material cost $34.65.

   Work out the cost of 12 metres of material at the same price.

2 3 kilograms of flour cost $10.20.

   Work out the cost of 5 kilograms of flour at the same price.

3 William is paid $41.25 for $5\frac{1}{2}$ hours' work.

   How much is he paid for working 8 hours?

4 55 litres of fuel cost $52.80.

   Calculate the cost of 68 litres of fuel at the same price.

5 x and y are two variables.

   When $x = 12, y = 40$

   When $x = 30, y = 100$

   Is x directly proportional to y? Show your working.

6 Two jars of strawberry jam are being compared.

Fruity Jam
Large
720 g
Contains
324 g fruit

Fruity Jam
Small
500 g
Contains
225 g fruit

The small jar has 225 g of fruit in 500 g of jam.

The large jar has 324 g fruit in 720 g of jam.

Is the mass of fruit directly proportional to the mass of jam?

You must show your working.

7 The table shows the quantity of sugar in different-sized boxes of breakfast cereal.

	Pack A	Pack B	Pack C	Pack D
Mass of sugar	75 g	150 g	100 g	54 g
Mass of cereal	500 g	1 kg	700 g	360 g

a Calculate the percentage of sugar in each box of cereal.

b Is the mass of sugar directly proportional to the mass of cereal?

If they are directly proportional, the percentages will all be the same.

8 Antoine is paid $101.20 for 8 hours' work.

Claudie is paid $135.85 for 11 hours' work.

a Work out their rates of pay.

b Are they both on the same rate of pay?

9 The exchange rate between the US dollar and the Saudi riyal is $1 = 3.75 riyals.

a Convert $120 into Saudi riyals.

b Convert 200 riyals into dollars. Give the answer to the nearest cent.

10 Warren takes a holiday from the USA to Chile.

He changes $400 into Chilean pesos at the start of his holiday.

The exchange rate between the US dollar and the Chilean peso is $1 = 470 pesos.

a Calculate how many Chilean pesos he has.

At the end of his holiday, he still has 6500 pesos left.

He changes them back to dollars at the same exchange rate.

b Work out how many dollars he gets, to the nearest cent.

19 Ratio and proportion

11 Jack and Ella travel to South Africa.

Jack converts $250 into 2225 South African rands.

Ella converts $320 at the same rate.

Work out how many rands she receives.

12 At the start of a holiday to Thailand, Danut converts $400 into 11 840 Thai bahts.

At the end of his holiday he has 2500 bahts left.

He converts them back to dollars at the same exchange rate.

Work out how much he gets back, to the nearest cent.

Use the original amounts to work out the dollar to baht exchange rate.

13 This internet advertisement offers a computer for £240 or $375.

The exchange rate is $1 = £0.61

Work out whether the computer is cheaper in dollars or pounds.

You can work in either pounds or dollars.

Netbook computer

Buy it online in dollars or pounds

$375 OR £240

14 Freya travels from the USA to Great Britain.

She changes $360 to pounds at a rate of $1 = £0.65

She now travels to Germany and changes all of her money to euros.

The exchange rate from pounds to euros is £1 = €1.19

a Work out how many euros she now has.

b The exchange rate from dollars to euros is $1 = €0.76

Work out how much she would have if she had converted her dollars directly to euros.

Show your working clearly. First change $360 to pounds. Then change those pounds to euros.

Review

1 Write these ratios in their simplest form.

    a 12 : 21   **b** 24 : 56   **c** 72 : 27

    d 21 : 33 : 15   **e** 48 : 28 : 36

2 Simplify these ratios. Give the answers in their simplest form.

    a 4 days : 2 weeks   **b** 1.5 kg : 500 g

    c 750 ml : 2 l   **d** 8 mm : 2 cm

3 Divide these quantities in the given ratio.

   a $320 in the ratio 5 : 3   **b** 160 cm in the ratio 7 : 1   **c** $176 in the ratio 8 : 3

4 The angles of a quadrilateral are in the ratio 3 : 4 : 6 : 7

   Work out the size of the largest angle.

5 A 2.4 m piece of wood is cut into two pieces in the ratio 3 : 2

   Work out the length of each new piece.

   Give your answer in centimetres.

6 A bag contains red beads and black beads in the ratio 5 : 2

   What fraction of the beads are red?

7 On a bus, $\frac{3}{4}$ of the people are male.

   Work out the ratio of males to females.

8 Ann and Graham share a prize in the ratio 5 : 2

   Ann receives $369 more than Graham.

   How much was the prize money altogether?

9 Simplify these pairs of ratios and state whether they are equivalent.

   a 28 : 35 and 36 : 45   **b** 54 : 36 and 72 : 36   **c** 63 : 27 and 28 : 12

10 a Write these ratios in the form n : 1. If necessary, round your answers to 1 decimal place.

      i 12 : 8   **ii** 5 : 3   **iii** 5 : 4   **iv** 9 : 6   **v** 8 : 3

   b Which two of the ratios in part **a** are equivalent ratios?

11 Barney mixes 375 ml of fruit juice with 3 l of water.

   Harriet mixes 600 ml of fruit juice with 5 l of water.

   a Work out the percentage of fruit juice in each mixture.

   b Who has the higher proportion of fruit juice in their mixture?

12 A school skiing trip has 51 students and 4 teachers.

   A sailing trip has 87 students and 7 teachers.

   a Work out the student-to-teacher ratio for the skiing trip in the form n : 1

   b Work out the student-to-teacher ratio for the sailing trip in the form n : 1

   c Which party has the higher student-to-teacher ratio?

13 8 metres of material cost $39.60.

   Work out the cost of 15 metres of material at the same price.

14 Isabelle and Michel are both paid at the same rate of pay.

   Isabelle is paid $53.20 for 7 hours' work.

   Work out how much Michel is paid for 5 hours' work.

15 x and y are two variables.

When $x = 45, y = 12$

When $x = 30, y = 8$

Is x directly proportional to y? Show your working.

16 The exchange rate between the US dollar ($) and the British pound (£) is $1 = £0.65

   a Convert $300 into pounds.

   b Convert £275 into dollars. Give your answer to the nearest cent.

17 Jane, Gillian and Neil travel from the USA to Norway.

They all exchange some dollars into Norwegian krone at the same exchange rate.

Jane converts $175 into 1015 krone.

   a Gillian converts $250 into krone.

      Work out how much she receives.

   b Neil converts his money and receives 2320 krone.

      How many dollars did he exchange?

Examination-style questions

1 Alan mixes red paint with white paint in the ratio 5 : 1

   a Work out how much red paint he mixes with 2 litres of white paint. [1]

   b Work out how much white paint he mixes with 2.5 litres of red paint. [1]

2 On a railway ride, the ratio of adults to children is 4 : 3

There are 120 children on the ride.

Work out how many people there are on the ride altogether. [2]

3 Write the ratio 63 : 35 : 49 in its simplest form. [1]

4 Samarah mixes 2.4 kg of gravel, 1.5 kg of sand and 600 g of cement together.

Write the ratio gravel : sand : cement in its simplest form. [2]

5 The exchange rate between the US dollar ($) and the Indian rupee (INR) is $1 = 53.8 INR.

   a Convert $250 into rupees. [1]

   b Convert 7500 INR into dollars. Give your answer to the nearest cent. [1]

6 This diagram shows a pattern of black and white squares.

Write the ratio of black squares : white squares in its simplest form. [1]

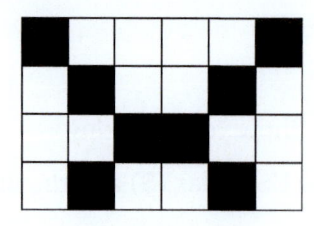

7 Here are the ingredients for making shortbread biscuits.
These quantities are enough to make 12 biscuits.

Shortbread biscuits
180g plain flour
60g sugar
120g butter

Makes 12 biscuits

  a Pedro wants to make 24 biscuits.

     How much sugar does he need? [1]

  b Jacques is making biscuits for 6 people.

     How much butter does he need? [1]

  c Gretel is making biscuits for 20 people.

     Work out how much flour she needs. [2]

8 A drink is made by mixing lemon juice and water in the ratio 2 : 7

  a Work out how much water is mixed with 800 ml of lemon juice. [1]

  b Gabriela wants to make 36 litres of the drink for a party.

     Work out how many litres of lemon juice are needed. [1]

9 Here are four ratios:

    9 : 21   30 : 80   24 : 56   60 : 140

Three of these ratios are equivalent.

Work out which one is different from the others.

You must show your working. [2]

10 The exchange rate between the US dollar ($) and the euro (€) is $1 = €0.79

  a Convert $420 into euros. [1]

  b Convert €160 into dollars. Give your answer to the nearest cent. [1]

20 Time and rates of change

Learning outcomes

- Solve problems involving average speed.
- Use compound measures to make comparisons in real-life contexts, e.g. travel graphs and value for money.

Up to speed

In December 1898 the first world land speed record was set.

Gaston de Chasseloup-Laubat of France became the first record holder.

His speed was measured at 63.15 km/h.

A record of 1227.986 km/h was set in the Black Rock desert in the USA in 1997.

It was the first time a vehicle on land had gone faster than the speed of sound.

20.1 Average speed

Constant speed

Speed is the rate at which something moves.

At a constant speed, the distance moved is directly proportional to the time.

You can use proportionality to solve problems involving speed.

Worked example 1

A car travels at a constant speed.

It goes 165 km in 3 hours.

Work out the speed of the car in km/h.

Mathematics for Cambridge Secondary 1

Speed in km/h means the distance travelled in 1 hour.

In 3 hours it goes 165 km.

In 1 hour it goes 165 ÷ 3 = 55 km   *divide by 3 to find the distance travelled in 1 hour*

The speed is 55 km/h.

This is the same as the unitary method for direct proportion.

Average speed

From the previous example, you can see that the relationship between speed, distance and time is:

$$\text{speed} = \frac{\text{distance}}{\text{time}}$$

But it is rare for anything to travel at a constant speed.

Instead we use **average speed**, which is based on a whole journey:

$$\text{average speed} = \frac{\text{total distance travelled}}{\text{time taken}}$$

Using s = average speed, d = total distance travelled and t = time taken, the formula is:

$$s = \frac{d}{t}$$

This formula can be re-arranged to make d or t the subject. The two new formulae are:

$$d = s \times t \quad \text{and} \quad t = \frac{d}{s}$$

This triangle will help you to remember the formulae.

To find the formula for speed, cover up the letter s.

The triangle now shows $\frac{d}{t}$. This tells you that $s = \frac{d}{t}$

If you cover t, the triangle shows $\frac{d}{s}$, so $t = \frac{d}{s}$

If you cover d, the triangle shows st, so $d = st$ *$d = st$ is the same as $d = s \times t$*

Worked example 2

A toy car takes 12 seconds to travel 96 cm.

Work out its average speed in cm/s.

Using letters, $t = 12$ and $d = 96$

The formula to find the average speed is $s = \dfrac{d}{t}$

$s = \dfrac{96}{12}$

$= 8$

The average speed is 8 cm/s.

Converting between hours and minutes

The times 3 hours 30 minutes, 3.5 hours and $3\tfrac{1}{2}$ hours are all the same.

Sometimes you need to convert between them.

You know how to change fractions to decimals and decimals to fractions.

To change minutes into hours, you divide by 60.

To change hours into minutes, you multiply by 60.

> You only divide or multiply the decimal or minutes part of the time. The whole number of hours stays the same.

Worked example 3

a Convert 1 h 18 min to decimal form.

b Convert 4.7 hours to hours and minutes.

a The whole number of hours will stay the same.

18 minutes $= \dfrac{18}{60}$ hour

$= 0.3$ hour

Add the 1 hour back on, so 1 h 18 min = 1.3 hours

b 4.7 hours is 4 hours and 0.7 hour.

0.7 hour $= 0.7 \times 60$ minutes

$= 42$ minutes

4.7 hours = 4 h 42 min

Mathematics for Cambridge Secondary 1

Worked example 4

A train travels 192 km from London to Bristol at an average speed of 80 km/h.

Work out the journey time in hours and minutes.

$d = 192$ and $s = 80$

Use the formula $t = \frac{d}{s}$

$t = \frac{192}{80}$

$= 2.4$

Journey time $= 2.4$ hours

$\qquad = 2$ hours and 0.4 hour

$\qquad = 2$ hours and 0.4×60 minutes

$\qquad = 2$ hours and 24 minutes

Exercise 20.1

1. Find the average speeds of objects that move:
   - **a** 150 km in 3 hours. Answer in km/h.
   - **b** 100 m in 10 seconds. Answer in m/s.
   - **c** 2475 km in 9 hours. Answer in km/h.
   - **d** 27 km in 6 hours. Answer in km/h.
   - **e** 16.5 km in 3 h 18 min. Answer in km/h.
   - **f** 23 km in 20 min. Answer in km/h.

2. Find the distance travelled by objects that travel:
   - **a** 85 km/h for 4 hours
   - **b** 435 km/h for 6.5 hours
   - **c** 23 m/s for 24 seconds
   - **d** 96 km/h for $3\frac{1}{2}$ hours
   - **e** 75 cm/s for 0.8 second
   - **f** 125 km/h for 4 h 54 min.

3. Calculate the time taken for the following journeys.
   - **a** 300 km at 75 km/h
   - **b** 160 km at 64 km/h
   - **c** 84 m at 28 m/s
   - **d** 22.5 km at 3.75 km/h
   - **e** 3750 km at 600 km/h
   - **f** 19 km at 4 km/h

4. A train travels 130 km in $2\frac{1}{2}$ hours. Work out its average speed.

5. The distance from London to Madrid is 1260 km.

   A plane flies from London to Madrid in 2 h 15 min.

   Work out the average speed of the plane.

6 On a holiday, Jane cycles 27.3 km at an average speed of 6.5 km/h.

Work out her journey time in hours and minutes.

7 The winner of a cycle race covers 209 km in 3 hours 48 minutes.

What is the average speed?

8 An aeroplane flies for 2 hours 24 minutes at 360 km/h.

Work out how far it flies in that time.

9 A car travels at 84 km/h for $1\frac{1}{2}$ hours and then at 66 km/h for 2 hours.

Work out how far it has travelled altogether.

10 A train travels 144 km at 80 km/h and then a further 130 km at 50 km/h.

Work out the average speed for the entire journey.

Give the answer to 1 decimal place.

> Work out the total distance and total time first.

11 A car and a bicycle set off together from the same point.

They follow the same route for $3\frac{1}{2}$ hours.

> Work out how far they have each travelled.

The car's average speed is 56 km/h and the bicycle's average speed is 16 km/h.

Work out the distance between them after $3\frac{1}{2}$ hours.

12 Roy walks 12 km to his friend's house at an average speed of 6 km/h.

His average speed for the journey to his friend's house and home again is 4 km/h.

How many hours does he take to walk home?

> Work out how long the walk to his friend's house takes. Then work out how long it takes altogether for the combined journey there and back.

20.2 Compound measures

A **compound measure** is one that involves two different quantities.

Average speed is an example of a compound measure as it is the distance travelled in a certain time.

Other examples include:

- pressure, measured in grams per square centimetre
- rate of flow of a liquid, measured in litres per second
- price of a commodity, measured in dollars per gram or dollars per litre.

There are many others.

Many compound measures involve the word 'per'.

This tells you that it is one quantity divided by another.

289

Mathematics for Cambridge Secondary 1

Worked example 1

Detergent comes in two sizes.

Bottle A holds 0.75 litre and costs $2.55.

Bottle B holds 1.6 litres and costs $5.25.

Work out which is the better value for money.

Method 1 Find the lower cost per litre.

Bottle A is $2.55 for 0.75 l.

　Cost per litre = 2.55 ÷ 0.75 = 3.4

　Bottle A costs $3.40 per litre.

Bottle B is $5.25 for 1.6 l.

　Cost per litre = 5.25 ÷ 1.6 = 3.28125

　Bottle B costs $3.28 per litre (to the nearest cent).

Bottle B costs less per litre, so it is better value for money.

Method 2 Find the higher number of litres per dollar.

Bottle A is 0.75 l for $2.55.

　Litres per dollar = 0.75 ÷ 2.55 = 0.294 (to 3 decimal places)

　Bottle A gives 0.294 litre per dollar.

Bottle B is 1.6 l for $5.25.

　Litres per dollar = 1.6 ÷ 5.25 = 0.305 (to 3 decimal places)

　Bottle B gives 0.305 litre per dollar.

Bottle B gives more litres per dollar, so it is better value for money.

Using graphs

Here is a **travel graph**. It is also called a **distance–time graph**.

It shows a cycle ride taken by Sue

She left home at 08 00 and returned at 14 00.

From the graph you can work out her speed between 08 00 and 09 00.

She travels 15 km in 1 hour. Her speed at that time is 15 km/h.

But what is her average speed?

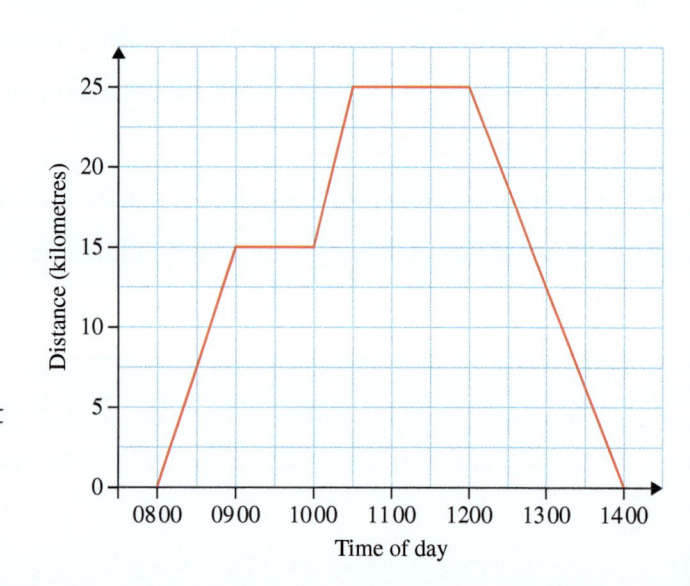

20 Time and rates of change

She rides 25 km and then back again. Her total distance is 50 km.

During the day she rests for $2\frac{1}{2}$ hours and rides for $3\frac{1}{2}$ hours.

Her average cycling speed is 50 ÷ 3.5 = 14.29 km/h (rounded to 2 decimal places).

Worked example 2

Two mobile phone companies are advertising special offers.

Fun Phones
Flat rate tariff
Just $15 per month
Unlimited calls

Phone Home
Special Offer
30 minutes free per month then only $0.10 per minute

a Show both these rates on a graph.

The horizontal axis has call times up to 200 minutes per month.

The vertical axis has cost up to $20.

b After how many minutes per month is Fun Phones better value for money?

a Here are the axes described in the question.

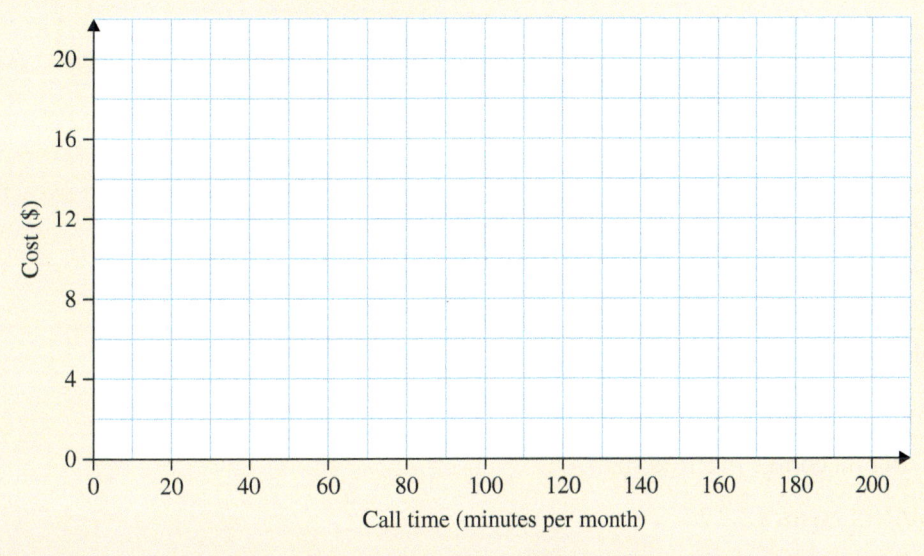

This table shows the cost for each company for different numbers of minutes per month.

Time (minutes)	0	30	60	100	150	200
Fun Phones cost ($)	15	15	15	15	15	15
Phone Home cost ($)	0	0	3	7	12	17

Plot these points on the graph and join them up with straight-line sections.

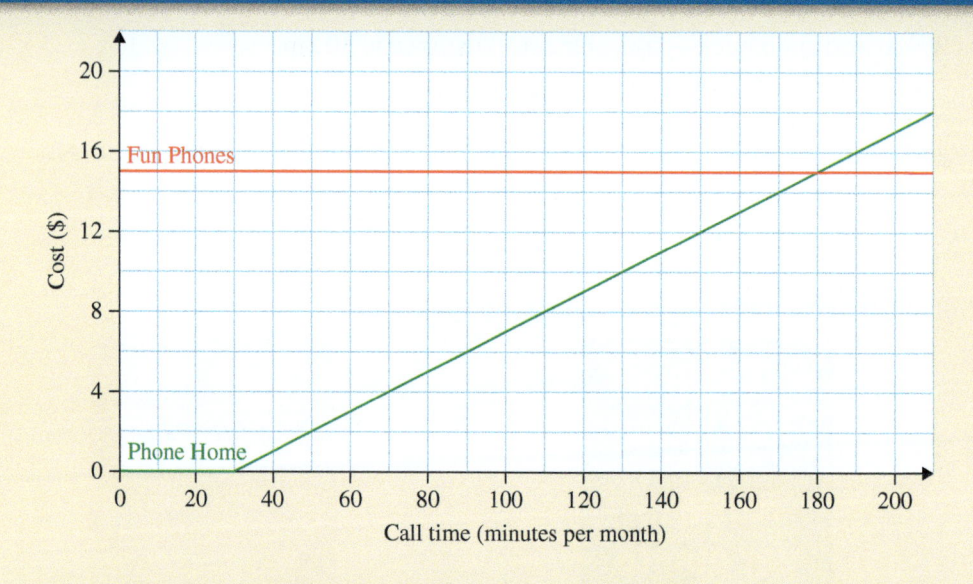

b The graph shows the lines crossing at 180 minutes per month.

To the left of this point the green line is lower, so Phone Home is cheaper.

To the right of this point the red line is lower, so Fun Phones is cheaper.

Fun Phones is better value if you use more than 180 minutes per month.

Exercise 20.2

1 Kitties cat food is sold in two different sizes.

A 420 g pack costs $1.26 and the larger 600 g pack costs $1.98.

Work out which is the better value for money.

Show all your working.

2 Cooking oil comes in three sizes.

   500 ml costs $3.19

   1 litre costs $5.99

   2.4 litres cost $14.49

Work out which is the best value for money.

You must show all your working.

3 Rafael leaves home at 09 30 to visit a village 80 km from his home town.

This graph shows his day out.

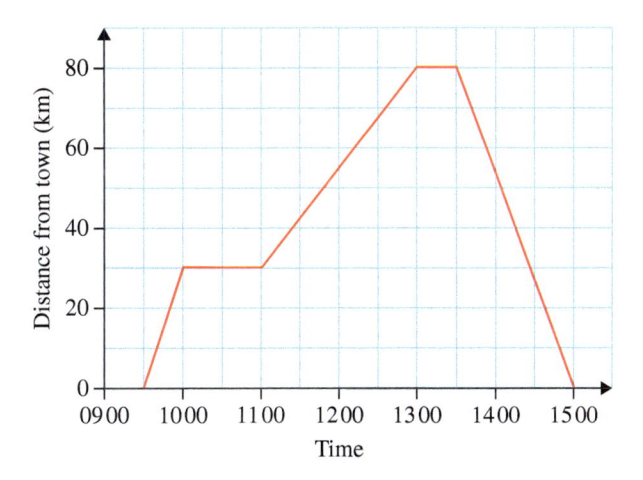

a Work out his speed between 11 00 and 13 00.

b On which part of his journey is he travelling fastest? How can you tell?

c Work out his average driving speed for the whole day.

4 A supermarket has a special offer on bread flour of 1.5 kg flour for $0.89.

At the flour mill they sell a 25 kg sack of flour for $15.

Which is the better value for money?

5 This graph shows the amount of water in a bath as it fills and then empties.

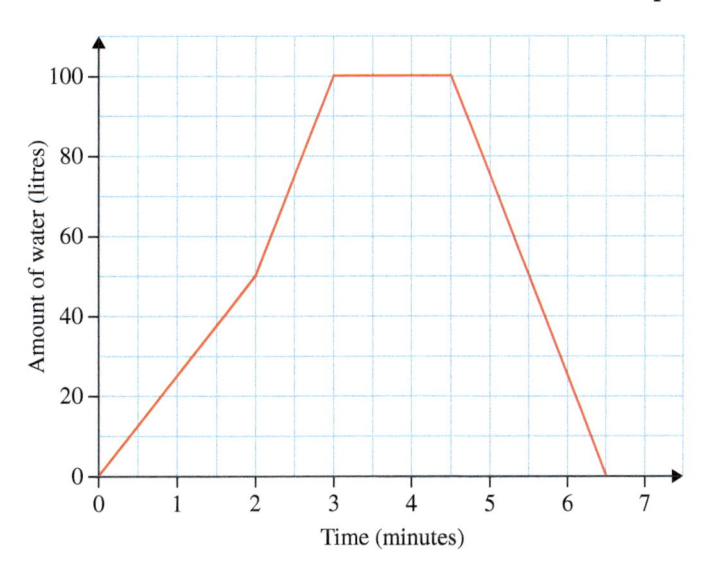

a How much water is in the bath when it is full?

b How much water flows into the bath in the first 2 minutes.

c Work out the rate of flow in the first 2 minutes in litres per minute.

d Work out the rate of flow as the bath empties in litres per minute.

6 The graph shows the price of posting parcels with company X.

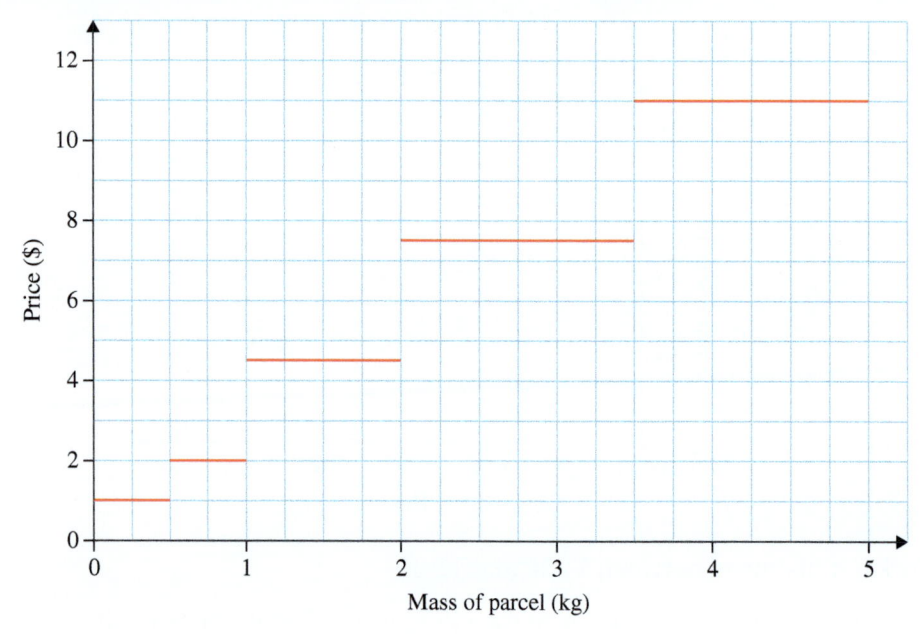

It shows, for example, that parcels up to 500 g cost $1 to send.

a Write down the cost of sending parcels of the following masses with company X.

  i 900 g   ii 1.4 kg   iii 3.4 kg   iv 3.75 kg

Company Y offers a flat rate of $2.50 per kilogram.

Sending a parcel of mass 3.1 kg would cost 3.1 × 2.50 = $7.75

b Work out the cost of sending the four parcels in part a with company Y.

c For each parcel, write down which company is cheaper and by how much.

7 The density of a solid is a compound measure of relative mass.

It is calculated by dividing the mass in grams by the volume in cm^3.

The units of density are grams per cm^3 and are written as g/cm^3.

a Calculate the density of these three cuboids in g/cm^3.

> Use the formula volume = length × width × height

i Mass = 420 g, 5 cm × 4 cm × 2 cm

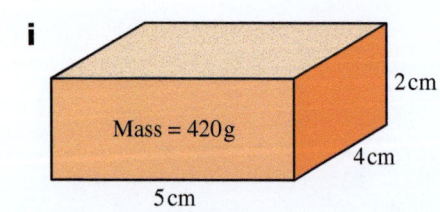

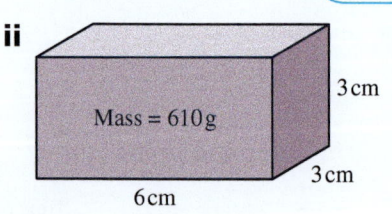

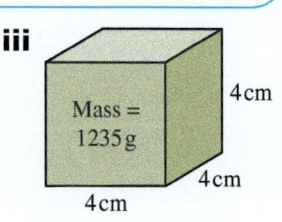

ii Mass = 610 g, 6 cm × 3 cm × 3 cm

iii Mass = 1235 g, 4 cm × 4 cm × 4 cm

Here is a table showing the density of some common metals.

Aluminium	Chromium	Copper	Gold	Iron	Lead	Silver	Titanium
2.7 g/cm³	7.15 g/cm³	8.96 g/cm³	19.3 g/cm³	7.86 g/cm³	11.3 g/cm³	10.49 g/cm³	4.51 g/cm³

b Write down what type of metal each of the cuboids is made of.

> Decide which metal has a density closest to the answer for each cuboid in part a.

Review

1 Work out the average speed in km/h of vehicles that move:

　a 210 km in $3\frac{1}{2}$ hours

　b 168 km in 4 hours

　c 17 km in 3 hours 24 minutes

　d 285 km in 3 hours 48 minutes.

2 Find the distance moved by objects that travel:

　a 56 km/h for 6 hours　　**b** 12 km/h for $4\frac{1}{2}$ h

　c 65 m/s for 12 seconds　**d** 250 km/h for 3 h 18 min.

3 Work out the time taken for the following movements. Write the answers in hours and minutes.

　a 400 km at 50 km/h　　**b** 135 km at 27 km/h

　c 240 km at 50 km/h　　**d** 23.5 km at 2.5 km/h

4 A snail moves 23.4 metres in 30 minutes.

　a Work out its average speed in metres/hour.

　b Convert the speed to metres/second.

5 A train travels for 2 h 48 min at an average speed of 140 km/h.

　How far does it travel in that time?

6 A bus travels at an average speed of 55 km/h.

　Calculate how long it takes the bus to go 22 km.

　Give the answer in minutes.

7 Motor oil is available in two sizes.

The small container holds 2 litres and costs $4.59.

The large container holds 4.5 litres and costs $9.99.

Work out which is the better value for money.

You must show your working.

8 Baked beans come in three different sizes.

Baked Beans 240 g — $0.69
Baked Beans 410 g — $1.20
Baked Beans 750 g — $2.05

Work out which size is the best value for money.

Show all your working.

9 This graph shows the distance travelled by a train after leaving the main station.

A short while after leaving the main station it stopped at a signal.

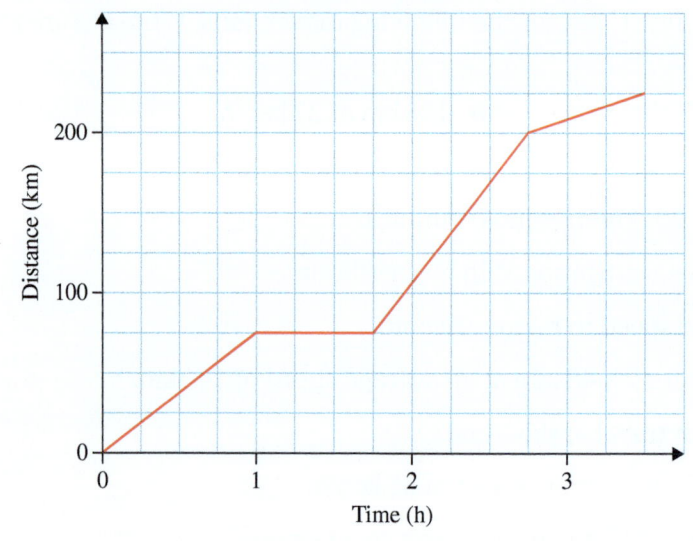

a How far did the train travel altogether?

b Work out the average speed for the whole journey, including the time stopped at the signal.

Give your answer to 1 decimal place.

c Work out the average speed the train travelled at, without including the time spent at the signal.

Give your answer to 1 decimal place.

Examination-style questions

1 This sign appears outside a local cinema.

    a Convert 19 45 to the 12-hour clock. [1]

    b Work out the finishing time of the film.

       Give the answer in 24-hour clock format. [2]

Tonight's film

Starts at 19 45

The film is:
2 hours 25 minutes long

2 A car travels 400 metres in 21.5 seconds.

   Work out the average speed, correct to 1 decimal place. [2]

3 Here is part of a bus timetable.

	Departures		
Bentown	13 45	14 54	16 01
The Village	14 08	15 17	16 24
Ascot Hill	14 22	15 31	16 38
Chesterville	14 58	16 07	17 14
Dranscot	15 20	16 29	17 36
Parkham	16 08	17 17	18 24

    a At what time does the 14 54 bus from Bentown reach Dranscot? [1]

    b Convert your answer to part **a** into the 12-hour clock. [1]

    c Estella finishes work in Ascot Hill at 3:50pm.

       How long does she need to wait for the next bus to Parkham? [2]

4 Fruit juice comes in three different sizes.

    A 600 ml bottle costs 69 cents.

    A 1 litre bottle costs $0.99

    A 2 litre bottle costs $1.89

    a Work out the price per litre for each size of bottle. [3]

    b Which is the best value for money? [1]

5   This table shows the flight times and distances between London, Kuala Lumpur and Auckland.
    The distance between Kuala Lumpur and Auckland is missing.

	Flight time	Distance (km)
London to Kuala Lumpur	13 hours 30 minutes	10 557
Kuala Lumpur to Auckland	11 hours 24 minutes	

   a   Work out the average speed between London and Kuala Lumpur. [2]

   b   The average speed between Kuala Lumpur and Auckland is 765 km/h.

       Work out the distance between Kuala Lumpur and Auckland. [2]

6   The green line on this graph shows a car journey between Plymscott and Queenstown.
    The two towns are 45 kilometres apart.

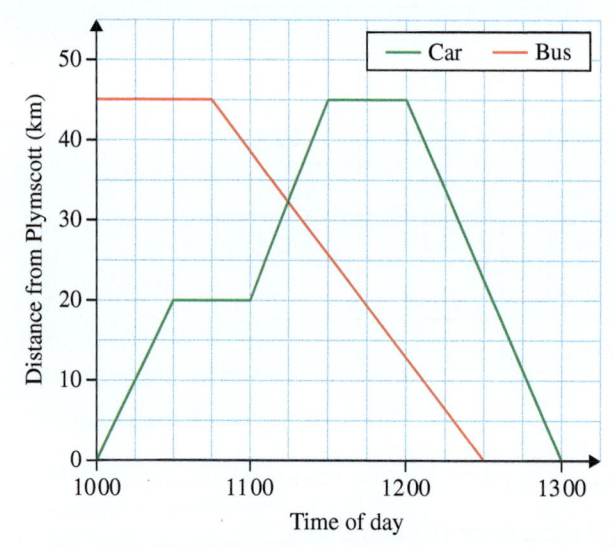

    The car left Plymscott at 10 00.

    a   At what time did it reach Queenstown? [1]

    b   For how long did it stop in Queenstown? [1]

    The red line shows a bus travelling from Queenstown to Plymscott.
    The bus left Queenstown at 10 45.

    c   Work out the average speed of the bus. [2]

    d   Write down the time at which they passed each other. [1]

7   Convert 1.1 hours to hours and minutes. [1]

21 Pythagoras' theorem

> **Learning outcomes**
>
> - Know and use Pythagoras' theorem to solve two-dimensional problems involving right-angled triangles.

Squaring up

Pythagoras was a Greek mathematician who lived about 2500 years ago.

His work involved different right-angled triangles.

Even before his time it was known that a triangle with sides in the ratio 3 : 4 : 5 was right angled.

Babylonian, Chinese and Egyptian mathematicians are believed to have used the so-called 3 4 5 triangle to make sure that the corners of their buildings were at right angles.

21.1 Pythagoras' theorem

The **theorem** with Pythagoras' name is one of the best known in mathematics.

It concerns right-angled triangles.

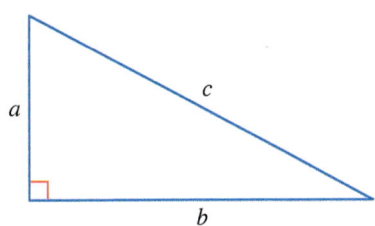

The sides of this right-angled triangle are labelled a, b and c.

The side labelled c is the **hypotenuse**, the longest side of the triangle.

The hypotenuse is the side opposite the right angle.

Pythagoras' theorem tells us that:

$a^2 + b^2 = c^2$

In words 'the square of the hypotenuse is equal to the sum of the squares of the other two sides'.

> Note that this works only if the three sides are given in the same units.

Worked example 1

Work out the length of the side marked x.

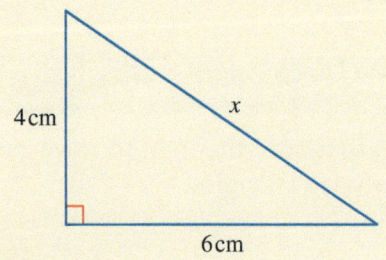

Always start by checking which side is the hypotenuse.

The side x is the hypotenuse. The two shorter sides are 4 cm and 6 cm.

Using Pythagoras' theorem:

$a^2 + b^2 = x^2$ state Pythagoras' theorem

$4^2 + 6^2 = x^2$ substitute the values 4 and 6 for a and b

$16 + 36 = x^2$ work out 4^2 and 6^2

$52 = x^2$

$x = \sqrt{52}$ square root is the inverse of squaring

$x = 7.21$ cm (to 2 d.p.)   give the answer to a sensible level of accuracy

Worked example 2

Work out the length of the side marked h.

Give your answer to 1 decimal place.

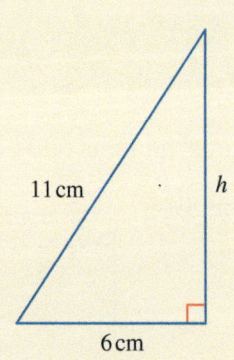

The hypotenuse is 11 cm and the other two sides are 6 cm and h.

Using Pythagoras' theorem:

$a^2 + b^2 = c^2$ state Pythagoras' theorem

$6^2 + h^2 = 11^2$ substitute the values

$h^2 = 11^2 - 6^2$ subtract 6^2 from both sides

$h^2 = 121 - 36$ work out 11^2 and 6^2

$h^2 = 85$

$h = \sqrt{85}$

$h = 9.2$ cm (to 1 d.p.)   round to 1 decimal place

> Always look at the diagram at the end and check the answer.
>
> In this case, 9.2 is less than 11 so it looks like a sensible answer.

21 Pythagoras' theorem

Exercise 21.1

Where, necessary, give your answers to 3 significant figures unless otherwise stated in the question.

1 Calculate the lengths of the sides marked with letters.

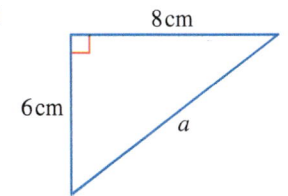

a (6 cm and 8 cm legs, hypotenuse a)

b (3 cm and 5 cm legs, hypotenuse b)

c (4.5 cm and 5 cm legs, hypotenuse c)

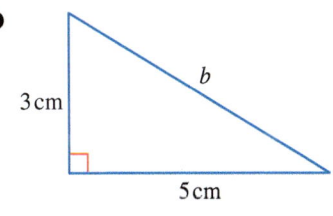

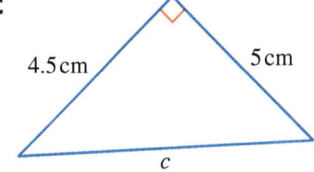

2 Calculate the lengths of the sides marked with letters.

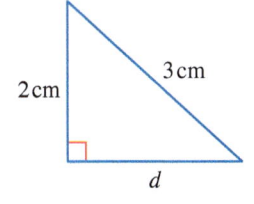

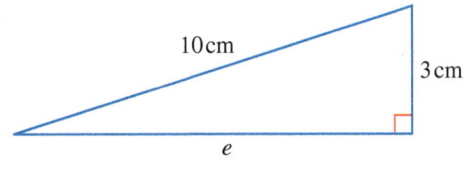

a (2 cm leg, 3 cm hypotenuse, other leg d)

b (10 cm hypotenuse, 3 cm leg, other leg e)

c (7.5 cm hypotenuse, 4.8 cm leg, other leg f)

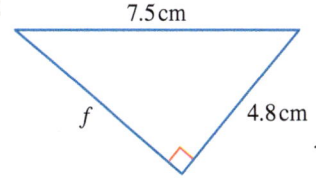

3 Calculate the lengths of the sides marked with letters.

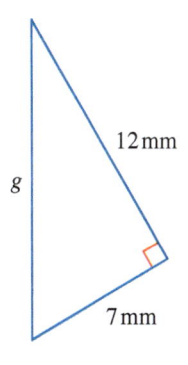

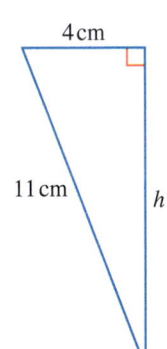

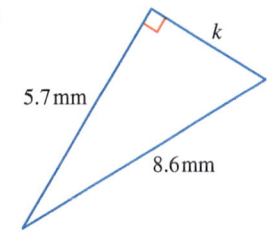

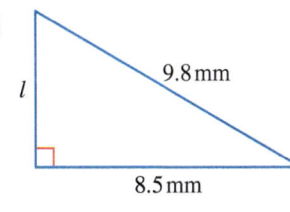

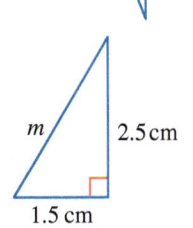

a (12 mm and 7 mm legs, hypotenuse g)

b (4 cm and 11 cm legs, hypotenuse h)

c (5.7 mm and 8.6 mm legs, hypotenuse k)

d (9.8 mm hypotenuse, 8.5 mm leg, other leg l)

e (2.5 cm hypotenuse, 1.5 cm leg, other leg m)

f (4 mm and 3.2 mm legs, hypotenuse n)

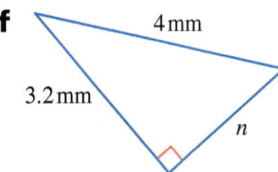

4 ABC is an isosceles right-angled triangle.

$BC = 6$ cm

a Write down the length of the side marked x.

b Calculate the length of the side marked y.

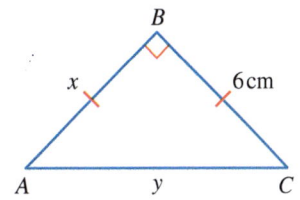

5 The diagram shows a right-angled isosceles triangle.

Calculate the length of the sides marked w.

> Start by making an equation using w.

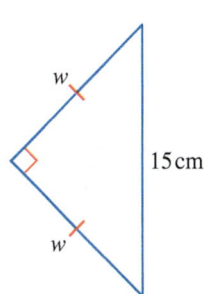

301

Mathematics for Cambridge Secondary 1

6 Charlie has used Pythagoras' theorem to find the length marked *p* in this triangle.

$7^2 + 9^2 = p^2$
$49 + 81 = p^2$
$130 = p^2$
$p = \sqrt{130} = 11.4\,cm$ (to 1 d.p.)

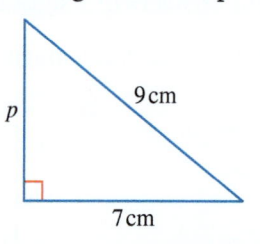

a How can you easily tell that his answer is wrong?

> Look at the lengths of the sides. Is this a sensible answer?

b Work out the correct answer.

7 Spirals can be built up by joining together a sequence of right-angled triangles.

Work out the lengths of the sides marked *x* and *y* in these diagrams.

a

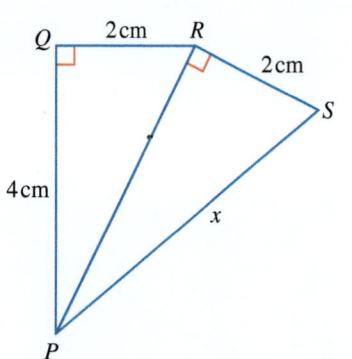

b

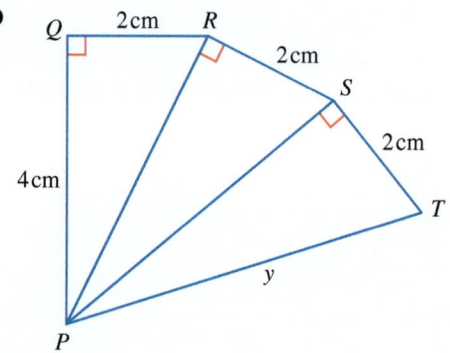

> You will need to work out the third side of the left-hand triangle first.

21.2 Applications of Pythagoras' theorem

Isosceles triangles

Any isosceles triangle can be made into two congruent right-angled triangles.

> Remember that congruent triangles are identical.

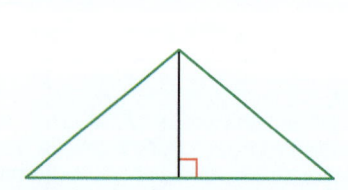

This allows you to use Pythagoras' theorem in isosceles triangles.

21 Pythagoras' theorem

Worked example 1

ABC is an isosceles triangle with $AB = AC$.

AM is a line perpendicular to *BC*.

$AB = 7.2$ cm and $BC = 10$ cm

a Calculate the length *AM*.

b Work out the area of triangle *ABC*.

Give your answers to 3 significant figures.

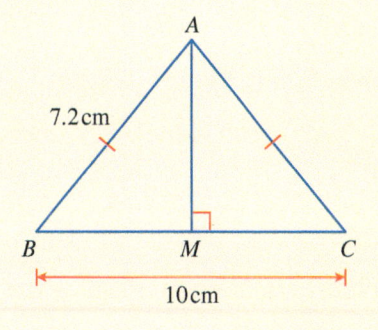

a It is a good idea to sketch one of the right-angled triangles.

This is triangle *ABM*.

$BM = 5$ cm as it is half the length of the base, *BC*.

The hypotenuse of the triangle is 7.2 cm.

Let $AM = x$

Using Pythagoras' theorem:

$5^2 + x^2 = 7.2^2$

$x^2 = 7.2^2 - 5^2$

$x^2 = 26.84$

$x = \sqrt{26.84} = 5.18$ cm (to 3 s.f.)

Length $AM = 5.18$ cm (to 3 s.f.)

b Area of triangle $= \frac{1}{2} \times$ base $\times$ height

Area of triangle $ABC = \frac{1}{2} \times 10 \times 5.1807\ldots$

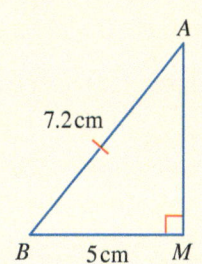

$= 25.9$ cm^2 (to 3 s.f.)

In part **b** do not use the 3 significant figure value from part **a**. Use at least 4 significant figures in the calculation so you do not lose accuracy.

If your calculator has an **Ans** button, you can type $\frac{1}{2} \times 10 \times$ **Ans** to use the previous answer.

Applying Pythagoras' theorem

There are many problems that can be solved by using Pythagoras' theorem.

For example, a ladder standing on horizontal ground leaning against a vertical wall makes a right-angled triangle.

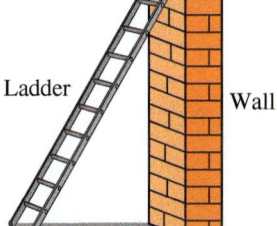

When solving problems it is a good idea to draw a diagram.

Give your answers to a sensible level of accuracy.

303

Worked example 2

A ladder stands on a horizontal floor leaning against a vertical wall.

The ladder is 5 metres long.

The base of the ladder is 2.2 metres from the wall.

How far up the wall does the ladder reach?

First draw a diagram to show the information.

The diagram shows a right-angled triangle.

The hypotenuse is 5 metres long.

Using Pythagoras' theorem:

$2.2^2 + h^2 = 5^2$

$\quad h^2 = 5^2 - 2.2^2$ subtract 2.2^2 from each side

$\quad h^2 = 20.16$

$\quad h = \sqrt{20.16} = 4.49\,\text{m}$ (to 2 d.p.)   round to 2 decimal places

The ladder reaches 4.49 metres up the wall.

Worked example 3

A helicopter flies 120 kilometres due north from an air base.

It then flies 45 kilometres due east.

How far is it now from the air base?

First draw a diagram to show the information.

Label the third side of the triangle x.

The diagram shows a right-angled triangle with hypotenuse x.

Using Pythagoras' theorem:

$45^2 + 120^2 = x^2$

$\quad 16\,425 = x^2$

$\quad\quad x = \sqrt{16\,425} = 128.2\,\text{km}$ (to 1 d.p.)   round to 1 decimal place

The helicopter is now 128.2 kilometres from the base.

21 Pythagoras' theorem

Exercise 21.2

For questions 1 to 3, give your answers to 3 significant figures where necessary.

For questions 4 to 10, choose a sensible level of accuracy if it is not given in the question.

1 Work out the lengths of the sides marked with letters in these isosceles triangles.

a 6 cm, h, 4.8 cm

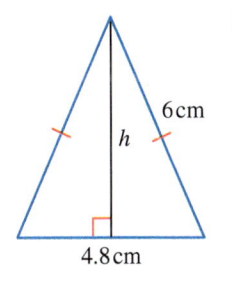

b x, 9.6 mm, 4 mm, x

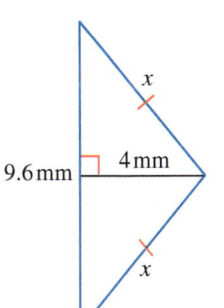

c y, 5.6 cm, 7 cm

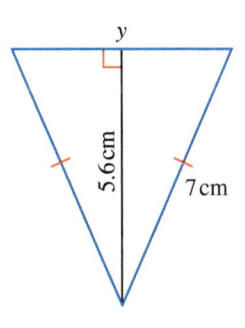

d w, 12 cm, 13 cm

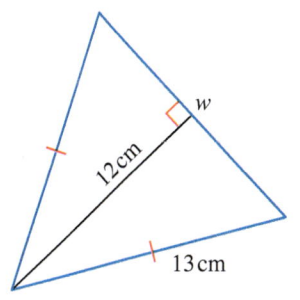

2 The diagram shows an isosceles triangle.

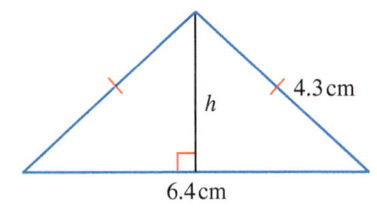

h, 4.3 cm, 6.4 cm

a Work out the height h.

b Calculate the area of the triangle.

3 Work out the area of each isosceles triangle.

a 8.4 mm, 4.2 mm

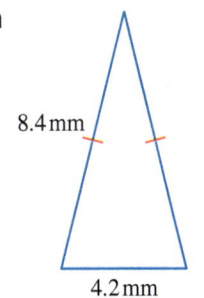

b 11 cm, 7.6 cm

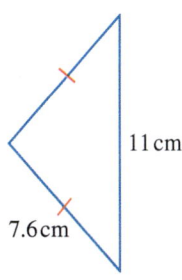

c 6.9 cm, 8.8 cm

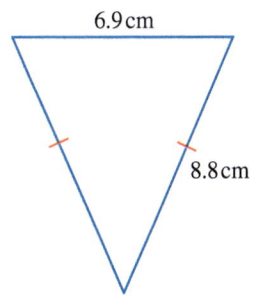

4 A ship sails 60 kilometres due south and then 48 kilometres due west.

How far is the ship from its starting point?

5 A ladder rests on horizontal ground leaning against a vertical wall.

The ladder is 6 metres long.

The top of the ladder is 5.35 metres up the wall.

How far is the foot of the ladder from the wall?

Give your answer correct to the nearest centimetre.

305

6 This is a sketch of the plan of a farmer's field.

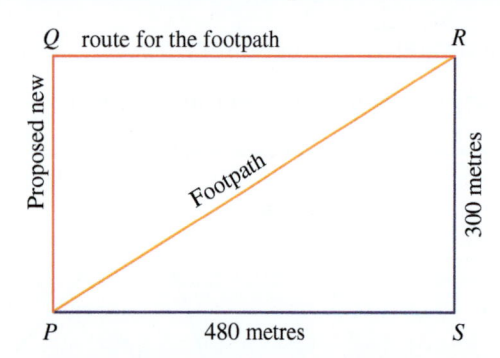

There is a footpath diagonally across the field from P to R.

a Calculate the length of the footpath.

The farmer wants to move the footpath to follow the edge of the field, from P to Q to R.

b How much longer is the proposed new route for the footpath?

7 The diagram shows a trapezium ABCD. BC is parallel to AD.

Angle DAB = angle ABC = 90°

Work out the length CD.

> Draw a perpendicular line from C to AD to make a right-angled triangle.

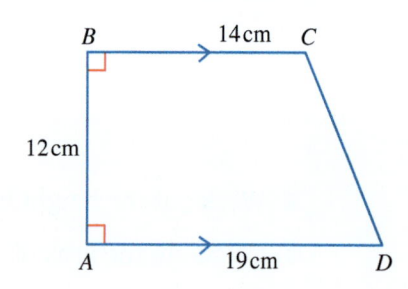

8 The diagram shows an isosceles triangle.

The area of the triangle is 72 cm².

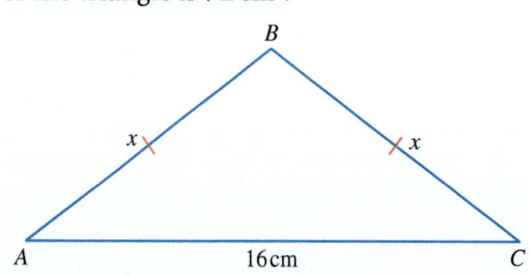

Calculate the length of the side marked x.

> Work out the perpendicular height first.

9 Work out the surface area of this prism.

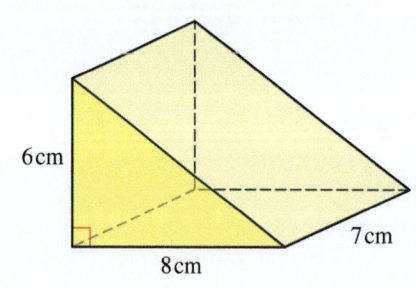

> Work out the area of a triangle using the formula area = $\frac{1}{2}$ × base × height

21 Pythagoras' theorem

10 The diagram shows a right-angled triangle with a semicircle on each side.
   The two shorter sides of the triangle are 3 cm and 4 cm. 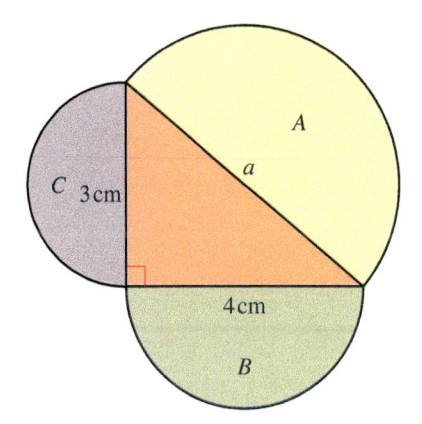
   a Work out the length of the hypotenuse, a.
   b Calculate the area of the semicircle labelled C.
     Give your answer to 3 decimal places.
   c Calculate the area of the semicircle labelled B.
     Give your answer to 3 decimal places.
   d Calculate the area of the semicircle labelled A.
     Give your answer to 3 decimal places.
   e Add together the areas of the two smaller semicircles. What do you notice?

Review

Give your answers to 3 significant figures unless stated otherwise in the question.

1 Work out the lengths marked with letters in these right-angled triangles.
   a m, 18 mm, 12 mm
   b 4.4 cm, 3.6 cm, n
   c 11 cm, 8 cm, p
   d 11.5 cm, 15 cm, q
   e r, 12 cm, 20 cm
   f 24 cm, 26 cm, s

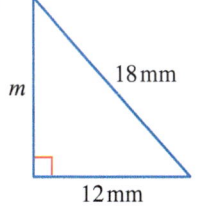

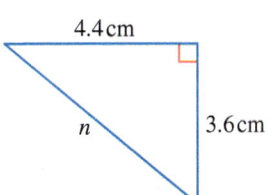

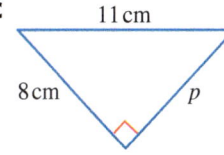

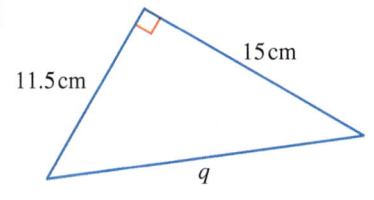

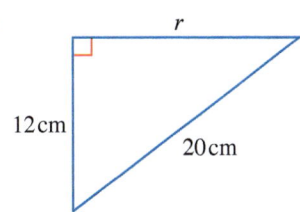

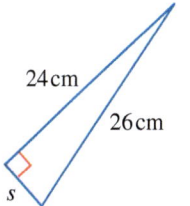

2 Work out the lengths marked with letters in these isosceles triangles.
   a a, 5.8 mm, 7.6 mm
   b b, 2.3 cm, 3.9 cm
   c c, 7 cm, 8.8 cm

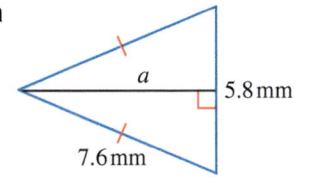

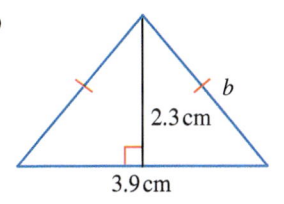

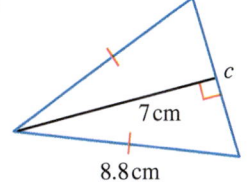

3 Calculate the areas of these isosceles triangles.
   a 8.4 cm, 5 cm, 5 cm
   b 6.1 cm, 6.1 cm
   c 3 cm, 6.8 cm

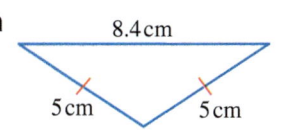

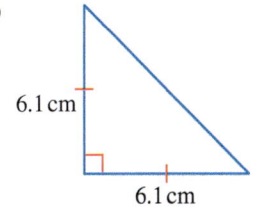

307

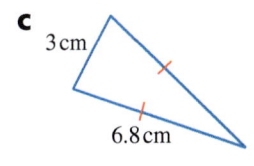

4 *ABC* is an equilateral triangle.

   a Calculate the height *h*.

   b Work out the area of triangle *ABC*.

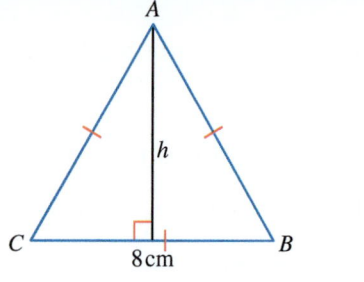

5 *PQRS* is a rhombus.

   The diagonals *QS* and *PR* have been drawn.

   QS = 8 cm and *PR* = 12 cm

   Work out the length of the side of the rhombus.

   > The diagonals of a rhombus intersect at 90°.

6 A boat sails 75 kilometres due north and then 150 kilometres due west.

   How far from the starting point is it now?

7 The diagram shows the end of a garden shed.

   The shed is 2.6 metres high at the front and 2 metres high at the back.

   The sloping edge of the roof is 1.9 metres.

   Work out the width of the shed, marked *w* on the diagram.

   Give the answer correct to the nearest centimetre.

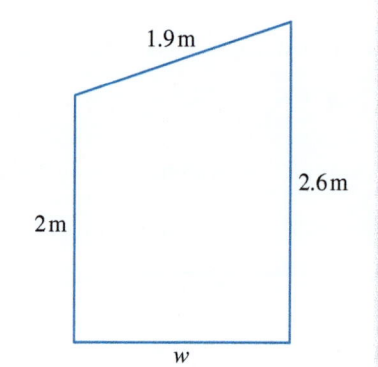

8 A ship sails 154 kilometres due north, then 76 kilometres due east, then 87 kilometres due south.

   How far from the starting point is it now?

   Give the answer to the nearest kilometre.

Examination-style questions

Give your answers correct to 3 significant figures unless stated otherwise in the question.

1 A rectangular TV screen measures 33 cm by 27 cm.

   Work out the length of the diagonal, *d*, across the screen.

   Round the answer to the nearest millimetre.

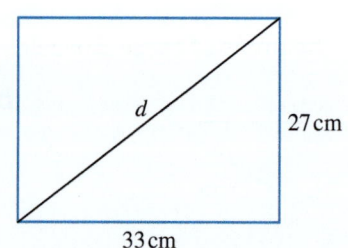

   [2]

21 Pythagoras' theorem

2 The diagram shows a 5.6 metre ladder leaning against a vertical wall.

The base of the ladder is 2.1 metres from the wall.

The floor on which the ladder stands is horizontal.

Work out how high the top of the ladder is above the floor.

Round the answer to the nearest centimetre. [2]

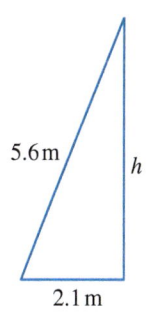

3 ABC is an isosceles triangle.

$AB = BC = 7$ cm and $AC = 11$ cm

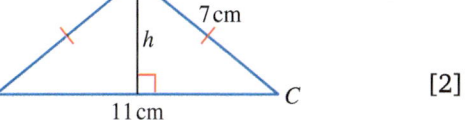

a Work out the perpendicular height of the triangle, h. [2]

b Calculate the area of triangle ABC. [1]

4 PQRS is a square with sides 5 centimetres.

Calculate the length of a diagonal of the square. [2]

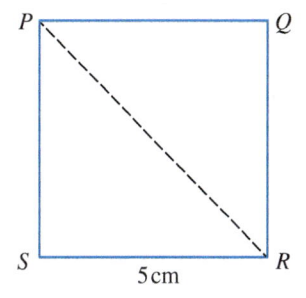

5 A vertical flagpole is 28 metres tall.

It stands on horizontal ground.

It is secured by cables from the top of the flagpole to the ground.

One of the cables is shown in the diagram.

Each cable is secured 12.5 metres from the base of the flagpole.

Work out the length of each cable. [2]

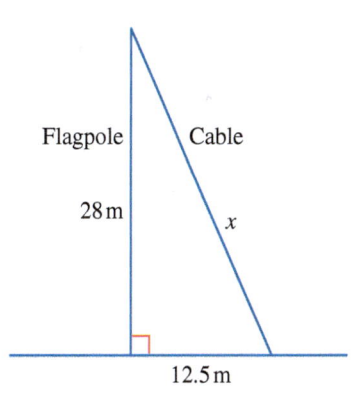

6 The diagram shows the sail of a yacht.

The sail is a right-angled triangle.

Work out the length marked h.

Round the answer to the nearest centimetre. [2]

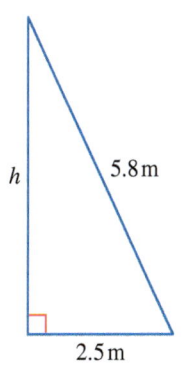

22 Trigonometry

> **Learning outcomes**
> - Use the tangent, sine and cosine ratios to calculate sides and angles.

How far? How high?

Trigonometry is used in many scientific fields, such as architecture (designing and making buildings), astronomy (study of stars and space), engineering, oceanography (study of oceans) and optics (study of light).

In your everyday life you are surrounded by objects where trigonometry has been used at some stage in their research, development or manufacture. For example, trigonometry is used to build cars and homes.

When police use triangulation to trace the location of cell phones they are using trigonometry.

Navigation and surveying also depend heavily on trigonometry. It enables us to send men into space, since trigonometry is needed to compute launch, orbital and re-entry paths.

You can use trigonometry to calculate the height of a mountain or the width of a river.

22.1 Naming the sides in a right-angled triangle

Trigonometry is used to find lengths and angles in a right-angled triangle.

Before you learn how to do trigonometry, you must learn the special names for the three sides in a right-angled triangle.

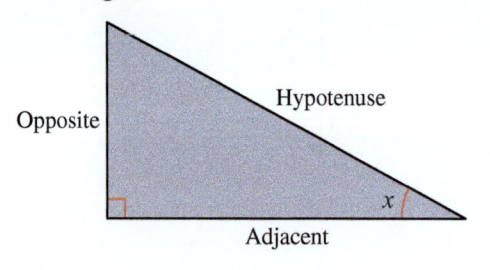

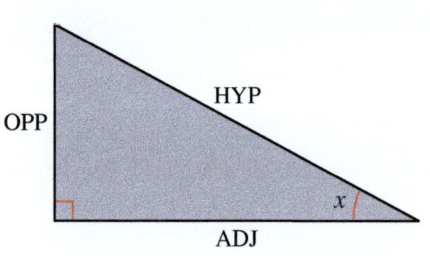

22 Trigonometry

The longest side, which is always opposite the right angle, is called the **hypotenuse** (HYP).

The side opposite angle x is called the **opposite** (OPP).

The third side, which touches angle x, is called the **adjacent** (ADJ).

It is important that you can give the correct name to each side.

Worked example

The sides of this right-angled triangle are p, q and r.

For the angle of 35°, identify which side is the:

a hypotenuse   **b** opposite   **c** adjacent.

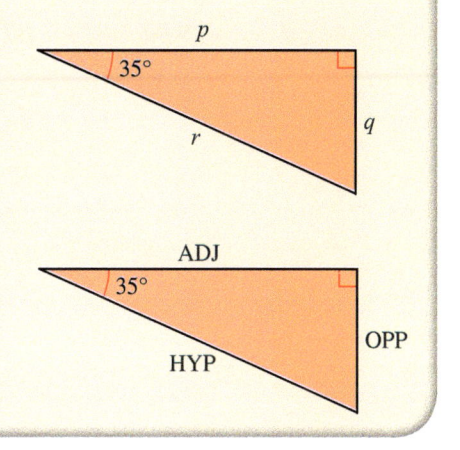

a The longest side = the hypotenuse = r
b The side opposite the 35° angle = the opposite = q
c The side touching the 35° angle = the adjacent = p

Exercise 22.1

1 Copy each diagram and label each side with the correct name for the angle shown.

The first one has been done for you.

a OPP, ADJ, HYP, 32°

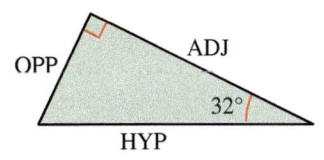

b 56°

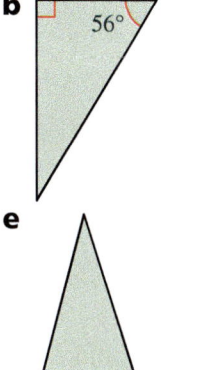

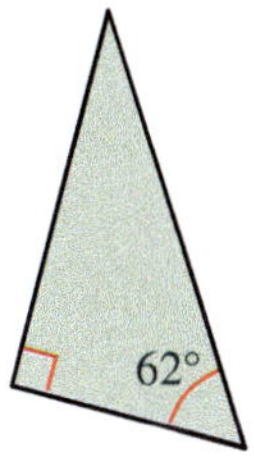

c 27°

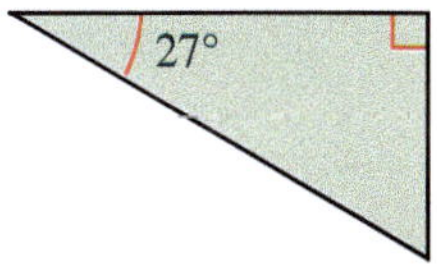

d 60°

e 62°

f 55°

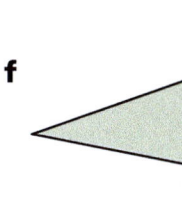

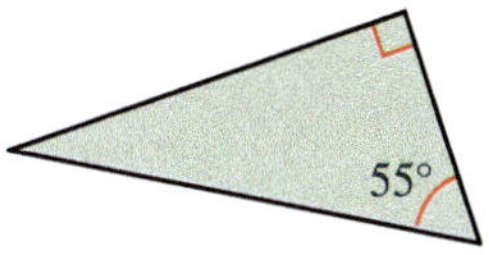

g 34°

h 40°

i 72°

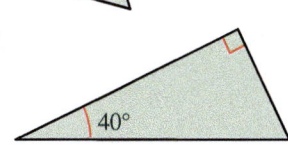

311

2 Measure the length of the opposite side and the length of the adjacent side for each of these right-angled triangles.

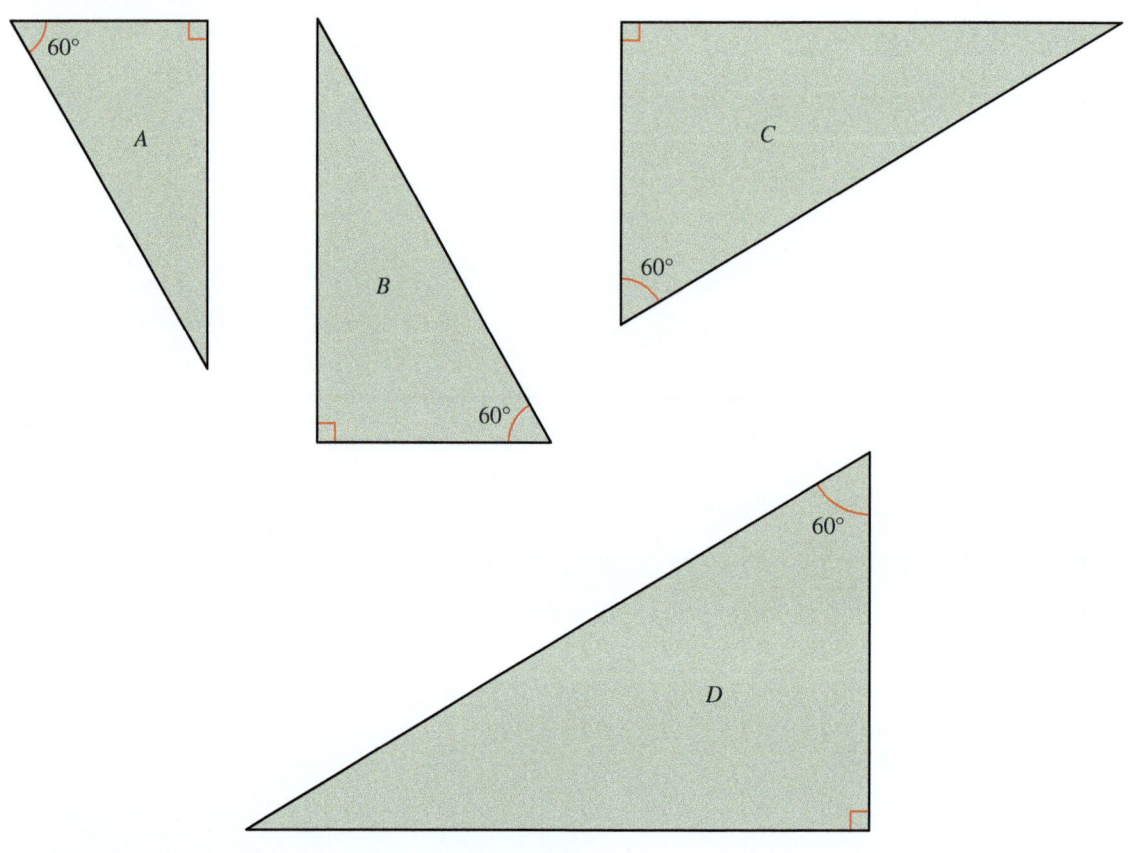

a Copy and complete the table for the triangles.

Triangle A has been done for you.

Triangle	OPP	ADJ	$\frac{OPP}{ADJ}$
A	4.6 cm	2.7 cm	4.6 ÷ 2.7 = 1.7
B			
C			
D			

b Comment on your answers for the ratio $\frac{OPP}{ADJ}$.

3 Draw some right-angled triangles that have an angle of 35°.

Calculate the ratio $\frac{OPP}{ADJ}$ for each triangle.

Comment on your results.

22.2 Finding a side using the tangent ratio

The ratio $\frac{\text{OPP}}{\text{ADJ}}$ is known as the **tangent ratio**.

The tangent ratio is usually abbreviated to tan.

$$\tan x = \frac{\text{OPP}}{\text{ADJ}}$$

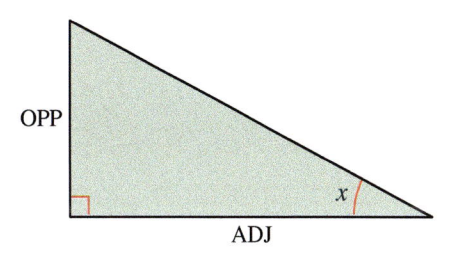

There is a tan key on your calculator, which gives the exact tangent ratios for any angle.

> Make sure your calculator is in degree mode. If it is, 'd', 'D', 'DEG' or 'deg' will show on the display.

The next example shows you how to use the tan key on your calculator to find the length of the opposite side in a right-angled triangle when you know an angle and the adjacent side.

Worked example 1

Find the value of x.

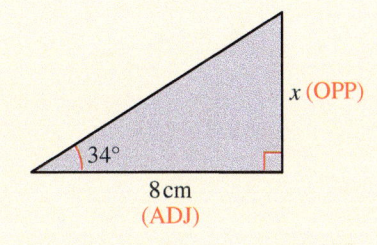

$\tan 34° = \dfrac{\text{OPP}}{\text{ADJ}}$ replace OPP by x and ADJ by 8

$\tan 34° = \dfrac{x}{8}$ multiply both sides by 8

$x = 8 \times \tan 34°$ use your calculator to find the value of $8 \times \tan 34°$

$x = 5.40\,\text{cm}$ (to 3 s.f.)

> Answers for lengths are usually given to 3 significant figures.

Worked example 2

Find the value of y.

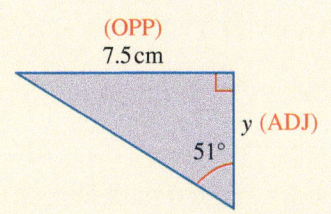

$\tan 51° = \dfrac{\text{OPP}}{\text{ADJ}}$ replace OPP by 7.5 and ADJ by y

$\tan 51° = \dfrac{7.5}{y}$ multiply both sides by y

$y \tan 51° = 7.5$ divide both sides by $\tan 51°$

$y = \dfrac{7.5}{\tan 51°}$ use your calculator to find the value of $\dfrac{7.5}{\tan 51°}$

$y = 6.07\,\text{cm}$ (to 3 s.f.)

Mathematics for Cambridge Secondary 1

Exercise 22.2

1 Calculate the lengths of the sides marked with letters.

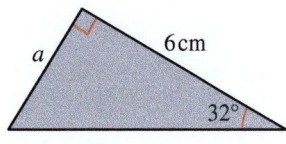

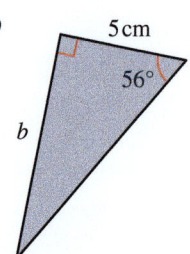

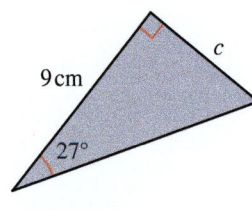

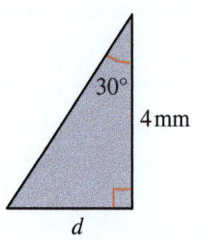

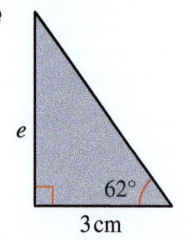

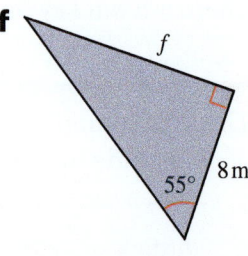

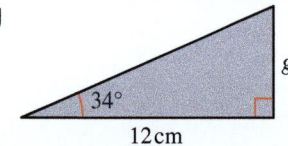

a) 6 cm, 32°, side a

b) 5 cm, 56°, side b

c) 9 cm, 27°, side c

d) 30°, 4 mm, side d

e) 3 cm, 62°, side e

f) 8 m, 55°, side f

g) 12 cm, 34°, side g

h) 10 cm, 40°, side h

i) 4 mm, 45°, side i

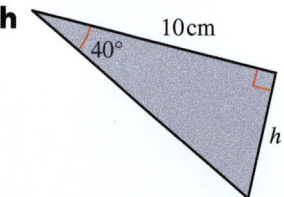

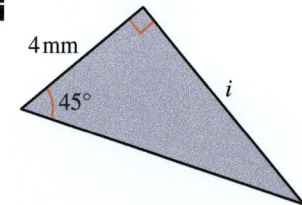

2 The diagram shows a tree standing on horizontal ground.
Calculate the height, h metres, of the tree.

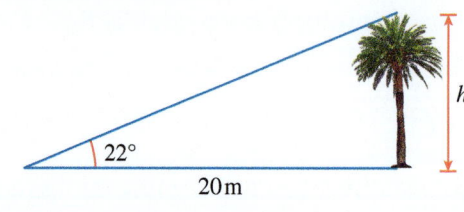

(22°, 20 m, height h)

3 a Calculate the height of this isosceles triangle.
  b Calculate the area of the isosceles triangle.

> Split the isosceles triangle into two right-angled triangles.

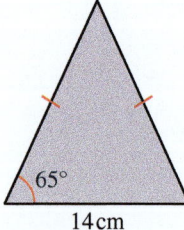

(65°, 14 cm)

4 Calculate the lengths of the sides marked with letters.

a) 6 mm, 52°, side x

b) 12 cm, 58°, side x

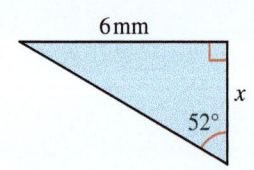

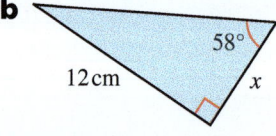

> In this question, x will be the denominator of the tangent ratio.

5 Find the lengths of the sides marked with letters.

a

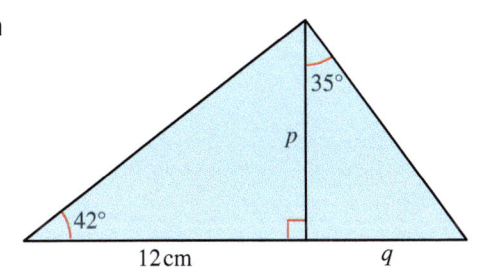

b

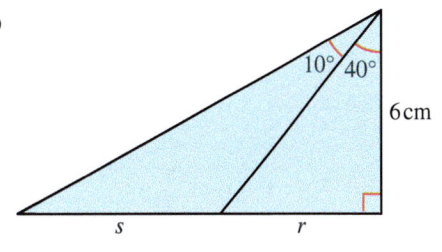

You will need to calculate the length $r + s$.

22.3 Finding an angle using the tangent ratio

You can use the value of the tangent to find the size of an angle in a right-angled triangle if you know the opposite and adjacent sides.

To do this you must use the 'inverse tan' button on your calculator.

(On most calculators this is labelled $\tan^{-1}$.)

Worked example

Find the size of angle x.

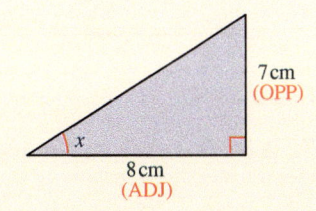

$\tan x = \dfrac{\text{OPP}}{\text{ADJ}}$ replace OPP by 7 and ADJ by 8

$\tan x = \dfrac{7}{8}$ to find x use the $\tan^{-1}$ button on your calculator

$x = \tan^{-1}\left(\dfrac{7}{8}\right)$ Remember to use brackets when you put the expression into your calculator.

$x = 41.2°$ (to 1 d.p.)   Answers for angles are usually given to 1 decimal place.

Exercise 22.3

1 Calculate the lengths of the sides marked with letters.

a

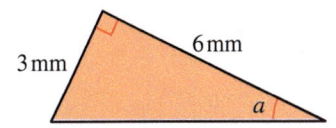

b

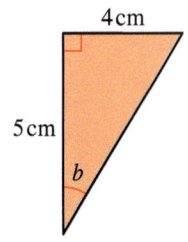

c

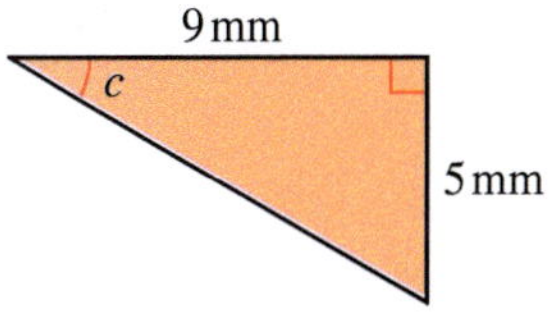

315

d
7 m
3 m
d

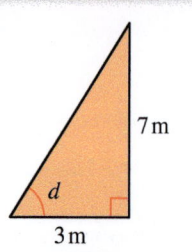

e 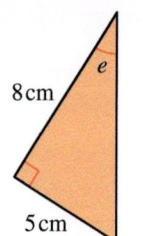
8 cm
5 cm
e

f
10 cm
8 cm
f

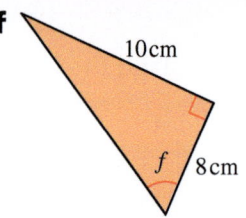

g
12 m
10 m
g

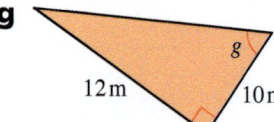

h 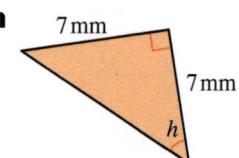
7 mm
7 mm
h

i
8 cm
4 cm
i

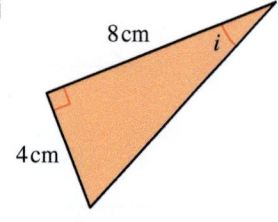

2 *ABCD* is a rectangle.

  AB = 12 cm and *BC* = 5 cm

  a Calculate angle *BAC*. **b** Calculate angle *BEC*.

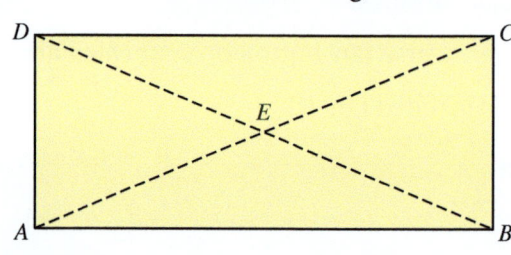

3 *ABCD* is a quadrilateral.

  Show that angle *ABC* is 83.6° correct to 1 decimal place.

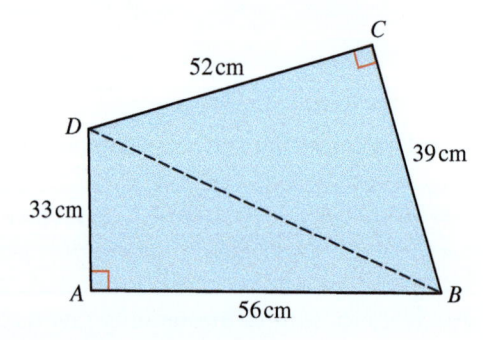

  52 cm, 39 cm, 33 cm, 56 cm

4 Find the size of the angles marked with letters.

  a 6 cm, 4 cm, 5 cm, *x*, *w*

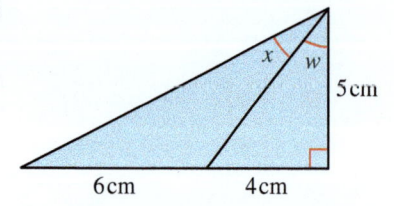

  b 5 mm, 5 mm, 7 mm, *y*, *z*

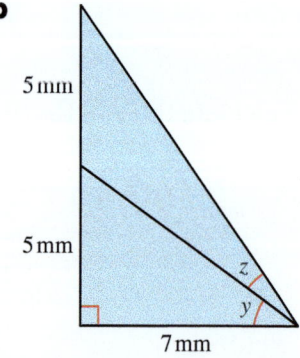

22.4 Finding a side using sine, cosine or tangent

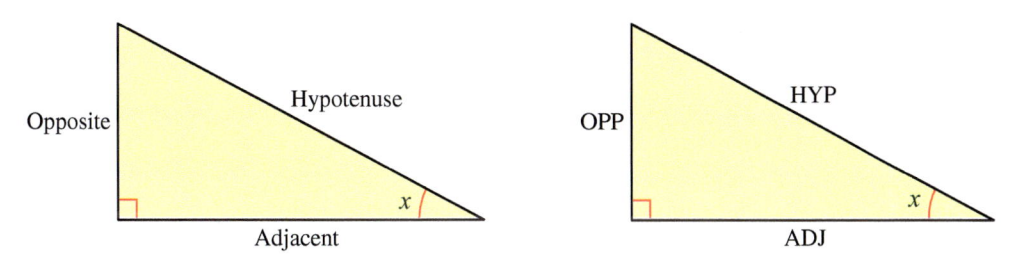

You have learnt how to use the trigonometric ratio $\frac{OPP}{ADJ}$, which is known as the tangent ratio.

There are two more ratios that are used in trigonometry.

These are the **sine ratio** and the **cosine ratio**.

The sine ratio is usually abbreviated to sin and the cosine ratio is abbreviated to cos.

The complete list of trigonometric ratios is:

$$\sin x = \frac{OPP}{HYP} \qquad \cos x = \frac{ADJ}{HYP} \qquad \tan x = \frac{OPP}{ADJ}$$

A memory aid is useful for learning these three ratios.

One possible memory aid is to learn the word: **SOHCAHTOA**

SOH represents **S** for sin, **O** for opposite and **H** for hypotenuse.

CAH represents **C** for cos, **A** for adjacent and **H** for hypotenuse.

TOA represents **T** for tan, **O** for opposite and **A** for adjacent.

So sin of x equals opposite over hypotenuse, cos of x is adjacent over hypotenuse and tan of x is opposite over adjacent.

Worked example 1

Find the value of x.

$\sin 31° = \frac{OPP}{HYP}$ replace OPP by x and HYP by 7

$\sin 31° = \frac{x}{7}$ multiply both sides by 7

$x = 7 \times \sin 31°$ use your calculator to find the value of $7 \times \sin 31°$

$x = 3.61$ cm (to 3 s.f.)

Mathematics for Cambridge Secondary 1

Worked example 2

Find the value of y.

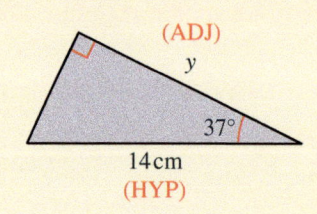

(ADJ) y

37°

14 cm (HYP)

$\cos 37° = \dfrac{\text{ADJ}}{\text{HYP}}$ replace ADJ by y and HYP by 14

$\cos 37° = \dfrac{y}{14}$ multiply both sides by 14

$y = 14 \times \cos 37°$ use your calculator to find the value of $14 \times \cos 37°$

$y = 11.2\,\text{cm}$ (to 3 s.f.)

Exercise 22.4

1 Calculate the lengths of the sides marked with letters.

a a, 41°, 4 mm

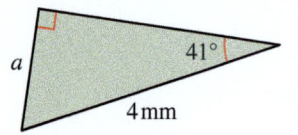

b 5 cm, 63°, b

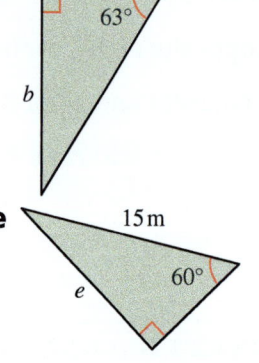

c 8 cm, 50°, c

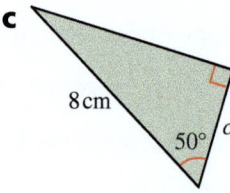

d 26°, 12 cm, d

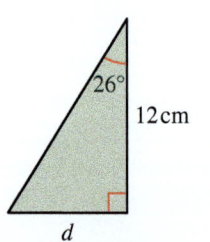

e 15 m, 60°, e

f f, 35°, 12 mm

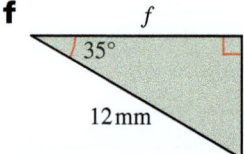

g g, 39°, 18 cm

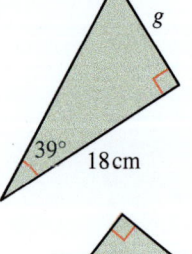

h h, 30°, 20 cm

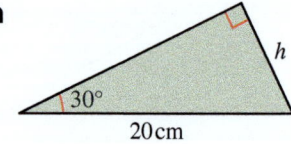

i i, 25 m, 47°

j j, 18°, 45 mm

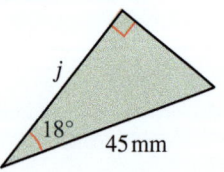

k 9 cm, 28°, k

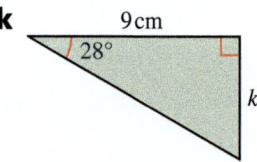

l l, 3.8 cm, 64°

318

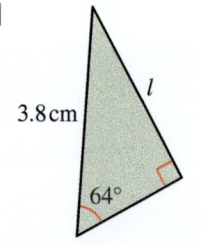

2 *ABCD* is a parallelogram.

BC = 7 cm and DC = 9 cm

Angle *DAB* = 50°

Calculate:

  a the perpendicular height, *h* cm, of the parallelogram

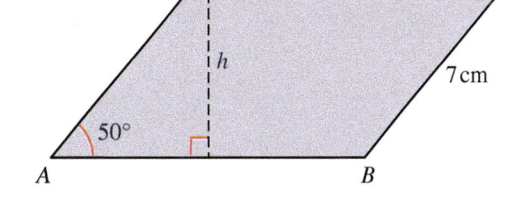

  b the area of the parallelogram.

3 Show that the area of the isosceles triangle is 96.5 cm² correct to 3 significant figures.

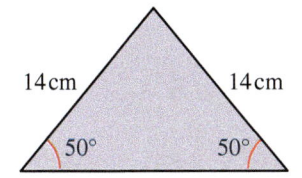

Split the isosceles triangle into two right-angled triangles.

4 *ABCD* is an isosceles trapezium.

AB = 15 mm and DC = 9 mm

Angle *DAB* = angle *CBA* = 55°

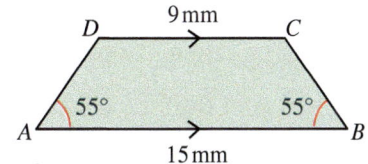

Show that the area of the trapezium is 51.4 mm² correct to 3 significant figures.

5 *ABCD* is a rhombus.

AC = 12 cm and angle *ABC* = 130°

Find the perimeter of the rhombus.

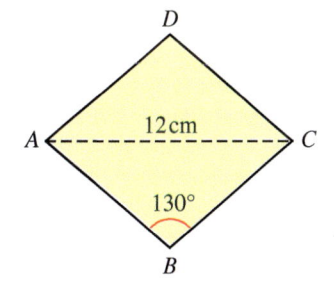

6 *ABC* is an equilateral triangle with sides of length 5 cm.

  a Calculate the perpendicular height of the triangle.

  b Calculate the area of the triangle.

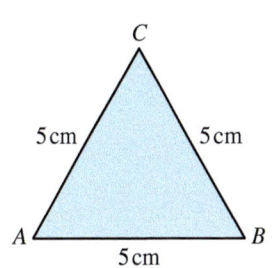

7 The diagram shows a bridge over a river.

A canoe *C* is 65 metres vertically below the bridge.

Angle *CXY* = 42° and angle *XYC* = 45°

Calculate the length, *XY*, of the bridge.

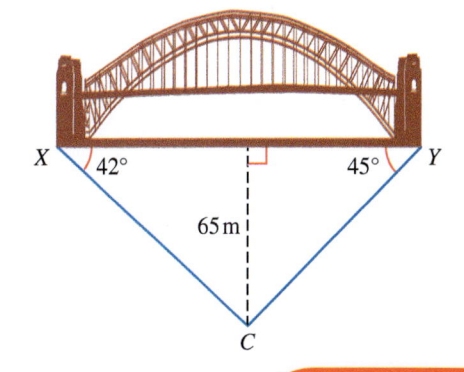

319

Mathematics for Cambridge Secondary 1

8 A ski lift travels for 150 metres at 48° to the horizontal and then for 200 metres at 32° to the horizontal.

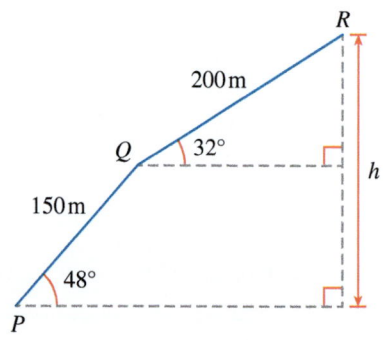

Calculate the vertical height, h metres, gained when travelling from P to R.

22.5 Finding an angle using sine, cosine or tangent

You learned how to find an angle using the value of the tangent in Section 22.3.

You can also use the value of the sine or cosine to find angles.

Worked example 1

Find the size of angle x.

(HYP) 10 cm   6 cm (OPP)

$\sin x = \dfrac{\text{OPP}}{\text{HYP}}$ replace OPP by 6 and HYP by 10

$\sin x = \dfrac{6}{10}$

$x = \sin^{-1}\left(\dfrac{6}{10}\right)$ to find x, use the $\sin^{-1}$ button on your calculator

$x = 36.9°$ cm (to 1 d.p.)

Worked example 2

Find the size of angle y.

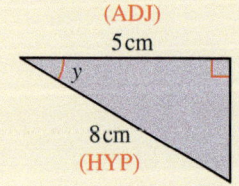

(ADJ) 5 cm   8 cm (HYP)

$\cos y = \dfrac{\text{ADJ}}{\text{HYP}}$ replace ADJ by 5 and HYP by 8

$\cos y = \dfrac{5}{8}$

$y = \cos^{-1}\left(\dfrac{5}{8}\right)$ to find y use the $\cos^{-1}$ button on your calculator

$y = 51.3°$ (to 1 d.p.)

22 Trigonometry

Exercise 22.5

1 Calculate the sizes of the angles marked with letters.

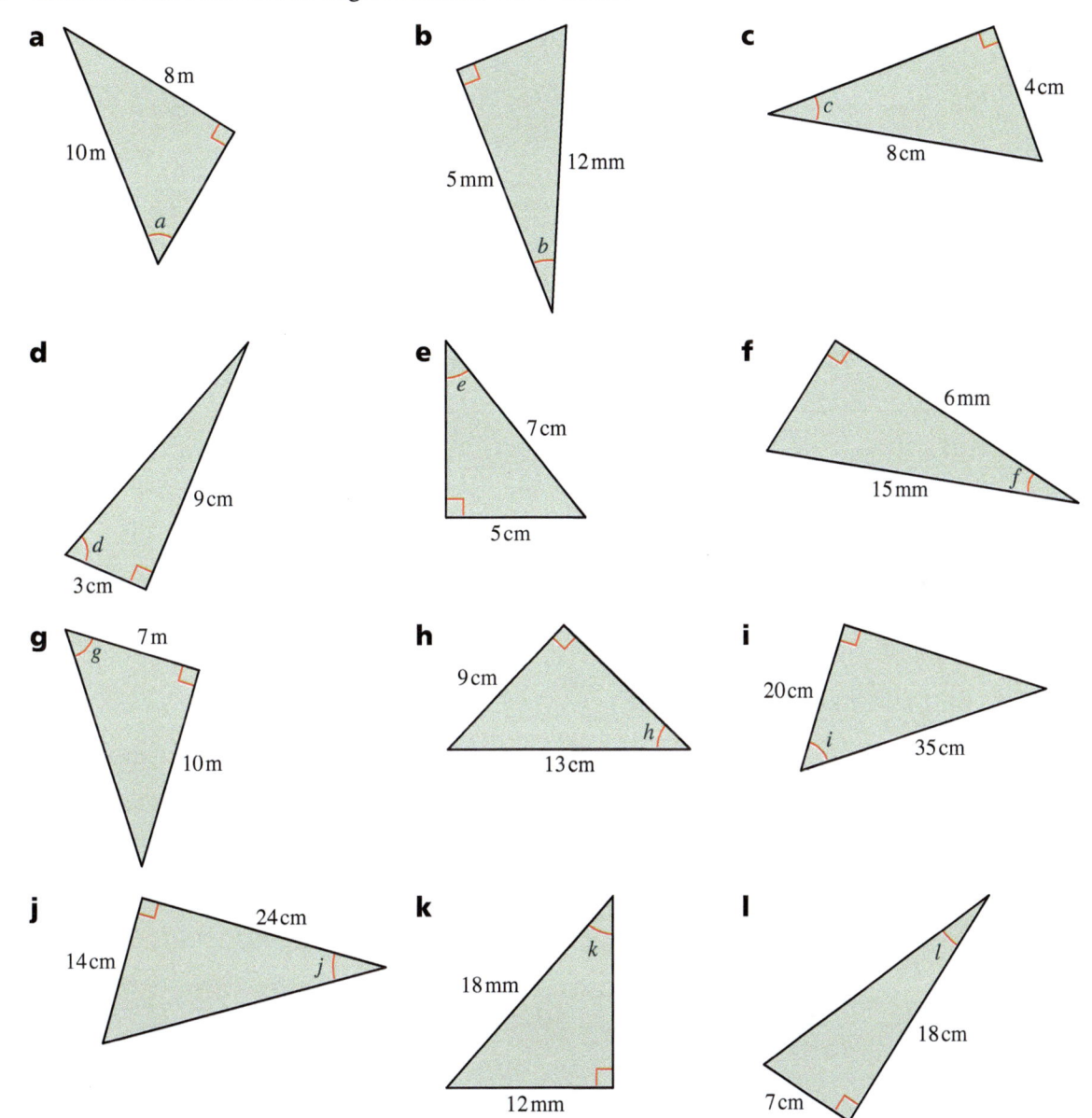

a 8 m, 10 m, angle a

b 5 mm, 12 mm, angle b

c 8 cm, 4 cm, angle c

d 9 cm, 3 cm, angle d

e 7 cm, 5 cm, angle e

f 15 mm, 6 mm, angle f

g 7 m, 10 m, angle g

h 9 cm, 13 cm, angle h

i 20 cm, 35 cm, angle i

j 24 cm, 14 cm, angle j

k 18 mm, 12 mm, angle k

l 18 cm, 7 cm, angle l

2 The diagram shows a flagpole standing on horizontal ground.

A rope, AB, supports the flagpole.

The flagpole is 6 metres high and the rope is 12 metres long.

Calculate the angle r that the rope makes with the ground.

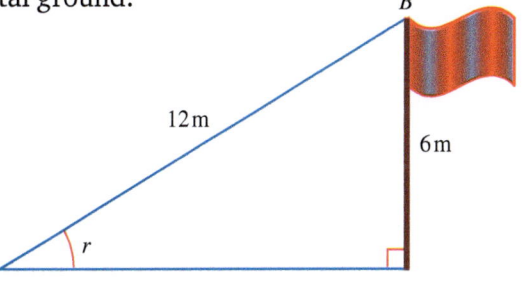

321

3 Triangle *ABC* is an isosceles triangle.

  a Calculate angle *ABC*.

  b Calculate angle *ACB*.

> Split the isosceles triangle into two right-angled triangles.

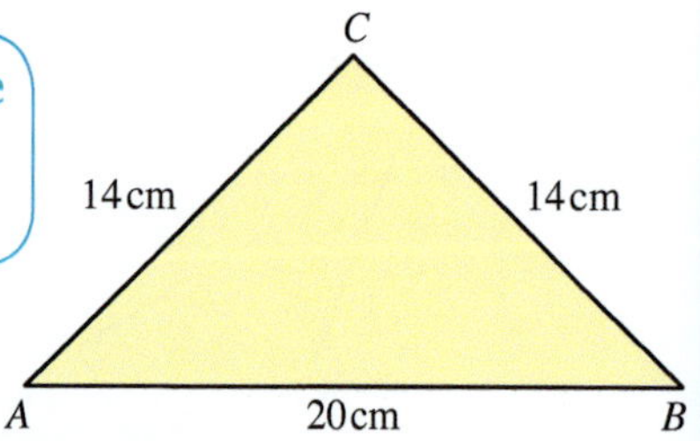

4 *ABCD* is an isosceles trapezium.

  AB = 20 mm and *DC* = 13 mm

  AD = *BC* = 14 mm

  a Calculate angle *DAB*.

  b Write down the size of angle *ADC*.

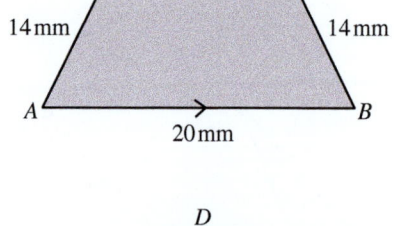

5 *ABCD* is a rhombus.

  AC = 30 cm and *BD* = 18 cm

  Calculate:

  a angle *DAB*

  b angle *ADC*.

  c the perimeter of the trapezium.

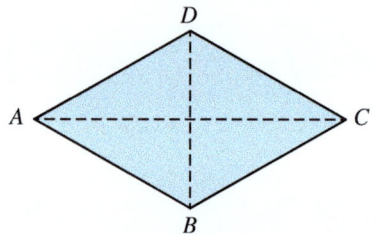

Review

1 Calculate the lengths and angles marked with letters.

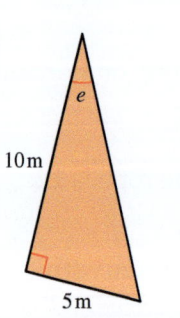

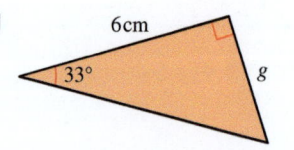

a 20°, 8 cm, *a*

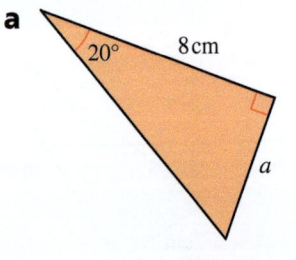

b 5 mm, 11 mm, *b*

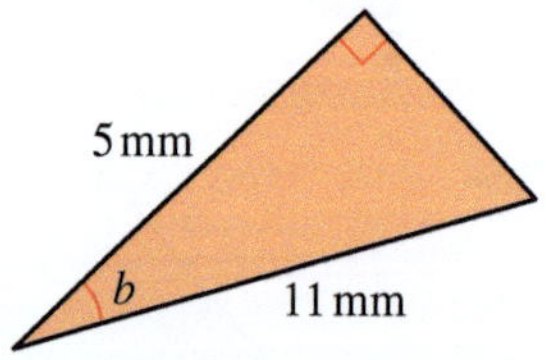

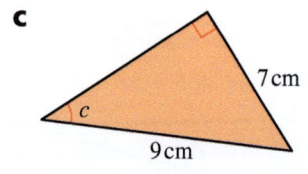

c 7 cm, 9 cm, *c*

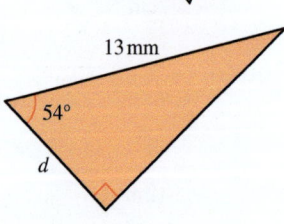

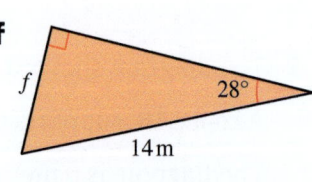

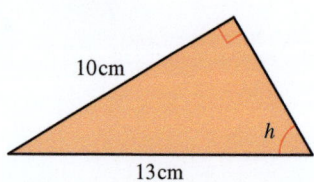

d 13 mm, 54°, *d*

e 10 m, 5 m, *e*

f 14 m, 28°, *f*

g 6 cm, 33°, *g*

h 10 cm, 13 cm, *h*

i 15 cm, 64°, *i*

322

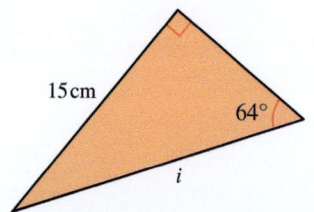

22 Trigonometry

2 a Calculate the value of *x*.

b Work out the area of the trapezium.

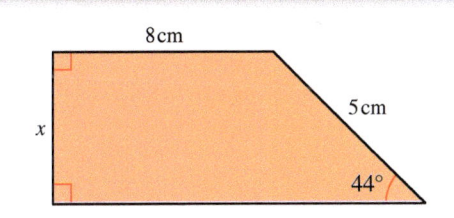

3 Calculate the lengths and angles marked with letters.

a

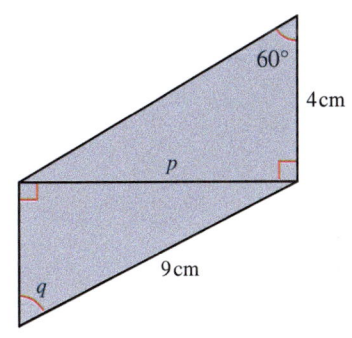

b

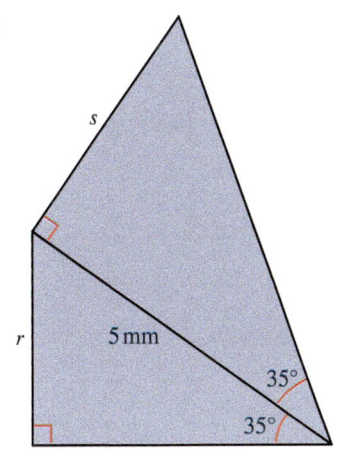

c

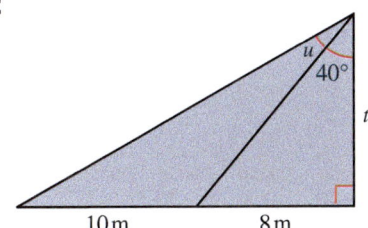

4 Calculate the lengths and angles marked with letters.

a

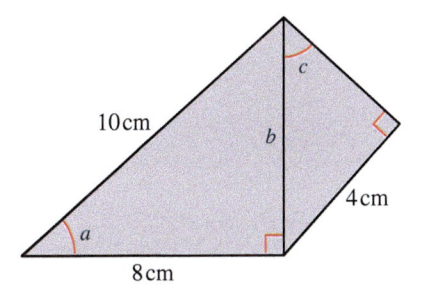

b

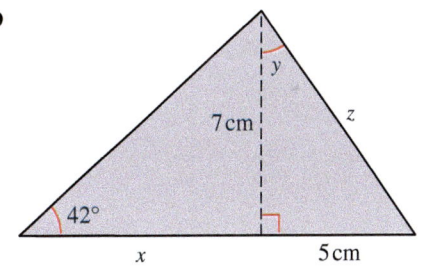

5 *ABCDE* is the cross section of a tent.

$AE = BC = 1.8\,\text{m}$

$DE = DC = 1.2\,\text{m}$

Angle *EAB* = angle *ABC* = 70°

Angle *DEC* = angle *DCE* = 20°

a Calculate the length of:

  i *EC*   **ii** *AB*.

b Calculate the height, *h* m, of the tent.

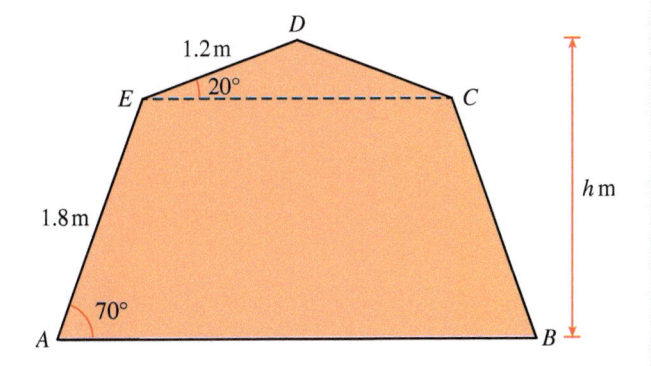

323

6 The diagram shows a vertical radio mast, PQ, which is 34 m high.

The mast stands on horizontal ground and two wires, PA and PB, support the mast.

PA = 45 m and PB = 50 m

 a Calculate the size of:

   i angle PAB   ii angle PBA.

 b Calculate the distance AB.

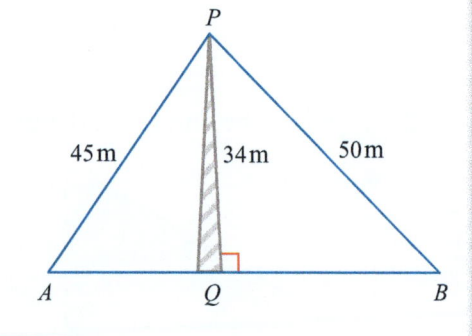

7 ABCD is a field.

 a Calculate the length of:

   i BD   ii CD.

 b Calculate the size of:

   i angle ADC   ii angle ABC.

 c Work out the area of the field.

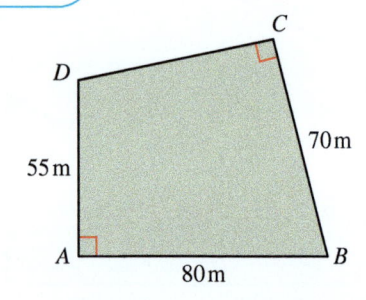

> You need to use Pythagoras' theorem in part a.

23 Matrices and transformations

Learning outcomes

- Transform a shape using a matrix.

Playing with transformations

Matrix transformations turn up almost everywhere within the world of computer graphics and computer games, where they have become an invaluable tool in a programmer's toolkit.

All software systems and hardware graphic processors use matrices to perform operations such as translating, reflecting, rotating and enlarging.

In this chapter you will learn how to use matrices to transform shapes in 2-D.

23.1 Multiplying matrices

In your Stage 8 Student Book you learnt about the **order** of a **matrix** and also how to multiply matrices.

It is important that you revise the rules for multiplying matrices before you learn how to transform a shape by a matrix.

The order of a matrix is the size of the matrix.

The matrix $\begin{pmatrix} 5 & 7 & 0 \\ -8 & 1 & 4 \end{pmatrix}$ has 2 rows and 3 columns.

You write the order of the matrix as 2×3.

> The number of rows is always written first.

You can only multiply two matrices when the number of columns in the first matrix is the same as the number of rows in the second matrix.

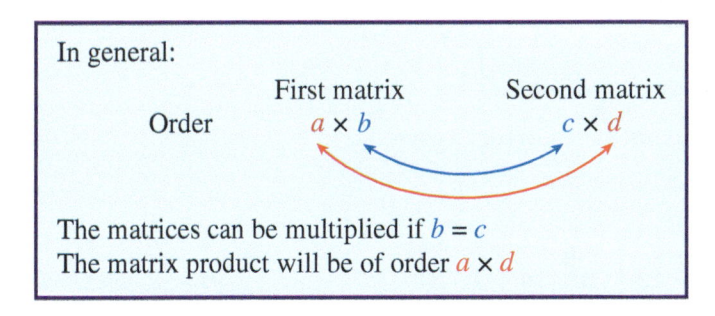

In general:

	First matrix	Second matrix
Order	$a \times b$	$c \times d$

The matrices can be multiplied if $b = c$
The matrix product will be of order $a \times d$

Mathematics for Cambridge Secondary 1

For the matrix product $\begin{pmatrix} 1 & 4 \\ 0 & 3 \\ 2 & 8 \end{pmatrix} \begin{pmatrix} 5 & 7 \\ 9 & 6 \end{pmatrix}$:

order $\quad 3 \times 2 \quad 2 \times 2 = 3 \times 2$

So $\begin{pmatrix} 1 & 4 \\ 0 & 3 \\ 2 & 8 \end{pmatrix} \begin{pmatrix} 5 & 7 \\ 9 & 6 \end{pmatrix} = \begin{pmatrix} * & * \\ * & * \\ * & * \end{pmatrix}$

> Multiply each row in the first matrix with each column in the second matrix to find the unknown numbers.

$1 \times 5 + 4 \times 9 = 41 \qquad 1 \times 7 + 4 \times 6 = 31$

$0 \times 5 + 3 \times 9 = 27 \qquad 0 \times 7 + 3 \times 6 = 18$

$2 \times 5 + 8 \times 9 = 82 \qquad 2 \times 7 + 8 \times 6 = 62$

$\begin{pmatrix} 1 & 4 \\ 0 & 3 \\ 2 & 8 \end{pmatrix} \begin{pmatrix} 5 & 7 \\ 9 & 6 \end{pmatrix} = \begin{pmatrix} 41 & 31 \\ 27 & 18 \\ 82 & 62 \end{pmatrix}$

Worked example

$A = \begin{pmatrix} 2 & 1 \\ 3 & 0 \end{pmatrix} \qquad B = \begin{pmatrix} 7 & 8 & 5 \\ 6 & 4 & -1 \end{pmatrix}$

Work out AB.

First write down the order of each matrix: $\quad 2 \times 2 \quad 2 \times 3$

It is possible to multiply the two matrices because the numbers in the middle are the same.

The answer will be a 2×3 matrix.

$\begin{pmatrix} 2 & 1 \\ 3 & 0 \end{pmatrix} \begin{pmatrix} 7 & 8 & 5 \\ 6 & 4 & -1 \end{pmatrix} = \begin{pmatrix} * & * & * \\ * & * & * \end{pmatrix}$

> Multiply each row in the first matrix with each column in the second matrix to find the unknown numbers.

$2 \times 7 + 1 \times 6 = 20 \qquad 2 \times 8 + 1 \times 4 = 20 \qquad 2 \times 5 + 1 \times -1 = 9$

$3 \times 7 + 0 \times 6 = 21 \qquad 3 \times 8 + 0 \times 4 = 24 \qquad 3 \times 5 + 0 \times -1 = 15$

$\begin{pmatrix} 2 & 1 \\ 3 & 0 \end{pmatrix} \begin{pmatrix} 7 & 8 & 5 \\ 6 & 4 & -1 \end{pmatrix} = \begin{pmatrix} 20 & 20 & 9 \\ 21 & 24 & 15 \end{pmatrix}$

Exercise 23.1

1 $A = \begin{pmatrix} 3 & 2 \\ 1 & 9 \end{pmatrix} \qquad B = \begin{pmatrix} -1 & 5 \\ 0 & 7 \end{pmatrix}$

Helen is asked to calculate AB. She writes:

$AB = \begin{pmatrix} 3 & 2 \\ 1 & 9 \end{pmatrix} \times \begin{pmatrix} -1 & 5 \\ 0 & 7 \end{pmatrix} = \begin{pmatrix} 3 \times -1 & 2 \times 5 \\ 1 \times 0 & 9 \times 7 \end{pmatrix} = \begin{pmatrix} -3 & 10 \\ 0 & 63 \end{pmatrix}$

Explain why Helen is wrong and calculate the correct answer.

2 Work out:

a $\begin{pmatrix} 1 & 3 \\ 4 & 0 \end{pmatrix} \begin{pmatrix} 2 & 1 \\ 3 & 2 \end{pmatrix}$
b $\begin{pmatrix} 5 & -1 \\ 2 & 3 \end{pmatrix} \begin{pmatrix} -1 & 1 & 7 \\ 4 & 0 & 3 \end{pmatrix}$
c $\begin{pmatrix} 6 & 2 \\ 0 & -2 \end{pmatrix} \begin{pmatrix} 1 \\ -2 \end{pmatrix}$

d $(-2 \ \ 4)\begin{pmatrix} -1 \\ 3 \end{pmatrix}$

e $\begin{pmatrix} 3 & 0 & 2 \\ 0 & -2 & 0 \end{pmatrix}\begin{pmatrix} -5 & 2 \\ 4 & 0 \\ -1 & 1 \end{pmatrix}$

f $\begin{pmatrix} 2 \\ 4 \end{pmatrix}(3 \ \ -3)$

g $\begin{pmatrix} -1 & 2 & 1 \\ 1 & 2 & -2 \\ -1 & 4 & 3 \end{pmatrix}\begin{pmatrix} 0 \\ 2 \\ 1 \end{pmatrix}$

h $\begin{pmatrix} 0 & 5 & 1 & -1 \\ 2 & 7 & 5 & -6 \end{pmatrix}\begin{pmatrix} 2 & 0 \\ 0 & -2 \\ 2 & 0 \\ -2 & 0 \end{pmatrix}$

3 Work out:

a i $\begin{pmatrix} 1 & 0 \\ 0 & 1 \end{pmatrix}\begin{pmatrix} 5 & -2 \\ 4 & 8 \end{pmatrix}$

ii $\begin{pmatrix} 1 & 0 \\ 0 & 1 \end{pmatrix}\begin{pmatrix} 2 & 3 & -1 \\ 9 & -4 & 4 \end{pmatrix}$

iii $\begin{pmatrix} 1 & 0 \\ 0 & 1 \end{pmatrix}\begin{pmatrix} 54 \\ -126 \end{pmatrix}$

iv $\begin{pmatrix} 1 & 8 \\ -4 & 5 \\ 0 & 2 \end{pmatrix}\begin{pmatrix} 1 & 0 \\ 0 & 1 \end{pmatrix}$

v $\begin{pmatrix} 3 & -1 \\ 7 & 10 \end{pmatrix}\begin{pmatrix} 1 & 0 \\ 0 & 1 \end{pmatrix}$

vi $(-7 \ \ 15)\begin{pmatrix} 1 & 0 \\ 0 & 1 \end{pmatrix}$

b Look at your answers to part **a**. What effect does multiplying by the matrix $\begin{pmatrix} 1 & 0 \\ 0 & 1 \end{pmatrix}$ have?

4 $A = \begin{pmatrix} 5 & 0 \\ 1 & 2 \end{pmatrix}$ $B = \begin{pmatrix} 3 & 1 \\ 5 & 0 \end{pmatrix}$

Simone says that AB = BA.

Calculate AB and BA to find out whether Simone is correct.

5 $A = \begin{pmatrix} 2 & -3 \\ 7 & 1 \end{pmatrix}$

Nathan is asked to find A^2. He writes:

$A^2 = \begin{pmatrix} 2^2 & (-3)^2 \\ 7^2 & 1^2 \end{pmatrix} = \begin{pmatrix} 4 & 9 \\ 49 & 1 \end{pmatrix}$

Explain why Nathan is wrong and calculate the correct answer.

6 $B = \begin{pmatrix} 5 & 2 \\ 0 & -1 \end{pmatrix}$. Work out B^2.

7 $C = \begin{pmatrix} 1 & 5 & -1 \\ 3 & 2 & 0 \\ 8 & -2 & 4 \end{pmatrix}$. Work out C^2.

8 $P = \begin{pmatrix} -3 & 1 \\ -2 & 5 \end{pmatrix}$. Work out P^3.

> First work out P^2 and then multiply by P.

9 If $Q = \begin{pmatrix} 1 & 0 \\ 0 & 1 \end{pmatrix}$, find Q^8.

Explain how you worked out your answer.

23.2 Transforming a shape by a matrix

You can transform a shape using a **transformation matrix.**

When a shape is transformed you are often asked to describe the **transformation.** It is important that you describe the transformation fully:

Mathematics for Cambridge Secondary 1

- To describe a **reflection** fully you must name the mirror line.
- To describe a **rotation** fully you must state the direction of the rotation, the angle of the rotation and the centre of the rotation.
- To describe an **enlargement** fully you must state the scale factor of the enlargement and the centre of the enlargement.

Follow these steps to transform triangle PQR using the transformation matrix $\begin{pmatrix} 0 & -1 \\ 1 & 0 \end{pmatrix}$

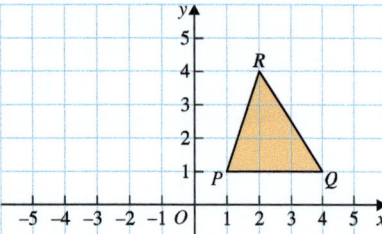

Step 1
Write the coordinates P, Q and R as the column vectors $\begin{pmatrix} 1 \\ 1 \end{pmatrix}, \begin{pmatrix} 4 \\ 1 \end{pmatrix}$ and $\begin{pmatrix} 2 \\ 4 \end{pmatrix}$

Step 2
Combine these column vectors to make the matrix $\begin{matrix} P & Q & R \\ \begin{pmatrix} 1 & 4 & 2 \\ 1 & 1 & 4 \end{pmatrix} \end{matrix}$

Step 3
Multiply the matrices: $\begin{pmatrix} 0 & -1 \\ 1 & 0 \end{pmatrix} \begin{matrix} P & Q & R \\ \begin{pmatrix} 1 & 4 & 2 \\ 1 & 1 & 4 \end{pmatrix} \end{matrix} = \begin{matrix} P' & Q' & R' \\ \begin{pmatrix} -1 & -1 & -4 \\ 1 & 4 & 2 \end{pmatrix} \end{matrix}$

Step 4
Plot the points $P'(-1, 1)$, $Q'(-1, 4)$ and $R'(-4, 2)$ on the grid. This is called the **image**.

Step 5
Describe the transformation that maps triangle PQR on to triangle $P'Q'R'$.

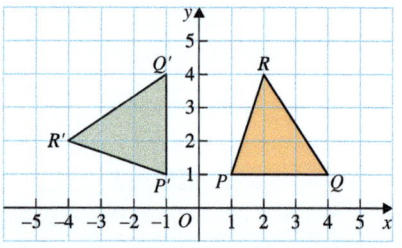

The transformation is a rotation of 90° anticlockwise about the origin.

Worked example

The coordinates of a quadrilateral are $A(2, 0)$, $B(3, 2)$, $C(4, 2)$ and $D(1, 4)$.
Transform quadrilateral ABCD using the matrix $\begin{pmatrix} -1 & 0 \\ 0 & 1 \end{pmatrix}$
Describe the transformation fully.

$\begin{pmatrix} -1 & 0 \\ 0 & 1 \end{pmatrix} \begin{matrix} A & B & C & D \\ \begin{pmatrix} 2 & 3 & 4 & 1 \\ 0 & 2 & 2 & 4 \end{pmatrix} \end{matrix} = \begin{matrix} A' & B' & C' & D' \\ \begin{pmatrix} -2 & -3 & -4 & -1 \\ 0 & 2 & 2 & 4 \end{pmatrix} \end{matrix}$

The transformation is a reflection in the y-axis.

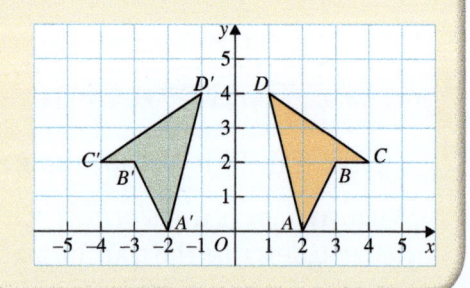

Exercise 23.2

1. Transform trapezium ABCD using the matrix $\begin{pmatrix} 2 & 0 \\ 0 & 2 \end{pmatrix}$

   Label the image $A'B'C'D'$.

   Describe fully the single transformation that maps trapezium ABCD on to trapezium $A'B'C'D'$.

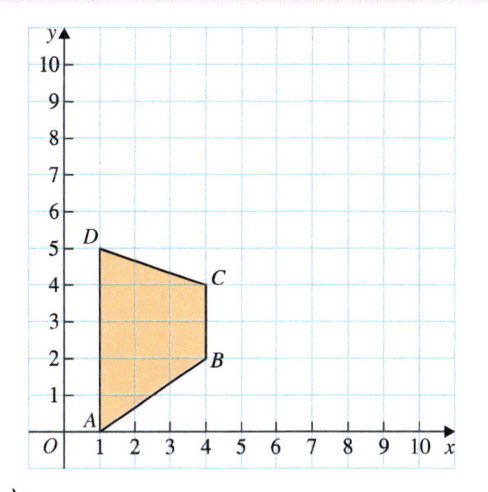

2. Transform quadrilateral ABCD using the matrix $\begin{pmatrix} 0.5 & 0 \\ 0 & 0.5 \end{pmatrix}$
   Label the image $A'B'C'D'$.

   Describe fully the single transformation that maps quadrilateral ABCD on to quadrilateral $A'B'C'D'$.

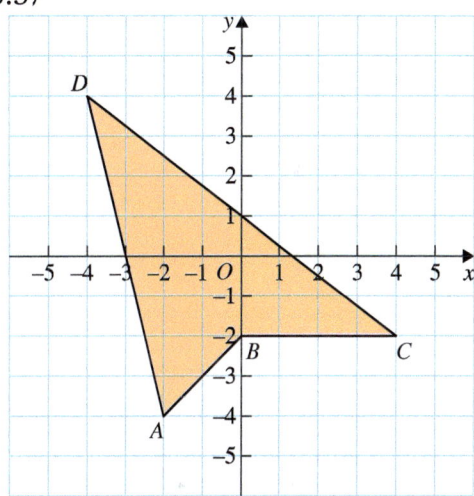

3. Transform triangle PQR using each of the matrices below.

   Label the image $P'Q'R'$.

   Describe fully the single transformation that maps triangle PQR on to triangle $P'Q'R'$.

   Draw a new diagram for each part of the question.

   a $\begin{pmatrix} 1 & 0 \\ 0 & -1 \end{pmatrix}$

   b $\begin{pmatrix} -1 & 0 \\ 0 & -1 \end{pmatrix}$

   c $\begin{pmatrix} 0 & 1 \\ 1 & 0 \end{pmatrix}$

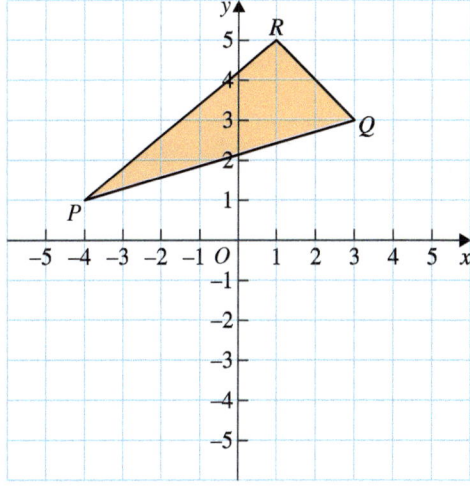

4. a Plot the points A(2, 1), B(4, 2), C(4, 4) and D(1, 5) on a grid.

   b Transform quadrilateral ABCD using the matrix $\begin{pmatrix} 0 & -1 \\ -1 & 0 \end{pmatrix}$

   Label the image $A'B'C'D'$.

   c Describe fully the single transformation that maps quadrilateral ABCD on to quadrilateral $A'B'C'D'$.

5 a Plot the points $A(1, 1)$, $B(2, 5)$ and $C(5, 5)$ on a grid.

  b Transform quadrilateral ABCD using the matrix $\begin{pmatrix} 0 & 1 \\ -1 & 0 \end{pmatrix}$
  Label the image $A'B'C'$.

  c Describe fully the single transformation that maps triangle ABC on to triangle $A'B'C'$.

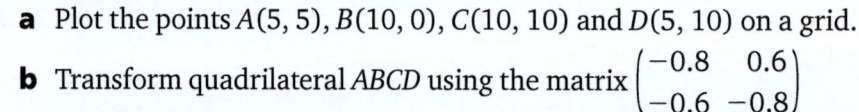

6 a Plot the points $O(0, 0)$, $P(2, 0)$, $Q(2, 4)$ and $R(0, 4)$ on a grid.

  b Transform quadrilateral OPQR using the matrix $\begin{pmatrix} 1 & 1 \\ 0 & 1 \end{pmatrix}$
  Label the image $O'P'Q'R'$.

  c Find the area of quadrilateral:

   i OPQR **ii** $O'P'Q'R'$.

  d Comment on your answers to part **c**.

7 a Plot the points $A(0, 0)$, $B(2, 0)$ and $C(1, 3)$ on a grid.

  b Transform triangle ABC using the matrix $\begin{pmatrix} 1 & 2 \\ 0 & 1 \end{pmatrix}$
  Label the image $A'B'C'$.

  c Find the area of triangle:

   i ABC **ii** $A'B'C'$.

  d Comment on your answers to part **c**.

8 The coordinates of a triangle are $A(0, 0)$, $B(4, 0)$ and $C(3, 4)$.

  a i Transform triangle ABC using the matrix $\begin{pmatrix} 2 & 0 \\ 0 & 1 \end{pmatrix}$
   Label the image $A'B'C'$.

   ii Find the areas of triangles ABC and $A'B'C'$.

  b Can you see a connection between what happens to the area and the transformation matrix?

  c Predict what will happen to the area of a shape if it is transformed by the matrix $\begin{pmatrix} 3 & 0 \\ 0 & 1 \end{pmatrix}$
  Check your prediction to see if you are correct.

9 a Plot the points $A(0, 0)$, $B(5, 10)$ and $C(0, 10)$ on a grid. *Measure angle CAC'.*

  b Transform triangle ABC using the matrix $\begin{pmatrix} 0.6 & -0.8 \\ 0.8 & 0.6 \end{pmatrix}$
  Label the image $A'B'C'$.

  c Describe fully the single transformation that maps triangle ABC on to triangle $A'B'C'$.

10 a Plot the points $A(5, 5)$, $B(10, 0)$, $C(10, 10)$ and $D(5, 10)$ on a grid.

  b Transform quadrilateral ABCD using the matrix $\begin{pmatrix} -0.8 & 0.6 \\ -0.6 & -0.8 \end{pmatrix}$
  Label the image $A'B'C'D'$.

  c Describe fully the single transformation that maps quadrilateral ABCD on to quadrilateral $A'B'C'D'$.

23.3 Combined transformations

Sometimes you can be asked to use more than one transformation.

More than one transformation is called a **combined transformation**.

Worked example

The coordinates of a triangle are $P(2, 1)$, $Q(4, 1)$ and $R(4, 4)$.

a Transform triangle PQR using the matrix $\begin{pmatrix} -1 & 0 \\ 0 & 1 \end{pmatrix}$

  Label the image $P_1Q_1R_1$.

b Transform triangle $P_1Q_1R_1$ using the matrix $\begin{pmatrix} 1 & 0 \\ 0 & -1 \end{pmatrix}$

  Label the image $P_2Q_2R_2$.

c Describe fully the single transformation that maps triangle PQR on to triangle $P_2Q_2R_2$.

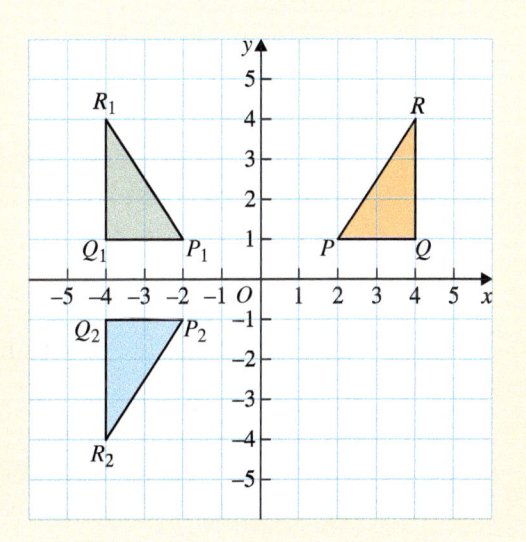

a $\begin{pmatrix} -1 & 0 \\ 0 & 1 \end{pmatrix} \begin{pmatrix} P & Q & R \\ 2 & 4 & 4 \\ 1 & 1 & 4 \end{pmatrix} = \begin{pmatrix} P_1 & Q_1 & R_1 \\ -2 & -4 & -4 \\ 1 & 1 & 4 \end{pmatrix}$

b $\begin{pmatrix} 1 & 0 \\ 0 & -1 \end{pmatrix} \begin{pmatrix} P_1 & Q_1 & R_1 \\ -2 & -4 & -4 \\ 1 & 1 & 4 \end{pmatrix} = \begin{pmatrix} P_2 & Q_2 & R_2 \\ -2 & -4 & -4 \\ -1 & -1 & -4 \end{pmatrix}$

c The transformation that maps triangle PQR on to triangle $P_2Q_2R_2$ is a rotation of 180° about the origin.

Exercise 23.3

1 a Transform triangle A using the matrix $\begin{pmatrix} 1 & 0 \\ 0 & -1 \end{pmatrix}$

  Label the image B.

b Transform triangle B using the matrix $\begin{pmatrix} 0 & -1 \\ -1 & 0 \end{pmatrix}$

  Label the image C.

c Describe fully the single transformation that maps triangle A on to triangle C.

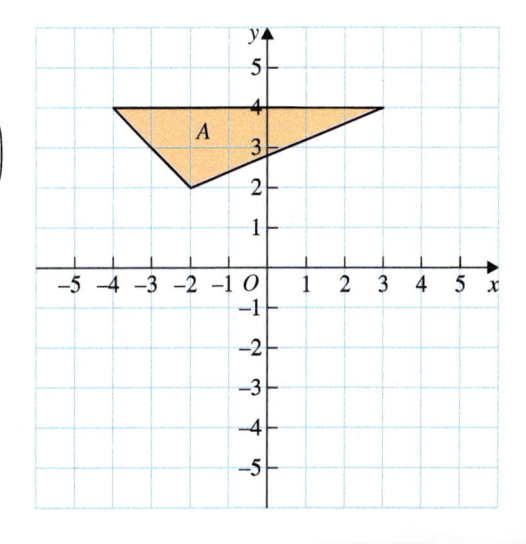

2 a Transform triangle P using the matrix $\begin{pmatrix} 1 & 0 \\ 0 & -1 \end{pmatrix}$
Label the image Q.

b Transform triangle Q using the matrix $\begin{pmatrix} 0 & 1 \\ 1 & 0 \end{pmatrix}$
Label the image R.

c Describe fully the single transformation that maps triangle P on to triangle R.

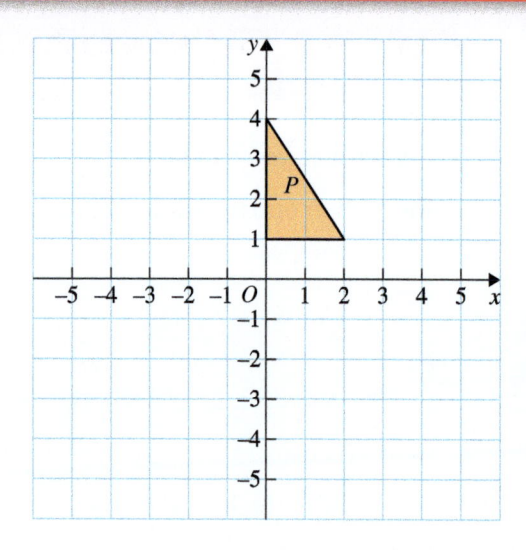

3 a Transform triangle E using the matrix $\begin{pmatrix} 0 & 1 \\ -1 & 0 \end{pmatrix}$
Label the image F.

b Transform triangle F using the matrix $\begin{pmatrix} -1 & 0 \\ 0 & 1 \end{pmatrix}$
Label the image G.

c Describe fully the single transformation that maps triangle E on to triangle G.

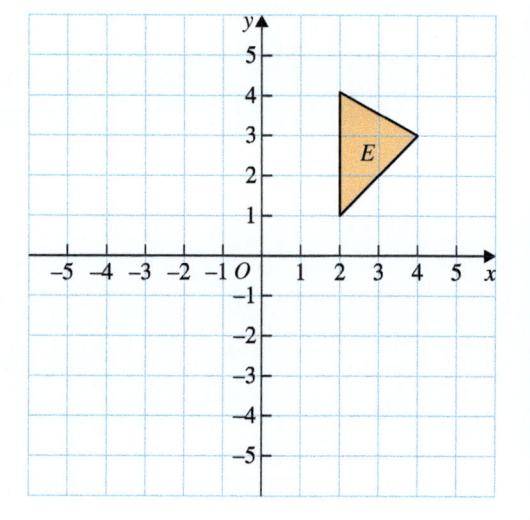

4 a Transform triangle X using the matrix $\begin{pmatrix} 0 & 1 \\ -1 & 0 \end{pmatrix}$
Label the image Y.

b Transform triangle Y using the matrix $\begin{pmatrix} 1 & 0 \\ 0 & -1 \end{pmatrix}$
Label the image Z.

c Describe fully the single transformation that maps triangle X on to triangle Z.

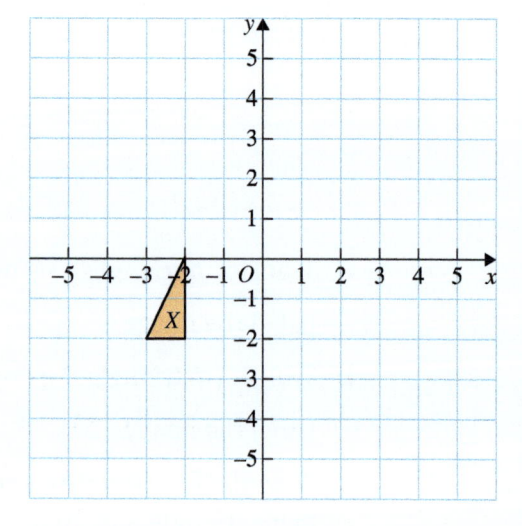

5 The coordinates of a triangle are $P(0, 2)$, $Q(1, 1)$ and $R(3, 2)$.

  a Transform triangle PQR using the matrix $\begin{pmatrix} 0 & 1 \\ -1 & 0 \end{pmatrix}$

  Label the image $P_1Q_1R_1$.

  b Transform triangle $P_1Q_1R_1$ using the matrix $\begin{pmatrix} -1 & 0 \\ 0 & -1 \end{pmatrix}$

  Label the image $P_2Q_2R_2$.

  c Describe fully the single transformation that maps triangle PQR on to triangle $P_2Q_2R_2$.

6 The coordinates of a triangle are $P(1, 1)$, $Q(2, 1)$ and $R(1, 3)$.

  a Transform triangle PQR using the matrix $\begin{pmatrix} 2 & 0 \\ 0 & 2 \end{pmatrix}$

  Label the image $P_1Q_1R_1$.

  b Transform triangle $P_1Q_1R_1$ using the matrix $\begin{pmatrix} 0 & 1 \\ -1 & 0 \end{pmatrix}$

  Label the image $P_2Q_2R_2$.

  c Transform triangle $P_2Q_2R_2$ using the matrix $\begin{pmatrix} 0.5 & 0 \\ 0 & 0.5 \end{pmatrix}$

  Label the image $P_3Q_3R_3$.

  d Describe fully the single transformation that maps triangle PQR on to triangle $P_3Q_3R_3$.

Review

1 Write down the order of the matrix $\begin{pmatrix} 2 & -1 & 0 & 4 & 0 \\ -1 & 2 & 3 & 4 & 2 \end{pmatrix}$

2 Work out:

  a $\begin{pmatrix} 2 & 1 \\ 0 & 1 \end{pmatrix}\begin{pmatrix} 0 & 5 \\ 1 & 4 \end{pmatrix}$

  b $\begin{pmatrix} 7 & 5 \\ 2 & -1 \end{pmatrix}\begin{pmatrix} 8 & -2 & 0 \\ 0 & 1 & -3 \end{pmatrix}$

  c $\begin{pmatrix} 6 & 1 \\ 0 & 4 \\ 1 & 3 \end{pmatrix}\begin{pmatrix} 1 & 0 & 8 \\ -2 & 7 & 1 \end{pmatrix}$

3 $A = \begin{pmatrix} 5 & -2 \\ 0 & 1 \end{pmatrix}$ $B = \begin{pmatrix} -3 & 1 \\ 1 & 7 \end{pmatrix}$

  Work out:

  a AB   **b** BA   **c** A^2 **d** B^2

4 Transform triangle A using the matrix $\begin{pmatrix} 0 & 1 \\ 1 & 0 \end{pmatrix}$

  Label the image B.

  Describe fully the single transformation that maps triangle A on to triangle B.

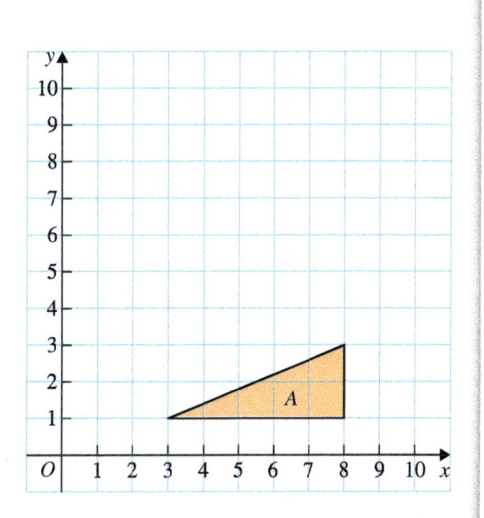

5 Transform shape P using each of the matrices below.

Draw a new diagram for each part of the question.

Describe fully the single transformation for each matrix.

a $\begin{pmatrix} 0 & -1 \\ 1 & 0 \end{pmatrix}$ **b** $\begin{pmatrix} -1 & 0 \\ 0 & 1 \end{pmatrix}$ **c** $\begin{pmatrix} -1 & 0 \\ 0 & -1 \end{pmatrix}$

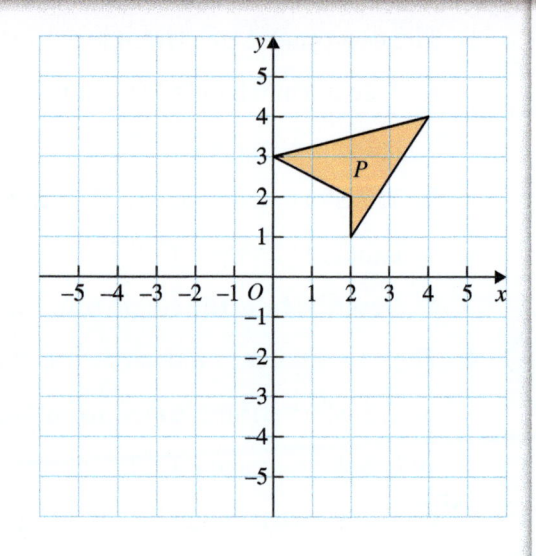

6 a Transform shape X using the matrix $\begin{pmatrix} 0 & 1 \\ -1 & 0 \end{pmatrix}$
Label the image Y.

b Transform shape Y using the matrix $\begin{pmatrix} 0 & 1 \\ 1 & 0 \end{pmatrix}$
Label the image Z.

c Describe fully the single transformation that maps shape X on to shape Z.

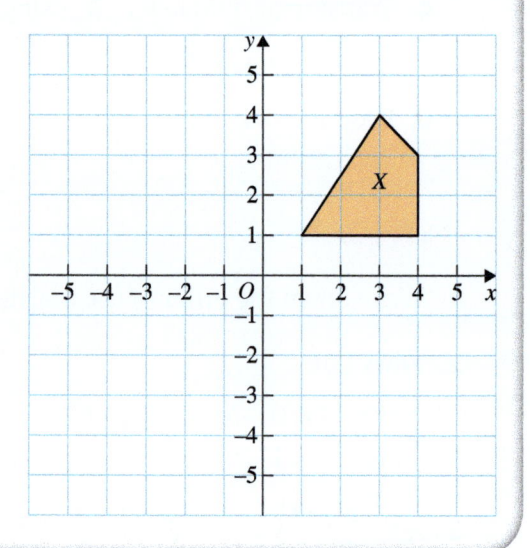

Glossary

A

adjacent side The side that touches the angle in a right-angled triangle.

average speed Speed calculated over a period of time.

B

back to back stem-and-leaf diagram Used to organise two sets of data in order of size.

base In the expression 3^5, the number 3 is called the base number.

base number The number that is raised to a power.

bearing An angle, measured clockwise from north, used for finding directions.

C

cancelling Another word for simplifying.

centi- Prefix denoting $\frac{1}{100}$ of the base unit.

circumference The perimeter, or distance around the edge, of a circle.

combined transformation More than one transformation.

compass An instrument for finding directions, usually by lining up due north.

compound measure A measure, like speed, that is calculated from two other measures.

concave A polygon with at least one interior angle greater than 180°.

congruent Congruent shapes have the same lengths and the same angles.

convex A polygon whose interior angles are all less than 180°.

correlation A connection between two sets of data.

cosine ratio The ratio $\frac{adjacent}{hypotenuse}$

cross section The shape produced by cutting across a prism at any point.

D

decimal places These follow the decimal point. 1 decimal place (1 d.p.) means one digit after the decimal point.

degree of accuracy The number of decimal places or significant figures a number is given.

denominator The number on the 'bottom' of a fraction.

diagonal A line across a polygon joining two vertices.

direct proportion Two quantities y and x are in direct proportion if $y = kx$ i.e. a constant ratio.

distance–time graph Another name for a travel graph.

dual bar chart A multiple bar chart that shows two sets of data.

E

elevation The view of an object from the side.

equation An equation is a mathematical statement that says that two expressions are equal.

equivalent calculations Calculations that give the same answer. 13×0.03 has an equivalent calculation $13 \times 3 \div 100$

equivalent fractions When two fractions have the same value.

equivalent ratios Ratios showing that the same relative sizes are equivalent.

estimate of the mean This is found using the mid-class values. It is an estimate as the actual values are unknown.

expand brackets Multiply each term inside the brackets by the term outside the brackets.

experiment A series of trials.

experimental probability
$= \frac{\text{number of successful outcomes}}{\text{total number of trials}}$

expression A collection of terms that do not include an equals sign.

exterior angle The angle outside a polygon created by extending a side.

F

factorise The 'reverse' of expanding brackets. For example $12x + 8$ factorised is $4(3x + 2)$

formula A rule that connects two or more variables.

frequency polygon A type of frequency diagram used for grouped data. Points are plotted at mid-class values.

front elevation The elevation from the front of an object.

G

gradient A measure of how steep a line is.

gram The metric base unit for mass.

H

hectare A unit of area equal to 10 000 square metres.

hypotenuse The longest side of a right-angled triangle.

335

I

image The final shape after a transformation.

imperial units Old, non-decimal system of weights and measures.

improper fraction Any fraction in which the numerator is greater than the denominator.

index (power) A number used to show a power: e.g. in 3^5, the number 5 is the index (power).

indices The plural of index.

inequality An inequality is a statement such as $x > 5$ or $y \leq -2$

inscribe Construct one diagram with vertices on the perimeter of another.

integer A whole number.

interior angle An angle inside a polygon at one vertex.

interview A conversation where one person asks another person questions.

inverse function A function that undoes what another function has done.

isometric paper Paper with a triangular array of dots lined up in rows at 120° to each other.

K

kilo- Prefix denoting 1000 times the base unit.

L

linear sequence The differences between the terms are all the same.

litre The metric base unit for capacity.

loci The plural of locus.

locus A set of points that satisfy a particular rule.

loss When an item is sold for less than its initial cost, the difference is called the loss.

M

matrix A rectangular array of numbers.

metre The metric base unit for length.

metric system A system of weights and measures based on decimal numbers.

milli- Prefix denoting $\frac{1}{1000}$ of the base unit.

mixed number A number that has a whole number part and a fraction part.

multiple bar chart A bar chart that shows more than one set of data.

mutually exclusive Outcomes are called mutually exclusive if only one of them can happen at the same time.

N

negative correlation As one quantity increases, the other decreases.

no correlation The points on a graph do not show any trend.

nth term A general term in the sequence.

numerator The number on the 'top' of a fraction.

O

object The original shape before a transformation.

observation Primary data that is recorded when it is seen, such as the number of cars passing a point.

opposite side The side opposite the angle in a right-angled triangle.

order The number of rows and columns that a matrix has. For example, 2×3 means 2 rows and 3 columns.

P

per annum Interest rates are given 'per annum'. Per annum means for each year.

percentage reduction The percentage by which something is reduced. It is the same as the percentage decrease.

perpendicular At right angles.

pi (π) The ratio of the circumference to the diameter of a circle. Approximately equal to 3.141 592 654

plan The view of an object from vertically above.

plane of symmetry A plane, or flat surface, which acts as a mirror in a solid with reflection symmetry.

polygon A two-dimensional shape with straight sides.

population The whole set of things you are investigating.

position-to-term rule A rule connecting the term with its position in the sequence.

positive correlation As one quantity increases, so does the other.

power A power tells you how many times a number is multiplied by itself.

primary data Data that you collect yourself.

Glossary

prism A three-dimensional solid with a constant cross section.

profit When an item is sold for more than its initial cost, the difference is called the profit.

Q

questionnaire A set of written questions.

R

ratio The relative size of two or more quantities.

reflection symmetry Symmetry where a line or plane acts like a mirror, cutting a shape or solid into congruent, but reflected, parts.

regular polygon A polygon with equal sides and equal angles.

regular tessellation A tessellation made from one type of regular polygon.

relative frequency $\frac{\text{number of successful outcomes}}{\text{total number of trials}}$

S

sample The part of a population you ask when you do not ask the whole population.

sample space diagram A table showing all possible outcomes in a trial.

scatter diagram A diagram that shows two sets of data as points on a graph.

secondary data Data that someone else has collected.

sequence A list of numbers or diagrams that are connected by a rule.

side elevation The elevation from the side of an object.

significant figures The digits of a number. The first non-zero digit is the first significant figure, after that all digits are significant.

simple interest This is when the interest is paid in one payment at the end of the time period.

simplest form When a fraction cannot be simplified it is in its simplest form.

simplifying Dividing the numerator and denominator by the same number to make them smaller.

simultaneous equations A set of equations in two (or more) variables for which there are values that can satisfy all the equations simultaneously.

sine ratio The ratio $\frac{\text{opposite}}{\text{hypotenuse}}$

speed The rate of movement. Distance travelled in one unit of time.

square number A square number is a number formed by multiplying any integer by itself. The second power of the integer.

stem-and-leaf diagram Used to organise data in order of size.

subject The subject of a formula is the term on its own in front of the equal sign.

substitution Replacing the letters in an expression by the given numbers.

survey A method of collecting data by asking questions or making observations.

T

tangent ratio The ratio $\frac{\text{opposite}}{\text{adjacent}}$

term-to-term rule A rule that connects a term with the previous term in the sequence.

tessellation A pattern of shapes covering a flat surface with no overlaps or gaps.

theorem A mathematical rule that has been proved to be true.

three-dimensional Having length, width and thickness.

transformation Movement or change of one shape into another.

transformation matrix A matrix that transforms a shape.

travel graph A graph showing distance travelled over a period of time.

trial A single experiment, such as the roll of a dice.

trial and improvement A method for finding approximate answers to equations where you make an estimate for the answer and then try to improve on the estimate.

trigonometry A branch of mathematics that studies lengths and angles in triangles.

two-dimensional Having only length and width but no thickness.

U

unitary ratio A ratio in the form $n:1$ or $1:n$

V

variable A quantity that can take different values.

vertex The point where two sides of a polygon meet.

vertices The plural of vertex.

Index

Note: key/glossary terms are in **bold**.

3-D shapes	107–15
congruent	113
cubes	114–15
elevations	107–13
front elevation	108–13
isometric paper/drawings	109–13
planes of symmetry	113–15
plans	107–13
prisms	114
pyramids	114
reflection symmetry	113–15
side elevation	108–13

A

adding	
algebraic fractions	24–6
combining signs	2
directed numbers	1–2
fractions	24–6, 49–53
adjacent side, right-angled triangles	310–11, **310**
algebra origins	18
algebraic expressions *see* expressions	
algebraic fractions	
see also fractions	
adding	24–6
subtracting	24–6
angles and lines	42–7
area	150–68
converting	150–1
hectares	151–2
average speed	286–7, **286**
compound measures	289
averages	71–80

B

back to back stem-and-leaf diagrams	140–2,**140**
bar charts, comparing distributions	143–6
base numbers	**8**
expressions	**18**
indices	8–9
multiplying	8–12
bearings, compass	123–30,**123**
BIDMAS	
expressions	26–7
order of operations	14–15, 26–7
substitution	26–7
brackets, expanding *see* expanding brackets	

C

cancelling, fractions	49–50, **49**
capacity	81–90
volume	155–7
centi-, metric units	**82**
circles	157–61
circumference	157–61
pi (π)	157–61
circumference, circles	157–61, **157**
combined transformations	331–3, **331**

combining signs	
adding	2
dividing	3–4
multiplying	3–4
subtracting	2
compass, bearings	123–30, **123**
compound measures	289–94, **289**
concave, polygons	34–6, **34**
congruent	
3-D shapes	**113**
isosceles triangles	302–3
constructing	
expressions	30–2
lines	115–19
polygons	119–23
converting	
area	150–1
hours and minutes	287
imperial units	85–8
kilometres	85
metric units	81–4
miles	85
times	287
convex, polygons	34–6, **34**
correlation	133–9, **134**
negative correlation	135
no correlation	135
positive correlation	134–5
variables	134
cosine ratio, trigonometry	317–22, **317**
cross section, prisms	161–2,**161**
cube roots	6–8
cubes, 3-D shapes	114–15
cylinders	161–8

D

data collection	
data sources	261
methods	262–4
observation sheets	264–5
planning	260–4
questionnaires	262, 266–9
sampling	261
sheets	264–9
decimal places	65
rounding	65–9
significant figures	66–7
decimals	59–70
decimal places	65–9
degree of accuracy	67–8
dividing	60–3
equivalent calculations	63–5
equivalent fractions	62
integer powers of 10:	59–62
integers	62
multiplying	60–2, 63–5
rounding	65–9
degree of accuracy	67
data collection	**262**
decimals	67–8
rounding	67–8

denominator, fractions	**49**, 51
deriving formulae	171–3
design, mathematics in	34
diagonals, polygons	35–6, **35**
direct proportion	
functions and graphs	248–50, **248**
proportionality	277–8, **277**
directed numbers	
adding	1–2
dividing	3–4
integers	1–2
multiplying	3–4
subtracting	1–2
distance-time graphs	290–1, **290**
distributions	
back to back stem-and-leaf diagrams	140–2
comparing	142–8
dividing	
combining signs	3–4
decimals	60–3
directed numbers	3–4
expressions	18–20
fractions	56–8
index laws	10–12, 20
indices	10–12, 20–1
integer powers of 10:	60–2
dual bar charts,	
comparing distributions	143–6,**143**

E

elevations, 3-D shapes	107–13, **107**
enlargements	198–203
describing	200–1
drawing	198–9
equations	91–106, **91**
see also inequalities	
linear equations	91–3
simultaneous equations	100–5
straight line, general equation	242–3
trial and improvement	98–100
equivalent calculations	**63**
decimals	63–5
multiplying	63–5
equivalent fractions	**49, 62**
decimals	62
equivalent ratios	**274**
estimate of the mean	71–2, **71**
examination-style questions	
area	168–70
averages	79–80
comparing distributions	149
data collection	270
decimals	70
equations	106
expressions	33
formulae	182
fractions	58
functions and graphs	251
integers, powers, roots	17
length, mass capacity	90

338

Index

percentages 259
probability 236–7
proportionality 283–4
Pythagoras' theorem 308–9
ratios 283–4
sequences 221–2
shapes and geometric reasoning 47, 131–2
time and rates of change 297–8
transformations 206–7
volume 168–70
exchange rates, proportionality 271, 279–81
expanding brackets 22
double brackets 27–30
expressions 27–30
factorising 22–3
linear equations 92, 96
experimental probability 230–5, **230**
relative frequency 230
using probability 231–5
experiments 230
data collection 262
expressions 18–33, **19**
algebraic fractions 24–6
base numbers 18
BIDMAS 26–7
constructing 30–2
dividing 18–20
expanding double brackets 27–30
factorising 22–4
indices 11–13,18
multiplying 18–20
simplifying expressions, indices 11–13,18–21
substitution 26–7
exterior angles, polygons 36–8, **36**, 39–41

F

factorising 22–4, **22**
expanding brackets 22–3
Fibonacci sequence 210, 212
formulae 171–82, **171**
deriving formulae 171–3
fractions 177–8
negative x terms 176–7
square numbers 179–81
square roots 179–81
subjects 174–6
substituting numbers 171–3
fractions 49–58
adding 24–6, 49–53
algebraic fractions 24–6
cancelling 49–50
denominator 49, 51
dividing 56–8
equivalent fractions 49, 62
formulae 177–8
history 49
improper fractions 51
indices 10,13
integers 53–4
mixed numbers 50–1
multiplying 53–5
numerator 49, 51
percentages 254
simplest form 49–50
simplifying 49–50

subtracting 24–6, 49–53
frequency polygons, comparing distributions 142–8,**142**
front elevation, 3-D shapes 108–13,**108**
functions and graphs 238–51
direct proportion 248–50
gradients 238–9
inverse function 243–6
real-life problems 246–8
simultaneous equations 241–2
straight line, general equation 242–3
straight-line graphs 240–1

G

gradients, functions and graphs 238–9, **238**
gram, metric units 81
graphs
see also functions and graphs
using 290–4

H

hectares, area 151–2, **151**
hypotenuse, right-angled triangles **299**, 310–11, **310**

I

images
transformation matrix **328**
transformations **192**
imperial units, converting 85–8, **85**
improper fractions 51
index laws
dividing 10–12, 20
multiplying 10–12, 20
raising to a power 10–12, 21
index (power) 8
indices 8–12, **8**
base numbers 8–9
dividing 10–12, 20–1
expressions 11–13,18
fractions 10, 13
index laws 10–12, 20–1
index notation 8–9
integer powers of 10: 59–62
multiplying 8–12, 20–1
negative indices 12–14
raising to a power 10–12, 21
simplifying expressions 11–13,18–21
inequalities 93–6, **93**
see also equations
number lines 93–5
solving linear inequalities 95
inscribing, polygons 119–23, **119**
integer powers of 10:
decimals 59–62
dividing 60–2
multiplying 60–2
integers 2, 53, 62
decimals 62
directed numbers 1–2
fractions 53–4
number lines 1–2
interior angles, polygons 34–6, **35**, 39–42
interviews, data collection 262

inverse function,
functions and graphs 243–6, **244**
isometric paper/
drawings, 3-D shapes 109–13,**109**
isosceles triangles,
Pythagoras' theorem 302–3, 305–7

K

kilo-, metric units **82**
kilometres, converting 85

L

length 81–90
linear equations 91–3
expanding brackets 92, 96
linear sequences 208–9, **208**
lines and angles 42–7
lines construction 115–19
perpendicular lines 115–19
litre, metric units **81**
locus/loci 189–92, **189**
loss, percentages 252–3, **253**

M

maps 123–30
bearings 124–30
map scales 124–30
mass 81–90
matrices 325–34, **325**
see also transformations
multiplying 325–7
order of a matrix 325–7
transformation matrix 327–30
mean 71–9
advantages/disadvantages 75
estimate of the mean 71–2
median 71–9
advantages/disadvantages 75
metre, metric units **81**
metric system **81**
metric units
converting 81–4
prefixes 81–4
miles, converting 85
milli-, metric units **82**
mixed numbers, fractions 50–1, **50**
mode 71–9
advantages/disadvantages 75
multiple bar charts,
comparing distributions 143–6, **143**
multiplying
base numbers 8–12
combining signs 3–4
decimals 60–2, 63–5
directed numbers 3–4
equivalent calculations 63–5
expressions 18–20
fractions 53–5
index laws 10–12, 20
indices 8–12, 20–1
integer powers of 10: 60–2
matrices 325–7
mutually exclusive
outcomes, probability 224–6, **224**

N

negative correlation	**135**
negative indices	12–14
negative x terms, **formulae**	176–7
no correlation	**135**
non-linear sequences	208–9, **208**
*n*th term, sequences	213–20
number lines	
inequalities	93–5
integers	1–2
numerator, fractions	49, **49**, 51

O

objects, transformations	**192**
observation sheets, data collection	264–5
observations, data collection	**262**
opposite side, right-angled triangles	310–11, **310**
order of a matrix	325–7, **325**
order of operations, BIDMAS	14–15, 26–7

P

parallel lines, angles and lines	42–3
per annum, percentages	252–3, **253**
percentage reduction	252–3, **252**
percentages	252–9
calculating with	256–8
finding	254–6
fractions	254
percentage change	252–6
ratios	275–7
uses	253
perimeter	158–61
perpendicular lines,	
constructing	115–19, **115**
pi (π), circles	157–61, **157**
planes of symmetry,	
3-D shapes	113–15, **113**
planning, data collection	260–4
plans, 3-D shapes	107–13, **107**
polygons	34–42, **34**, 119
concave	34–6
constructing	119–23
convex	34–6
diagonals	35–6
exterior angles	36–8, 39–41
frequency polygons	142–8
inscribing	119–23
interior angles	34–6, 39–42
regular	39–42
vertices	35
populations, sampling	**261**
position-to-term rule, sequences	213–17, **213**
positive correlation	134–5, **134**
powers *see* indices; integer powers of 10;	
primary data, data source	**261**
prisms	161–8, **161**
3-D shapes	114
cross section	161–2
surface area	162–6
volume	161–2
probability	223–37

experimental probability	230–5
mutually exclusive outcomes	224–6
sample space diagrams	227–30
profit, percentages	252–3, **253**
proportionality	277–81
direct proportion	277–8
exchange rates	271, 279–81
pyramids, 3-D shapes	114
Pythagoras' theorem	299–309
applications	302–8
isosceles triangles	302–3, 305–7

Q

questionnaires, data collection	**262**, 266–9

R

raising to a power, index laws	10–12, 21
ratios	271–8, **271**
comparing ratios	274–7
equivalent ratios	274
percentages	275–7
unitary ratios	275
real-life problems, functions and graphs	246–8
reflection symmetry, 3-D shapes	113–15, **113**
regular polygons	39–42, **39**, 187
regular tessellations	186–7, **187**
relative frequency	**230**
see also experimental probability	
right-angled triangles	
Pythagoras' theorem	299–309
trigonometry	310–12
rounding	
decimal places	65–9
decimals	65–9
degree of accuracy	67–8
significant figures	66–7

S

sample space diagrams, probability	227–30, **227**
sampling	
data collection	**261**
populations	**261**
scales, map	124–30
scatter diagrams	133–9, **133**
secondary data, data source	**261**
sequences	208–22, **208**
Fibonacci sequence	210, 212
linear sequences	208–9
non-linear sequences	208–9
*n*th term	213–20
position-to-term rule	213–17
term-to-term rule	208–9
side elevation, 3-D shapes	108–13, **108**
significant figures	**66**
decimal places	66–7
rounding	66–7
simple interest, percentages	253, 255, 257–8
simplest form, fractions	49–50, **49**

simplifying expressions, indices	11–13, 18–21
simplifying fractions	49–50, **49**
simultaneous equations	100–5, **100**
functions and graphs	241–2, **241**
sine ratio, trigonometry	317–22, **317**
speed	285–98, **285**
average speed	286–7
constant speed	285–6
square numbers	5–8, **5**
formulae	179–81
square roots	5–6
formulae	179–81
stem-and-leaf diagrams	139–42, **139**
straight-line graphs	240–1
subjects, formulae	174–6, **174**
substituting numbers, formulae	171–3, **171**
substitution	**26**
BIDMAS	26–7
expressions	26–7
subtracting	
algebraic fractions	24–6
combining signs	2
directed numbers	1–2
fractions	24–6, 49–53
surface area, **prisms**	162–6
surveys, data collection	**262**

T

tangent ratio, trigonometry	313–22, **313**
term-to-term rule, sequences	208–9, **208**
tessellation	183–9, **183**
regular tessellations	186–7
theorem	**299**
three-dimensional	**107**
see also 3-D shapes	
time and rates of change	285–98
transformation matrix	327–30, **327**
transformations	192–7, **192**, 325–34, **327**
see also matrices	
combined transformations	331–3
images	192, 328
transformation matrix	327–30
travel graphs	290–1, **290**
trial and improvement, equations	98–100, **98**
trials, experiments	**230**
trigonometry	310–24, **310**
cosine ratio	317–22
right-angled triangles	310–12
sine ratio	317–22
tangent ratio	313–22
two-dimensional	**109**

U

unitary ratios	**275**

V

variables, correlation	**134**
vertex/vertices, polygons	**35**
volume	154–68
capacity	155–7
prisms	161–2